"Through compassionate listening and vulnerable conversation, Neal fashions a homiletic of place that calls for preaching God's presence even in the most desolate realities. Masterfully integrating creation theology and Ezekiel's exilic hope, Neal puts the truths of climate catastrophe in every pulpit as necessary proclamation. Preachers will be both challenged and empowered by Neal's proposal to view preaching as a matter of life and death for the sake of ecological justice."

—**Karoline M. Lewis**, *Professor and Marbury E. Anderson Chair of Biblical Preaching, Luther Seminary*

"This courageous book explores homiletical truth-telling in solidarity with climate refugees and communities enduring the consequences of ecological catastrophe. Attending to the prophetic witness of Ezekiel and the preaching of Fijian Methodist leaders, Jerusha Neal weaves a riveting narrative of resistance to globalizing exploitation and naïve environmentalism, crafting her homiletical approach through a holistic theology of place and an astute ethnographic sensibility. Neal calls practical theologians to a posture of radical relationality that honors diverse Indigenous perspectives on place and dislocation, spiritual and geopolitical causes of climate change, and covenantal hope. Generative and beautifully written!"

—**Carolyn J. Sharp**, *Professor of Homiletics, Yale Divinity School*

"Jerusha Neal has written a necessary book. The analysis of our time she provides is sharp and illuminating, and her writing is clear and compelling. If you have wondered what churches have to say to the multiple crises that climate change represents, read this book."

—**Norman Wirzba**, *Gilbert T. Rowe Distinguished Professor of Theology, Duke Divinity School*

"Jerusha Neal's *Holy Ground* transports readers to a place many of us living in North Atlantic regions have little knowledge of. It successfully demonstrates how listening to voices from Fiji and neighboring islands can teach those of us not from there about prophetic and ecologically sound preaching. Her commitment to the specificity of that place, her dedicated relationship-building with its dwellers, and her deep dive into the biblical interpretation of Ezekiel to shed light on Fijian contexts show her commitment to that place and its people. But her complexifying engagements with theological, postcolonial, and Indigenous topics on sovereignty, belonging, and loss powerfully resonate with global problems that all preachers must face today. Readers will be also equally delighted to see sermon examples regarding how to tackle the climate crisis. *Holy Ground* is a much-needed excellent addition to ecological homiletics."

—**HyeRan Kim-Cragg**, *Principal and Timothy Eaton Memorial Church Professor in Preaching, Emmanuel College, Toronto School of Theology, University of Toronto*

"A world of rising seas, burning forests, imperial extractions, forced migrations, and border walls requires a different kind of preaching. This book points the way to a genuinely apocalyptic homiletic. It is not just addressing issues. It is born of deep, personal encounters with these crises—and with preachers in Fiji who are already addressing them, and with the searing visions of Ezekiel. The result is a book unlike any other. It is a book we preachers need."

—**Ted A. Smith**, *Charles Howard Candler Professor of Divinity, Candler School of Theology, Emory University*

HOLY GROUND

Climate Change, Preaching, and the Apocalypse of Place

Jerusha Matsen Neal

BAYLOR UNIVERSITY PRESS

Cover and book design by Elyxandra Encarnación
Cover image: Painting by Wendy Michelle Davis, wendymichelledavis.com

Library of Congress Cataloging-in-Publication Data

Names: Neal, Jerusha Matsen, 1972- author.
Title: Holy ground : climate change, preaching, and the apocalypse of place / Jerusha Matsen Neal.
Description: Waco, Texas : Baylor University Press, [2024] | Includes bibliographical references and index. | Summary: "A conversation about the significance of place for practical theology given the experiences of pastors and congregations in Fiji forcibly displaced by climate change"-- Provided by publisher.
Identifiers: LCCN 2024025175 (print) | LCCN 2024025176 (ebook) | ISBN 9781481319072 (hardcover) | ISBN 9781481319102 (adobe pdf) | ISBN 9781481319096 (epub)
Subjects: LCSH: Climatic changes--Religious aspects--Christianity. | Ecology--Religious aspects--Christianity. | Environmental justice.
Classification: LCC BT695.5 .N393 2024 (print) | LCC BT695.5 (ebook) | DDC 261.8/8--dc23/eng/20240701
LC record available at https://lccn.loc.gov/2024025175
LC ebook record available at https://lccn.loc.gov/2024025176

For those who preach in worlds undone

O God . . .
Your way was through the sea,
your path through the mighty waters;
yet your footprints were unseen.

Psalm 77:16, 19

Contents

Foreword

Preaching is a matter of life and death. That is the conviction that animates all of Jerusha Matsen Neal's work. She has such a deep love for people and such a profound passion for God that she aches, literally aches, for everyone to encounter the living God, the God of life. Preaching is a matter of life and death. Jerusha appreciates the craft of preaching, and she practices that craft remarkably well. But like the Palestinian poet Noor Hindi, who writes for her displaced people, Jerusha understands that craft is secondary when people are dying.[1] So don't look to this book for simplistic, catchy "how-tos." Rather, expect a deep and profound engagement with human suffering and the living God.

In one of his many upbuilding discourses, which for all practical purposes are sermons, Søren Kierkegaard writes, "Woe to the upbuilding discourse that only wants to chat with people about all sorts of inconveniences in life but does not dare to venture out into the more terrible sufferings."[2] In *Holy Ground*, Jerusha ventures out into the more terrible sufferings, in particular the suffering of people deeply rooted in a place who are now being *displaced* by ecological devastation. Jerusha takes off her shoes, like Moses, to encounter the God of Exodus who hears the cries of a suffering people. She takes off her shoes because the holy ground about which she writes is a specific place where feet touch the land as rising seas lap over toes. And, simultaneously, this holy ground is a relationship with a mysterious, ambivalent God who encounters God's people in that particular place. There is no controlling or managing God in this place. When Moses seeks to do so by asking God's name, God reminds Moses, "I am

1 See Noor Hindi, "Fuck Your Lecture on Craft, My People Are Dying," https://www.poetryfoundation.org/poetrymagazine/poems/154658/fuck-your-lecture-on-craft-my-people-are-dying.

2 Søren Kierkegaard, "An Occasional Discourse," in *Upbuilding Discourses in Various Spirits*, ed. and trans. Howard V. Hong and Edna H. Hong (Princeton, N.J.: Princeton University Press, 1993), 33–77 [106].

who I am," "I will be who I will be." On this unsettled, unmanageable holy ground of human place and divine relationship, among people suffering displacement and seeking God's Word, Jerusha wrestles with the meaning and purpose of preaching for our time.

On such holy ground, as Moses discovers, one must listen. So Jerusha listens. She listens attentively to human lives, and she listens expectantly for God. Her listening is grounded in years of ministry in Fiji, where the deathly realities of colonialism, ecological devastation, and displacement impinge upon the everyday lives of the people, and where preachers are wrestling in profound ways to speak God's Word into this place. So Jerusha listens to these preachers. She converses with them, asks questions of them. She reflects on their sermons in their context. She doesn't attempt to impose the categories and standards of North American homiletics, which can often be myopic and full of hubris. Embracing diversity and complexity, she eschews all forms of universalizing rhetoric that deny the specificity of a place and a community.

Rather, she listens as these Fijian preachers challenge not only her, but also North American homiletics. And challenge they do, often in ways that will make readers uncomfortable. Unlike North American preachers, pastors in Fiji do not have the luxury of addressing the climate crisis at a safe and comfortable distance, without rising tides killing crops, destroying homes, and threatening a communal way of life. As a result these faithful preachers reveal the underlying colonialism in many of our Western assumptions, and they unmask the powers driving the climate catastrophe. Jerusha's work is humbling for a North American preacher, for she "invites settled scholars and preachers to take seriously the unsettling resonances" in these Fijian witnesses. She reminds us that we've done too much speaking and not enough listening. She confronts us with the ways we too often "chat with people about all sorts of inconveniences in life but [do] not dare to venture out into the more terrible sufferings."

There are enormous life-and-death issues facing people around the globe. And Jerusha reminds us that, while preaching is always contextual, in the face of ecological devastation and massive human displacement our context is now the world, and we dare not speak without the kind of cross-cultural engagement she models. Indeed, this vulnerable and possibly transformative cross-cultural listening is one of *Holy Ground*'s most important challenges and contributions to North American homiletics.

But Jerusha doesn't just listen to the preachers in Fiji. She finds another conversation partner in that most unsettling, displaced prophet, Ezekiel, who also understands that preaching is a matter of life and death. She takes off her shoes and listens intently and reverently for God's Word through this exilic prophet. Here again, she avoids easy resolutions and comfortable conclusions, but rather dares to confront the ambivalent, covenanting God whom Ezekiel proclaims. In her honest wrestling with God's Word as she stands on holy ground with the prophet Ezekiel and the Fijian preachers, Jerusha discerns God's unsettling, disruptive presence in the midst of the climate catastrophe.

What emerges from this attentive listening to the prophetic voices of the displaced is an apocalyptic theology of place, which seeks to help preachers proclaim "communal repentance and covenantal hope in the face of ecological catastrophe." In the language of the Fijian preachers, a "New Exodus" may be called for, one that resists colonial oppression and journeys toward something new with the God who covenants with displaced people and all of creation. Such an Exodus transforms and reframes our imaginations and requires us to confront loss and confess painful truths. It invites preachers to stand vulnerable on the holy ground of a groaning world and an ambivalent, covenanting God. It requires nothing less than the "apocalypse of proclamation," including both the radical interruption of old assumptions and the unveiling of new realities.

Preaching is a matter of life and death. And people are dying! What Jerusha ultimately calls for is proclamation grounded in a new relationality. Such preaching listens to and is changed by displaced others, whether they be in the South Pacific or on the southern border of the United States or, like Ezekiel, on the banks of the Chebar Canal. As Naomi Klein has noted, "Climate change isn't just about things getting hotter and wetter: under our current economic and political order, it's about things getting meaner and uglier."[3] In the face of climate catastrophe, Klein reminds us, displacement and dehumanization go together, as borders are secured and walls are built. In its ugliest form, displaced people simply disappear from the imagination; they are "imaginatively displaced" by rhetoric that makes them invisible. Before their literal displacement, they become "unimagined communities."[4] With passionate intensity, Jerusha cries out against this "meaner and uglier." She resists the economic and political colonialism that imaginatively displaces people and makes them invisible. She

[3] Naomi Klein, *On Fire: The Burning Case for a Green New Deal* (New York: Simon & Schuster, 2019), 166.

[4] Rob Nixon, *Slow Violence and the Environmentalism of the Poor* (Cambridge, Mass.: Harvard University Press, 2011), 150, 152.

calls for preachers, like the God of Exodus, to hear the cries of those who are suffering and to be transformed by them.

This transformation, Jerusha recognizes, can only come from a deep theological place. It's the place of holy ground before God, where we all stand as mere human beings, confessing that we are *not* divine, and we are *not* the norm, but simply creatures seeking to live in covenantal relationship with God, with others, and with all of creation. Standing in this very human place, we live with a humble "exilic hope," which laments what has been lost and refuses to appease with easy resolutions. In this very human place we listen attentively to others, engage vulnerably with Scripture, and encounter the sovereign, apocalyptic God, hoping against hope that this God "might yet remove our hearts of stone and give us hearts of flesh" (Ezek 36:26).

Charles L. Campbell

Preface

Valerusa: from "vale" (house) and "rusa" (destruction)

Valerusa is the third son of Taniela Balenaikorodawa, a fellow practical theologian who became a friend during the three years I taught in the South Pacific. Valerusa is also my *yaca*—my Fijian namesake.

In 2016, the summer following Cyclone Winston, Taniela told me his son's name, parsing its meaning in the middle of a crowded hallway at an ecotheology conference. He had combined the names of Valerie, a mission partner teaching at Pacific Theological College in Suva,[1] and myself, a missionary stationed at Davuilevu Theological College—the oldest pastoral training school in the country. Through the name of his child, Taniela embraced us both, extending familial kindness to cultural outsiders.

But the name was more than inclusive gesture. It was also holy lament. When Cyclone Winston hit Fiji in February of 2016, the storm was the strongest recorded cyclone in the Southern hemisphere. Entire communities were reduced to wind-blown debris. It was not lost on Taniela that the combination of "vale" and "rusha" sounded like "valerusa": the destruction of home. He had crafted no simple honorific. His son's name was a memorial for villages flattened and lives lost.

Taniela explained this to me without sentiment or irony. The scrap of my name embedded in the name of his son was intentionally ambivalent, and he was not afraid for me to know it. "In this difficult year of our homes' destruction, God sent you and Valerie to us," he said, implying a characteristic commitment to claims of God's providence in the midst of tragedy. But we both knew that providence was only part of this story. My home country had, in no small part, fueled the increasingly fierce storms that battered Fijian coastlines. My identity as a foreign missionary recalled the colonial destructions of home that theologians like Taniela were working to undo. My earnest desire to assist in that anti-colonial work only

1 Rev. Dr. Valerie Ogden served as Director of Theological Education by Extension at Pacific Theological College from 2014 to 2018.

got me so far. On any given day, my presence might reinforce that history rather than undermine it. While I taught at Davuilevu at the request of the Indigenous leaders of the Fijian Methodist Church, that request was embedded in systems of inequities and the expectations of Western-funded accrediting agencies. The line between subverting those systems and being co-opted by them was alarmingly thin.

In that hallway, Taniela gave me a sacred gift. He told me the truth. *Valerusa*. You are part of our family. *Valerusa*. You are destroying our homes. I groped at the intersection of that holy confession in 2016, and I am still groping today.

The years that I lived in Fiji were years of transformative stirrings in Fijian Methodist pulpits. Church leaders would call them "New Exodus" years. It was a season of naming and renaming, a season of reimagining church and reclaiming heritage. It was a season for confession and difficult truths. Much of that story is unfinished—and not mine to tell.

But part of it is.

In his final words to our family before our departure from Fiji in 2017, former Fijian Methodist president Tevita Nawadra Banivanua gave us this exhortation: "Tell your people that climate change is real." This book is a response to that charge.

As a teacher of preaching, I can think of few better ways to communicate that truth than through the sermons of South Pacific ministers who are presently facing down the crisis. Their sermons witness to a world unraveling, implicating the resignations and settled apathies of more insulated populations. They implicate me as outsider and insider, threat and family. These sermons stand with their feet planted in the soil of a particular place, even as they face the terror of displacement. They describe the apocalypse of empire, as well as an apocalypse (a revealing) of a different sort.

The goal of preaching in such a place is more than persuasive public service announcements or problem/solution exposition. Something bigger is at stake. These are sermons that ask the deepest of biblical questions: what does faithfulness look like after the "unthinkable"?[2] And what does the "unthinkable" say about the faithfulness of God? Such questions shape the Israelite experience of exile—and therefore the shape of the biblical canon. The crucible of displacement provides the historical and

2 Ellen Davis uses the word to describe the theological implications of exile for the prophet Ezekiel and the Israelite people: *Swallowing the Scroll: Textuality and the Dynamics of Discourse in Ezekiel's Prophecy* (Sheffield: Almond Press, 1989), 110.

theological context for the Hebrew Bible's compilation. They are questions that also shape the early church's ecclesiology. When Peter addresses his first epistle to "exiles," he is not only speaking in spiritual metaphors. The identity of the colonizing empire has changed, but the questions have not. What is God's relevance—and the relevance of God's covenant—to those dislocated by imperial violence? Is God's creation powerless in the face of voracious greed?

Biblical narratives of exilic displacement are particularly apt conversation partners for South Pacific preaching on the environment as they bring the connection between colonial empire and climate crisis into view. When Ezekiel stands displaced by the Chebar Canal, Babylon has removed the prophet from the land of his birth. The nation of Judah has been stripped of political sovereignty, theological grounding, and the place of God's promise. In Fiji, this imperial threat is no spiritual analogy or historical abstraction. To preach about climate change in Fiji is to tell the truth about the unthinkable histories of Western colonialism and the fraught history of Fiji's relationship to the British crown. Such preaching traces the links between climate change and foreign aggression, often masked as foreign apathy. Climate catastrophe surfaces the wounds of generations. Perhaps this is why Fijian preaching on the environment regularly makes a theological turn. Like Ezekiel, Fijian preachers know that the loss of place—be it through colonial fiat or a globalized, growth-obsessed economy—is more than a scientific problem to be solved. It is a problem of broken covenant. To address the depth of the crisis, prophetic vision, spiritual formation, and ecclesial reformation are needed. In Fiji, to preach about climate change is to preach about more than the weather. It is to preach about new hearts (Ezek 36:26).

The link between colonial asymmetries of power and climate catastrophe may seem theoretical to many preachers standing in Western pulpits. Colonialism may seem a thing of the past, having little to do with the shape of theological education, a preacher's reading of Scripture, or the function of preaching itself. But the 5:00 a.m. drumbeat of the Davuilevu *lali*, the drum that calls the community to morning prayer, is bound together with the 5:00 p.m. carillon of the gothic-style cathedral. These soundwaves do not exist in different worlds, despite the dateline that divides them. For those listening from the comfort of air-conditioned pews and settler-colonial theologies, Davuilevu's drum vibrates with a truth that names us and even claims us. *Valerusa*. You are part of our family. *Valerusa*. You are destroying our home.

Holy Ground: Climate Change, Preaching, and the Apocalypse of Place has more than a descriptive aim. It constructs a practical theology of place that equips preachers from various contexts to proclaim God's Word in the face of climate catastrophe. To do so, it centers the sermonic, prophetic performances of displaced people—in particular Fijian Methodists and the biblical prophet Ezekiel—to name the boundaries and responsibilities of covenantal relationship. For all of their differences, these two communities face resonant crises. Each has involuntarily lost land, and in losing land, each has lost access to places that give meaning to its identities and faith commitments. For Ezekiel, this means standing with exiles beside a Babylonian irrigation ditch. For Fijian preachers, it means standing in rising tides, watching the graves of village ancestors wash away. In response, both turn to the place of the biblical text. They search the Scriptures for holy ground, as the ground under their feet turns strange.

What truths do displaced preachers have to tell, and what difference does that witness make? This book will argue that preachers who are seeing their homes destroyed know something crucial about God's creation and God's promise. They know something crucial about God. In a world that is both deeply divided and inextricably linked, these preachers name a broken past and uncertain future. What will that naming require of those who perform it and those who receive it as God's proclamation?

Perhaps, to use the language of the Fijian Methodist Church, it will require a new exodus of our own—a season of transformative stirrings and reimaginings. A season of holy confessions and difficult truths. A season of loss. It will most certainly require renewed attention to the prophetic speech of the displaced, be they standing on the banks of a Babylonian canal or among the mangroves of Suva Harbor. It will require the apocalypse of proclamation, unveiling the holy ground of a groaning world and a holy God, in sermons preached by those who have seen their homes' destruction.

Acknowledgments

It is no small thing to trust one's sermons to another.

Sermons are woven together with the hearts of a community. They are precious offerings—labors of prayer, study, and love. They are also fragile. They are far too easy for an outsider to misread and undervalue. This project would not have been possible without fifteen Fijian Methodist ministers who decided that sharing their sermons with others was worth the risk. They extended the gift of their proclamation in a season of global crisis, trusting God's Word to speak beyond boundaries of language and culture to those willing to listen. In a moment when the scope of climate catastrophe threatens to render sermonic speech irrelevant at best, these ministers have invested in preaching's power to create social, ecclesial, and environmental change. Such proclamatory power has a decidedly different character than the powers that hold sway in much of the globalizing North. But this is, perhaps, the point. For those preachers who sent sermon outlines, edited descriptions, and at times traveled across the open sea for interviews, I am deeply grateful. I am more grateful still for the courage and commitment that guides their pastoral leadership to their communities.

The belief that preaching can change society fueled the leaders of the Methodist Church in Fiji's New Exodus. Qase Levu Vakacegu (president emeritus) Tevita Nawadra Banivanua and the Rev. James Bhagwan, current general secretary of the Pacific Council of Churches, did more than make their sermons accessible for this project. They gave of their time and wisdom. Our conversations unfolded over years, and their careful readings of chapter 4 were invaluable. Rev. Dr. Banivanua's sudden passing in the final months of this book's production process has made me all the more thankful for our hours together. I continue to be grateful for Radini Qase Levu Vakacegu Joana Waqairatu, who entrusted me with copies of Tuikilakila Waqairatu's sermon notes in 2016. Her conviction that a church's preaching history is a significant treasure amplifies my own. In 2019, Qase Levu Vakacegu Epineri Vakadewavosa gave this project his blessing, and in 2021, that blessing was extended by Qase Levu Vakacegu Ili Vunisuwai. The leadership of the Methodist Church in Fiji (MCIF) continues to prioritize the conversation about creation justice in its ministries and structures, and I pray that the U.S. church attends to this witness.

The historical research of Rev. Dr. Jolame Lasawa Sedra, former principal of Davuilevu Theological College and current MCIF general secretary, provided vital insights to my descriptions of Fijian ecclesial history, as did the careful memory of Livai Tuisaravere, trustee of the MCIF's Holding Trust. Any mistakes in this accounting are my own. The guidance and stalwart encouragement of Pacific Theological College principal Upolu Lumā Vaai also proved crucial to the project's completion—particularly in his recommendation of Rev. Marika Baleidelabaci (Bale) Tamanikaigavo as a research assistant. Bale invested in this effort from the start, strengthening both its grassroot and systemic import. He gave great care to the work of sermon collection and translation, traveling to areas of the country inaccessible by phone or Internet service in order to retrieve sermon notes and outlines. I am especially grateful for our conversations on what it means to listen well—and for the opportunity to drink tea and eat Radini Evelyn's delicious coconut buns on their family porch.

The completion of my on-site research in 2023 was made possible through the assistance of many. The inspiring office of the Fijian Council of Social Services facilitated my husband's and my visit to Cogea—a village that continues to braid tears and hope with great faithfulness. Special thanks to Rev. Esala Nacuva, who served as our *matanivanua* and translator on this visit. It was a gift to fellowship together with Rev. Nacuva's Kia Island Circuit and his family, who welcomed us with attentive care. Maelin Pickering-Bhagwan, Francisco, and Antonia opened their home over their Christmas holiday, making space at their dinner table and spreading their joy. Rev. Sung Il Lee and Rev. Jungae Lee, Global Ministries missionaries to the Methodist Lay Training Center, also opened their home and ministries to my husband and me. I am especially thankful for our morning prayer together. I am grateful for Aneeta Reuben's curry, the songs of the Dilkusha Church choir, the grounded wisdom of Akanisi Tarabe, and the prophetic insight of Lorini Tevi. The poem recited by Sitiveni Nabuka's daughter still rests in my heart. Mostly, I give thanks for the testimonies of former students who now serve the Fiji Islands with conviction and purpose. These voices have shaped me.

But not theirs alone. This is a project that has bridged the Pacific. I am grateful for Duke University's prioritization of its global climate commitment and the Josiah Charles Trent Memorial Foundation grant that funded this research. The university's Climate Research Innovation Seed Program (CRISP) was also instrumental in facilitating interdisciplinary conversations between theologians and natural and social scientists at the Duke Marine Lab over three years. These conversations strengthened this work, and I am grateful to Avery Davis Lamb and Liz DeMattia for their vision and collaborative skill in weaving these connections.

Charles "Chuck" Campbell championed the project from its inception. I remember an afternoon in 2019 when I told him that I had no idea how to layer the practical, ecological, theo-biblical, and ethnographic implications of this research in a single volume. He replied with a smile, "You don't know how . . . *yet*." I'm grateful for his confidence, his accompaniment, and his long-standing passion for climate justice. I am honored that he writes the foreword. Ellen Davis and Anathea Portier-Young have also given kind attention to this manuscript. Their careful readings have given a nonspecialist courage to wade into Ezekiel's murky depths! Norman Wirzba's guidance in the book's exploration of *place* has sharpened the book's argument. All mistakes, again, are mine. If nothing else, a project of this sort underscores the significance of cross-disciplinary scholarship in engaging problems that stretch beyond reductive classifications. I am grateful to work in an institution where such collaboration is possible. I am also grateful to Baylor University Press and David Nelson for courageously embracing such cross-disciplinary work. Dave had eyes to see the significance of this conversation, in spite of its particularity—and perhaps because of its particularity. When necessary, he also honored the need of this homiletician to use her sermonic voice. His editorial trust and attentive feedback have been a gift.

Ellen and I continue to teach our Hope for Creation? An Exilic Perspective course at Duke Divinity School and are regularly surprised by the new insights biblical texts yield when addressing ecological loss. The next-generation preachers in our classrooms are rising to this challenge. Through the course Preaching Place: The Challenge and Promise of a Global Gospel, these young preachers have helped me map the connections between Oceania's ecological crisis and crises of displacement here in Durham, North Carolina. I am grateful for the wisdom of Cullen McKenney, Melissa Florer-Bixler, and Bishop Clarence Laney in helping me attend to these resonances.

The road to this completed project has been long, and I have been sustained in the work by friends and family. I give particular thanks for the Porch Women—Christine, Megan, and Molly—for Asbury UMC's Bread Group, and for Annie; for my children, whose life choices are marked by a commitment to climate justice; and for my dear Wes, who has read every word of this book multiple times and who traveled with me in 2023 to the villages named. His preaching and teaching have nourished the soil of my perseverance, and his gentleness has seeded joy in that garden. It was his facilitation of a Bible study on Psalm 77 that inspired this book's opening dedication.

Introduction

Where Are We Standing?

I often tell my preaching classes that figuring out "where a congregation is standing" in relation to a biblical passage is one of the hardest tasks in sermon preparation. It's hard because individuals within congregations stand in all sorts of places. Caregivers bump shoulders with cancer patients. Comfortable retirees and idealistic teenagers sit beside the newly unemployed. But it's also hard because, for all a congregation's differences, there are significant connections that bind a preacher's listeners together in unrecognized ways. These connections are embedded in matrices of land, history, and community commitments—the relationships and memories that make up a "place."[1] These webs of connection are regularly the least exegeted voices in the sermonic process. Preachers don't know the rhythms of the ground under their feet, much less the story of the soil. More to the point, most white, North American congregations don't know their relation to that soil or how that soil's history places them in relation to their neighbor.[2] In the classrooms in which I teach, this lack of

1 Place, distinct from space, connects locations with their histories, relationships, and significations. For examples of thinkers who explore place as a social construct, see Yi-Fu Tuan, *Space and Place: The Perspective of Experience* (Minneapolis: University of Minnesota Press, 2001); Pierre Bourdieu, "Social Space and Symbolic Space," *Poetics Today* 12, no. 4 (1991): 627–38; Edward Casey, *Getting Back into Place* (Bloomington: Indiana University Press, 1993). Norman Wirzba describes "our need of each other and our dependence on local places" as a key component of creaturely existence: *This Sacred Life: Humanity's Place in a Wounded World* (Cambridge: Cambridge University Press, 2021), 109.

2 While rural communities may have a more pronounced relationship to soil, Indigenous activists note that stories of how communities came to own land and who was displaced in that process are regularly glossed over or muted, even in agrarian contexts. See Eve Tuck, Marcia McKenzie, and Kate McCoy, "Land Education: Indigenous, Post-colonial, and Decolonizing Perspectives on Place and Environmental Education Research," *Environmental Education Research* 20, no. 1 (2014): 1–23.

relationship presents a significant preaching challenge, making it difficult to locate a congregation in relation to a particular biblical text.

Albert Raboteau's landmark essay, "African Americans, Exodus, and the American Israel," testifies to what is at stake in that location. He notes that Puritan clergymen like John Winthrop imagined the North American landscape to be a promised land, with European settlers playing the part of the Israelites crossing the Jordan. But enslaved Christians perceived "where they were standing" quite differently. North America was no promised land. It was Egypt. And if it was Egypt, then a different sort of sermonic faithfulness was required.[3] For many white congregations in North America today, reframing place in relation to histories of displacement and enslavement continues to necessitate a painful reframing of identity and a costly repentance.

This speaks to why careful, comprehensive conversations about place are rare in the pulpit and in the church. They are hard. North America is particularly haunted by the histories of violence and genocide that accompanied settler colonialism.[4] Elaine Enns and Ched Meyers use the word "haunted" to describe realities that are "neither fully present nor fully absent,"[5] realities unacknowledged and yet formative for the dreams and fears of communal life. When HyeRan Kim-Cragg preaches on the repentance of Zacchaeus (Luke 19:1–10) at the 2018 Annual

[3] Albert J. Raboteau, "African Americans, Exodus, and the American Israel," in *Religion and American Culture: A Reader*, ed. David Hackett, 2nd ed. (New York: Routledge, 2003), 73–88.

[4] Tuck, McKenzie, and McCoy ("Land Education," 6) differentiate settler colonialism from exploitation colonialism. Settler colonialism is defined as the displacing of Indigenous people from their land and resettling that land as if it was the home of the colonial power. Exploitation colonialism focuses on the extraction of resources, often making the land unable to sustain the prior population. For an example of how exploitation colonialism stripped the world's places of resources, fueling the climate crisis, see Amitav Ghosh's discussion of the Dutch nutmeg monopoly in the Banda Archipelago: *The Nutmeg's Curse: Parables for a Planet in Crisis* (Chicago: University of Chicago Press, 2021).

[5] Elaine Enns and Ched Meyers, *Healing Haunted Histories: A Settler Discipleship of Decolonization* (Eugene, Ore.: Cascade, 2021), 11. Enns and Meyers draw on the work of Eve Tuck and C. Ree to describe the ways that "social life, settler colonialism and haunting are inextricably bound; each ensures there are always more ghosts to return," in "A Glossary of Haunting," in *Handbook of Autoethnography*, ed. Stacy Holman Jones, Tony Adams, and Carolyn Ellis (Philadelphia: Routledge, 2016), 639–58 [642].

Meeting of the Saskatchewan Conference of the United Church of Canada, she acknowledges that the gathering takes place on Treaty 6 land, and she names the pain that has shaped that place, including the exonerated killings of Native people that have marked the land's past and present. For Kim-Cragg, interrogating place is an "essential component of the preaching act."[6] Not only does it provide trajectories of relation in the biblical narrative and sharpen the text's prophetic edges; it also forecloses a congregation's co-opting of self-serving identities. Willie James Jennings argues that it was Western theology's supersessionist assumptions about where it stood in relation to biblical Israel that fueled colonialism's terror.[7] John Winthrop was not the first European to imagine his people standing in for God's chosen vessels, and he would not be the last. Sustained attention to the specific relationships between land and communities can unravel these dangerous projections. To know the place one stands is to know a place's particularity—and one's own. A place may resonate with the biblical story, but never substitute for it.

All of this underlines the significance of knowing "where one stands" in sermons addressing climate catastrophe.[8] What are the particulars of the land in this historical moment that necessitate and shape such sermons? What are the particulars of past and present facing preachers who address climate justice in their congregations? And what does the Bible have to say in the conversation?

This last question is more than rhetorical, as the biblical witness has often been characterized as environmentally dangerous. For decades,

6 HyeRan Kim-Cragg, *Postcolonial Preaching: Creating a Ripple Effect* (Lanham, Md.: Lexington Books, 2021), 8, 58–63.

7 Willie James Jennings, *The Christian Imagination: Theology and the Origins of Race* (New Haven: Yale University Press, 2011). "Western traditions," in this project, refers to Euro-American traditions that have engaged in settler-colonial and exploitative colonial relationships with Indigenous peoples. It is an umbrella term that includes those who benefit from this history in Australia, the European continent, and the Americas.

8 This question—"Where are we standing?"—was a pedagogical touchstone for the course Hope for Creation? An Exilic Approach. The class addressed biblical preaching in the face of climate change and was co-taught with Ellen Davis at Duke Divinity School in spring 2021. It informed and was informed by a related question: "Where are we standing in the biblical text?" I am grateful for joint reflection on these questions with Dr. Davis over two years of course planning, as well as her ongoing commitment to creation care, local preachers, and biblical scholarship.

historians,[9] feminists,[10] and postcolonial thinkers[11] have critiqued the Genesis creation account as hierarchical and anthropocentric. Robin Kimmerer draws a sharp line between the stories of biblical Scripture and the stories that shaped her own Potawatomi heritage. Contrasting the Skywoman story with the story of Eve, she notes that "one story leads to the generous embrace of the living world, the other to banishment. One woman is our ancestral gardener, a cocreator of the good, green world; . . . the other was an exile, just passing through an alien world on a rough road to her real home in heaven."[12] Kimmerer's assessment of Genesis rings true given the ways that much Western theology has used the Bible to subdue the voice of creation. But it ignores large swathes of Jewish scholarship and Indigenous interpretations from Christians in the global south[13] who

9 Lynn White Jr., "The Historical Roots of Our Ecologic Crisis," *Science* 155, no. 3767 (1967): 1203–7. Critiques of White's thesis, as well as a description of its significant impact, are outlined in Elspeth Whitney, "Lynn White's 'The Historical Roots of Our Ecologic Crisis' after Fifty Years," *History Compass* 13, no. 8 (2015): 396–410.

10 Sally McFague's influential framing argues that the Genesis account sets up polarities between humanity and creation that undergird patriarchal practice: *Models of God: Theology for an Ecological, Nuclear Age* (Minneapolis: Fortress, 1987). See also McFague, *The Body of God: An Ecological Theology* (Minneapolis: Augsburg Fortress, 1993).

11 For example, Peter Harrison describes the ways that Genesis 1:28 framed the colonial project in "'Fill the Earth and Subdue It': Biblical Warrants for Colonization in Seventeenth Century England," *Journal of Religious History* 29, no. 1 (2005): 3–24; HyeRan Kim-Cragg critiques the "anthropocentrism" in Scripture that underlies the commodification of the earth in *Interdependence: A Postcolonial Feminist Practical Theology* (Eugene, Ore.: Pickwick, 2018); and Arthur Walker-Jones describes Genesis 1's "dualistic separation of God and nature" as counter to traditional Fijian understandings of the divine: "Psalm 104: A Celebration of the *Vanua*," in *The Earth Story in the Psalms and the Prophets*, ed. Norman Habel and Shirley Wurst (Cleveland, Ohio: Pilgrim Press, 2001), 84–97.

12 Robin Wall Kimmerer, *Braiding Sweetgrass: Indigenous Wisdom, Scientific Knowledge, and the Teachings of Plants* (Minneapolis: Milkweed Editions, 2015), 7.

13 The relationship between Indigeneity and Christianity is understood by many grassroots, *iTaukei* communities in Fiji to be noncompetitive—and even mutually affirming. The Fijian-based Ecumenical Centre for Research, Education and Advocacy (ECREA) named significant ways in which the Bible had become interwoven with traditional, Indigenous self-understandings, particularly with regard to communal rights, rejection of secularism, and support of traditional *iTaukei* family structures. See Aisake Casimira, *Who Do You Say That I Am? Interaction between Bible and Culture* (Suva: ECREA Press, 2008), 65–157. Similarly, Matt Tomlinson discusses the sermonic uses of scriptures like Genesis 1:26 as a call to return to a perceived past purity of *iTaukei* Fijian life, mirroring "the likeness" of God: "Compelling Replication:

hear a very different account in the Genesis text.[14] For these readers, Eve's story is not about a garden that does not matter or a home found only in heaven. It is about a home that matters very much, to human creation and to God—a home that is lost. Kimmerer's description comes closest to this vision when she uses the word "exile."

For that is where Israel stands when its creation stories—the creation stories of the world and the creation stories of God's people—were compiled, edited, and preserved. They were standing as refugees on unfamiliar soil, facing horrific grief: the loss of land and home. There is a dialogical relationship between the lament-filled warnings of the prophets and the preservation of the Torah as we know it.[15] The biblical text is more than a collection of disinterested origin stories mapping doctrinal abstractions. It orbits around the searing question of an exiled people—a question with sharp, contemporary relevance: *Can the creation and covenant of a good God come undone?*

Genesis 1:26, John 3:16 and Biblical Politics in Fiji," *Journal of the Royal Anthropological Institute* 16 (2010): 743–60. The decolonial project of the Methodist Church in Fiji leadership described in chapter 4 complicates this view—while also affirming the significance of Indigenous biblical interpretation.

14 Rabbi Shai Held notes that for Maimonides, "creation is declared good prior to, and separate from, the arrival of human beings on the scene (*Guide of the Perplexed*, 3:13)": "Bere'shit #2: Created in God's Image," in *The Heart of Torah*, vol. 1, *Essays on the Weekly Torah Portion: Genesis and Exodus* (Omaha: University of Nebraska Press, 2017), 7–11. Also see Eric Katz, "Judaism and the Ecological Crisis," in *Worldviews and Ecology*, ed. Mary E. Tucker and John A. Grin (Maryknoll, N.Y.: Orbis Books, 1994), 55–70; and Ellen Bernstein, "Rereading Genesis: Human Stewardship of the Earth," in *Righteous Indignation: A Jewish Call for Justice*, ed. Or N. Rose, Jo Ellen Green Kaiser, and Margie Klein (Woodstock, Vt.: Jewish Lights, 2008), 55–59. Tongan biblical scholar Jione Havea notes the ways that the earth is invested with *mana* in the Genesis creation narrative: "The Land Has Colors," in *People and the Land: Decolonizing Theologies*, ed. Jione Havea (London: Lexington Books, 2020), 1–14. And Amadi Ahiamadu uses Genesis 1 to critique multinational and industrial oil operations in the Niger Delta: "A Postcolonial Critical Assessment of the *Imago Dei* in Gen. 1:26–28 in Nigerian Perspective," *Scriptura* 103 (2010): 97–106.

15 The Hebrew scriptures were the result of centuries of narrative streams, both oral and written, but the impact of the exilic experience on the Hebrew Bible's current form is clear. See David Carr's discussion of exilic trauma in *Holy Resilience: The Bible's Traumatic Origins* (New Haven: Yale University Press, 2014). J. Richard Middleton's *The Liberating Image: The* Imago Dei *in Genesis 1* (Grand Rapids: Brazos, 2005) describes the importance of the Babylonian exile for understanding the creation account of Genesis 1, in particular.

The 2021 Intergovernmental Panel on Climate Change (IPCC) Report describes environmental changes that are reshaping our planet in ways that are "unprecedented in thousands, if not hundreds of thousands of years." Some of those changes "already set in motion—such as continued sea level rise—are irreversible for hundreds to thousands of years."[16] Such data can present a crisis of meaning for preachers who see the primary goal of climate-care preaching as preventing climate catastrophe. If the goal of such preaching is persuading a congregation that it can save the world with bullet points of data and public-policy tweaks, the goal seems increasingly suspect. Public policy has its place in the struggle. But that place is rapidly shifting—and even shrinking—as sea levels rise and temperatures warm. Different trajectories of action are emerging. In the words of Jonathan Franzen, there is a new task for those who are looking clear-eyed into the crisis: "Accept that disaster is coming and begin to rethink what it means to have hope." Franzen is not advocating passive fatalism in the face of the planet's certain warming. He argues that "if you accept the reality that the planet will soon overheat to the point of threatening civilization, there's a whole lot more you should be doing."[17] And perhaps there are different actions that need prioritizing—actions focused less on preventing disaster and more on preparing communities to live just, meaningful, connected lives in the face of it.

Franzen's words are no surprise to the preachers whose sermons are gathered in this volume. For many in the South Pacific, the disaster is already here. Seforosa Carroll, World Council of Churches program executive for the Ecumenical Indigenous Peoples Network, describes the gut-wrenching decisions facing governments of low-lying atoll countries like Kiribati and Tuvalu. Forced relocation is a brutal reality as sea levels rise. The former prime minister of Tuvalu, Enele Sapoaga, emphasizes that such a move is no "migration with dignity." It is "forced displacement," borne by the countries least responsible for the catastrophe.[18] In Fiji, forty-two villages have presently been identified for relocation due to causes

[16] Intergovernmental Panel on Climate Change, "Climate Change Widespread, Rapid, and Intensifying," August 9, 2021, https://www.ipcc.ch/2021/08/09/ar6-wg1-20210809-pr/.

[17] Jonathan Franzen, "What If We Stopped Pretending?" *New Yorker*, September 8, 2019, https://www.newyorker.com/culture/cultural-comment/what-if-we-stopped-pretending.

[18] Carroll quotes Sapoaga in "Displacing Liturgy: A Pacific Exploration," *Liturgy* 36, no. 2 (2021): 26–35.

related to climate change.[19] For Indigenous communities, this land loss is more than a geographical crisis. The psychological, emotional tolls of "eco-anxiety and solastalgia"[20] are real, and for South Pacific Christians, spiritual riptides loom. *Where is God in the face of climate change?* Have the promises of this covenanting God also unraveled?

The scriptures of the Hebrew people, scriptures which shaped the witness of the Jew named Jesus, speak directly into that fear. To foreground the voices of displaced preachers grappling with ecological devastation is to glimpse a future that is on its way—indeed, a future that, for many, is upon us.[21] It is also to see again the relevance of the biblical text in addressing the crisis. It is to glimpse the Scriptures through the eyes of those who face the threat of cultural extinction, military occupation, and forced relocation and, in so doing, attend carefully to what may have been at stake for the biblical writers themselves.

A disclaimer is needed here. There are no simple categories in these tangled histories. Those who have studied colonialism know that it takes various guises. The genocidal settler colonialism that marked the North American continent is distinct from the exploitative colonialism that extracted natural resources from Fijian soil.[22] The driving concerns of Indigenous communities in these locations are different. Fijian *iTaukei* communities, for example, are proud of the fact that Indigenous villages retain ownership of 90 percent of the nation's land through their *mataqali* groupings (clans).[23] To be sure, the climate crisis is muddying

[19] Kate Lyons, "How to Move a Country: Fiji's Radical Plan to Escape Rising Sea Levels," *Guardian*, November 8, 2022, https://www.theguardian.com/environment/2022/nov/08/how-to-move-a-country-fiji-radical-plan-escape-rising-seas-climate-crisis. See also Michael Green, "Contested Territory," *Nature: Climate Change* 6 (2016): 817–20.

[20] Carroll, "Displacing Liturgy," 28. "Solastalgia" describes the environmentally induced distress experienced by those whose home environments are being transformed by climate change.

[21] Elizabeth Rush describes the ways that rising sea levels are impacting communities across North America in *Rising: Dispatches from the New American Shore* (Minneapolis: Milkweed Editions, 2018).

[22] For a helpful overview of evolving academic understandings of settler colonialism, where settlers occupy "empty land," displacing or destroying Indigenous populations, and exploitative colonialism, where natural resources are extracted from colonized lands for export, see Lorenzo Veracini, "'Settler Colonialism': Career of a Concept," *Journal of Imperial and Commonwealth History* 41, no. 2 (2013): 313–33.

[23] A concise description of the unique Fijian landholding structure can be found in the *iTaukei* Land Trust Board's "Land Ownership in Fiji," accessed July 30, 2022,

these theoretical distinctions. Extraction of resources and globalizing development are now actively displacing communities that are watching their land disappear, surfacing common currents in distinct Indigenous experiences.[24]

There is also the thorny categorization of Ezekiel and the Judean exiles in relation to Indigenous communities.[25] "Where a community stands" is no static point. Certain Indigenous scholars stress the primacy of Canaanite land claims in their readings of Judean history, problematizing the divine gift of the land to the Hebrew people.[26] Scholars also continue to debate Ezekiel's approach to ethnic and religious others.[27] Communities

---

https://www.tltb.com.fj/getattachment/Media/Brochures/Land-Ownership-in-Fiji-Booklet-(1).pdf.aspx?lang=en-US.

24 Using the lens of ecological devastation, Ghosh (*Nutmeg's Curse*) underscores the resonances between colonial histories across continents.

25 The definition of indigenous has itself been contested. Chris Cunningham, the director of health research at the School of Maori Studies at Massey University, notes that Indigeneity contains multiple valences in contemporary usage—including historic ties to land, lived experiences of oppression, and a worldview that emphasizes a relational ontology with the natural world: "Indigenous by Definition, Experience, or Worldview," *British Medical Journal* 327 (2003): 403–4. This project understands the term to describe an experience that is narrower than "born in a specific place," which is the definition given in "indigenous" in *The Concise Oxford Dictionary of Current English*, ed. R. E. Allen (London: Clarendon, 1990). My use of the term is more consistent with the background paper presented by the United Nations' Secretariat of the Permanent Forum on Indigenous Issues, "The Concept of Indigenous Peoples," Workshop on Data Collection and Disaggregation for Indigenous Peoples, New York, January 19–21, 2004. Their working definition reads, "Indigenous communities . . . are those which, having a historical continuity with pre-invasion and pre-colonial societies that developed on their territories, consider themselves distinct from other sectors of the societies now prevailing on those territories. They form, at present, non-dominant sectors of society and are determined to preserve, develop, and transmit to future generations their ancestral territories, and their ethnic identity, as the basis of their continued existence as peoples, in accordance with their own cultural patterns, social institutions, and legal system." This definition foregrounds historical experiences of displacement and marginalization, even as it describes these factors as part of a provisional "present." The definition continues to be negotiated.

26 Robert Allen Warrior, "A Native American Perspective: Canaanites, Cowboys, and Indians," in *Voices from the Margin: Interpreting the Bible in the Third World*, ed. R. S. Sugirtharajah, 25th anniversary ed. (Maryknoll, N.Y.: Orbis Books, 2016), 283–90, is a potent example of this critique.

27 In juxtaposition to scholars who emphasize Ezekiel's exclusion of foreigners from the temple in his final temple vision (Ezek 44:6–7), scholars like Stephen L. Cook

are multivocal, and the power dynamics that shape their relationships to others shift over time. A community can be displaced, and it can displace others. Those familiar with the simmering tensions between the Fijian *iTaukei* community and the country's Indo-Fijian minority—tensions that have bubbled over in a string of military coups—point to this painful complexity.[28] The goal of this cross-contextual comparison is not to idealize, conflate, or essentialize.[29] The goal is to attend to what diverse communities shaped by deep connection to land and experiences of forced displacement know about places under threat. In the context of the climate crisis, the goal is for settled scholars and preachers like myself to take seriously the unsettling resonances in these communities' witnesses.

These resonances point to colonialism's complicity in our present crisis and underscore the necessary uprooting of colonial philosophies and theologies in environmental advocacy. More constructively, they suggest guiding questions and orientations for settled communities who will be navigating environmental displacement themselves in the not-too-distant future. In centering displaced conversation partners, different framings of the climate crisis come into view, requiring deeper relational commitments in our research, theologies, and sermons.

---

and Mark Brett note the radical gift of land and citizenship to ethnic/cultural "aliens" in Ezek 47:21–23. See Stephen L. Cook, *Ezekiel 38–48* (New Haven: Yale University Press, 2018), 280–81; and Mark Brett, *Political Trauma and Healing: Biblical Ethics for a Postcolonial World* (Grand Rapids: Eerdmans, 2016), 108–9.

28 Jacqueline Ryle describes how the term "my God, my land," coined by revered statesman Ratu Sir Lala Sukuna to describe Indigenous Fijians' connection to land and faith, was manipulated by "divisive, nationalistic, ethno-religious rhetoric and attacks on Indo-Fijians in the coups of 1987 and 2000": *My God, My Land: Interwoven Paths of Christianity and Tradition in Fiji* (Surrey: Ashgate, 2010), xxxi. These tensions will be explored directly in chapter 4.

29 The capitalized form of Indigenous is not meant to suggest homogeneity within the category. The capitalized form is increasingly common in climate-activist literature and is closely related to the United Nation's definition of the term, referring broadly to communities of long connection to specific lands adversely affected or displaced by the political and historical realities of European colonialism. When using the capitalized form, I refer to contemporary, rather than biblical contexts. Resonances between these contemporary communities are striking, but so are their distinctions. I foreground the particularity of the *iTaukei* community in my methodology and analysis, as it is the *iTaukei* community to whom the project is accountable. For an accessible rationale of the capitalized form of Indigenous, see Christine Weeber, "Why Capitalize 'Indigenous'?" *Sapiens Anthropology Magazine*, May 19, 2020, https://www.sapiens.org/language/capitalize-indigenous/.

Centering the Witnesses of Displaced Communities

Through collaboration with the Methodist Church in Fiji and Pacific Theological College (Suva),[30] this book will bring the sermons of Fijian Methodist ministers, many of whom are seeing climate change's devastations play out in their local villages, into conversation with the writings of the displaced prophet Ezekiel. Ezekiel is not the only exilic text useful in illuming the sermons with Fijian preachers, nor is it the only exilic text usefully illumined by them. It is certainly not the only biblical text useful in constructing a homiletic response to the climate crisis. I have delimited my discussion, however, for the purpose of attentive listening. Targeted, textured conversations facilitate more careful relationships between preachers and the place of the biblical narrative. They also stretch preachers into new postures of relationship with present-day communities and the complex ecologies in which they live. In other words, they are deeply relevant to the work of climate justice. In a world where the local particulars of ecological diversity are threatened by globalization, even as the global community is recognizing the significance of those particulars for its survival, tightly framed conversations cannot be written off as provincial.

The words of these faith leaders, separated by millennia, witness to God's ambivalent presence in the midst of apocalypse. They are witnesses that will, no doubt, set certain teeth on edge. Proclamations of divine sovereignty, calls for religious purity, and prophetic fury seem far removed from many mainline North American pulpits—and even dangerous. Ezekiel's words and the sermons of these Fijian preachers point to other dangers. Their witnesses resonate at moments and diverge at others. Displaced peoples differ in their particular fears and hopes.[31] But in their particularity, a shared testimony is braided. The biblical text offers no pacifying panacea. It provides no theological justifications for washing one's hands of the global suffering that has resulted from imperial violence or industrial disregard. It brims with

[30] My deep appreciation to research assistant Rev. Marika Baleidelabaci (Bale) Tamanikaigavo for his expertise and translation assistance, to Principal Upolu Lumā Vaai for championing the project, and to the leadership of the Methodist Church in Fiji for their wise counsel and support. Three years of living in Fiji sharpened my awareness of my limits in engaging the theologies and practices of Fijian communities, and I am grateful for the ways that these partners have shaped this conversation and held me accountable in attentive listening.

[31] Eve Tuck and K. Wayne Yang, for example, describe the dangers of "ambiguating First Nations with Third World migrants" in "Decolonization Is Not a Metaphor," *Decolonization: Indigeneity, Education and Society* 1, no. 1 (2012): 1–40 [29].

lament and calls God's people to a fierce repentance that descends all the way to the bones of human personhood. In these pages, I hope that readers hear this call to repentance and respond in the particulars of their lives.

Reframing the Climate Crisis

But this call is more than a personal call for individual readers. To speak of climate change is to speak of power, influence, and ownership. *It is to speak of place*—its theological and historical import and its loss. The scope and urgency of climate change necessitates communal conversations about a common life, requiring repentance rooted in the complexities and limits of human belonging. It requires a repentance willing to see shared responsibility and local sovereignty as two sides of the same coin. Such repentance, evidenced through action, would impact more than ocean temperatures and sea-level rise. It would impact what we believe about human and nonhuman community, our covenantal commitments, and God.

The word *apocalypse* suggests an "unveiling" or "revealing" of hidden truth. Lisa Wells notes that climate change "reveals what we, on some level, have always known to be true: infinite growth is not possible, we are not above the laws of ecological reciprocity, and we are not as independent as we once believed."[32] Those truths do not only impact a South Pacific island where I used to live. They have present-day implications for "where I stand" now, at a divinity school embedded in an elite southern research university in the United States. They have implications for how research is conducted and whose concerns are centered. They have implications for the function and goals of that research—and the theological proposals to which it leads. What would it mean for climate change's apocalypse to descend to the bones buried in places of academic inquiry, places of theological formation, and places of sermonic speech? What histories would be revealed and whose questions rise?

[32] Lisa Wells, *Believers: Making a Life at the End of the World* (New York: Farrar, Straus and Giroux, 2021), 266. For biblical scholars, *apocalyptic* most properly signifies a literary genre that draws broadly on "the belief that God has revealed the imminent end of the ongoing struggle between good and evil in history." See John J. Collins, ed., *The Origins of Apocalypticism in Judaism and Christianity*, vol. 1 of *Encyclopedia of Apocalypticism*, ed. Bernard McGinn, John J. Collins, and Stephen J. Stein (New York: Continuum, 1998), vii. Christian preaching has long viewed the term through a "stereoscopic lens"; see James F. Kay, "The Word of the Cross at the Turn of the Ages," *Interpretation* 53, no. 1 (1999): 44–56. It has, in other words, held together the cataclysm of the world's end alongside the revelation of hidden, potentially generative, truths. Wells' observation makes a similar move from a secular perspective.

The conversation between the displaced prophet Ezekiel and displaced preachers from the South Pacific provides more than a descriptive map of colonial trauma. It does more than diagnose (and further marginalize) the theologies of displaced peoples.[33] The testimonies of communities that fully embrace the spiritual and phenomenological importance of place—and then lose land—reveal critical insights for settler-colonial communities responding to the current climate crisis. They uncover colonial understandings of place that continue to undergird many Western ecclesial and scientific approaches to climate justice,[34] as well as Western readings of Scripture. In so doing, these communal testimonies assert the critical importance of displaced voices in shaping theologies of place and creation. Fijian Methodism's recent reframing of exodus imagery is one example of this constructive unveiling. The church's rereading of the biblical exodus narratives from an experience of ecological exile illumines covenantal approaches to holiness that inform broad spectrums of Fijian preaching on climate change—as well as that preaching's critique of colonial power. This book is organized around these two trajectories of critique and covenantal commitment. My first three chapters explore unexpected sites of resistance to the colonial place-making project in Fijian sermons and in the biblical text, and my final three chapters draw on these same resources to chart a different relationship to emplaced life. I will argue that Fijian Methodism's constructive wrestling with the Israelite's exodus journey resonates with Ezekiel's own attention to the exodus narratives, his rereading of holiness from exilic ground, and his corresponding denouncement of false hopes and false gods. The book's final chapter takes

[33] Rebecca Copeland's exegesis of Ezek 47:1–12 worries that Ezekiel's traumatic displacement leads to the text's vision of a "magical restoration of the land," when "rational" approaches are preferred: "'Their Leaves Shall Be for Healing': Ecological Trauma and Recovery in Ezekiel 47:1–12," *Biblical Theology Bulletin* 49, no. 4 (2019): 214–22 [220–21].

[34] J. M. Bacon briefly outlines the historical connections between the U.S. conservation movement and discourses of "national expansion," noting more recent scholars like Roderick Nash who argue that "the gospel of ecology" should be seen as an "application of . . . American traditions of expansion," in "Settler Colonialism as Eco-social Structure and the Production of Colonial Ecological Violence," *Environmental Sociology* 5, no. 1 (2019): 59–69 [62], citing R. Nash, "Rounding Out the American Revolution: Ethical Extension and the New Environmentalism," in *Deep Ecology*, ed. Michael Tobias (New York: Pfeiffer, 1985), 170–81 [179]. More recently, Tuck, McKenzie, and McCoy describe the "seduction of claiming Indigenous land as [settlers'] special places" in place-based education, in "Land Education," 14.

these proposals seriously for the teaching of preaching and the preaching of climate-justice sermons in places that continue to accrue privilege from colonial history—places that include much of North America.

Risking Relationality in Practical Theology Method

This is, admittedly, a difficult task. There are two seductive cul-de-sacs in practical theologies that attend to differences of place and culture. The first is to ignore the boundaries of emplaced knowledge, attempting to universalize insights across contexts. Courtney Goto names practical theology's desire for broad, prescriptive relevance as the driving force behind this minimization of place's specificity.[35] It is a drive that Goto finds problematic, noting that insights from one place do not translate to another without remainders or shifts of meaning. Western climate science has increasingly recognized its shortcomings in this regard.[36] The danger, however, is not only seen in attempts to universalize culturally specific epistemologies. It is also seen in the instrumentalizing of Indigenous practices as consumables. Indigenous scholars have critiqued breezy translations of local practices that do not respect the "site specific" nature of their origins, treating place as if it did not matter at all.[37] Ecological relationships—similar to scientific exhortation—cannot be exported from one continent to another without careful attention to local soil.

There is, ironically, an inverted danger in cross-contextual practical theology: a descriptive insularity that exoticizes and essentializes places other than one's own. Amiria Salmond recognizes this tendency in her own fields of anthropology and ethnography, where a supposed scholarly respect for difference can mask a detached, neocolonial metaposition. "Declaring the existence of different ontologies is one thing," Salmond notes, summarizing this approach. "Subjectively entertaining the possibility of ontological

35 Goto argues that practical theologians "are invested in providing answers and solutions not simply for *particular* faith communities . . . but for *many*." Courtney Goto, *Taking on Practical Theology* (Leiden: Brill, 2018), 129 (emphasis original).

36 Willis Jenkins, Evan Berry, and Luke Beck Kreider note the "need for more . . . comparative scholarship that examines convergences and divergences across traditions" in "Religion and Climate Change," *Annual Review of Environment and Resources* 43 (2018): 85–108 [89].

37 Jace Weaver notes the "site specific" nature of Indigenous religious traditions and TEK (traditional ecological knowledge): "Misfit Messengers: Indigenous Religious Traditions and Climate Change," *Journal of the American Academy of Religion* 83, no. 2 (2015): 320–35 [333].

alternatives is quite another."[38] Practical theologies that never press descriptive work toward new possibilities of cooperation, relationship, and restorative justice undermine theology's moral telos. They refuse vulnerability and transformation. Such theologies imagine disengaged tolerance as largesse, when what is required is a riskier, "affectable" reflexivity.[39] Linda Tuhiwai Smith notes the long history of colonial knowledge production in which "the West came to see, to name, and to know" Indigenous communities without committing to the "visions, aspirations, and aims" of those communities.[40] Practical theologies that repent of that past cannot avoid the costly, often conflictual work of attentive relation that leads to fuller accountability and meaningful change.

The climate crisis crystalizes the stakes of this methodological catch-22. The current environmental upheaval makes visible the connections between places, the impacts of one place on another, and the need for collaboration in addressing a problem that does not respect borders. It also reveals stark differences in how particular places define the problem and strategize solutions, foregrounding imbalances of power in these negotiations. Climate scientists increasingly acknowledge the difficulty of mobilizing the world for action while honoring local sovereignty, even as they name the necessity of both.[41]

In so doing, climate science presses practical theology to imagine new methodologies of engagement—engagements that are humbler and more courageous. Can theologies that grow in different soils work together, grow together, and learn from each other in ways that are life-giving? Or does place finally separate us from each other, bounding our understanding and

[38] Amiria Salmond, "Re-building Ships at Sea: Ontological Innovation in Action," in *Environmental Transformations and Cultural Responses: Ontologies, Discourses, and Practices in Oceania*, ed. Eveline Dürr and Arno Pascht (New York: Palgrave Macmillan, 2017), 215–26 [223].

[39] My use of this term draws on Denise Ferreira da Silva's description of Western constructions of racialized others as "affectable others," as opposed to the supposed self-determination of European subjects; see *Toward a Global Idea of Race* (Minneapolis: University of Minnesota Press, 2007), xv.

[40] Linda Tuhiwai Smith, *Decolonizing Methodologies: Research and Indigenous Peoples*, 2nd ed. (London: Zed Books, 2012), 63, 218.

[41] Piérick C. M. Martin et al. note, for example, "To be effective and sustainable, adaptation strategies should acknowledge residents' worldviews and beliefs rather than try to uncritically substitute them," in "Responding to Multiple Climate-Linked Stressors in a Remote Island Context: The Example of Yadua Island, Fiji," *Climate Risk Management* 21 (2018): 7–15.

reinforcing our difference—perhaps even protecting our difference? These questions are less about hermeneutical philosophy and more about the privileges, resources, and perils of "where we stand" in the contemporary world and in relation to the biblical text. They are questions that surface the haunted histories of our past while tracking the fierce storms of our present. It is one thing to ask such questions as settlers who have forgotten place's claim on identity, living on land taken from others. It is different to ask them as Indigenous communities fully aware of their relationship with the soil, experiencing the "slow violence"[42] of displacement firsthand.

Sermons as Sites of Apocalyptic Witness

I have written elsewhere that sermons are "shifting performances of local theology," and in this way they are "placed and displaced, all at once."[43] Their hypercontextuality resists universal application, tethering sermons to particular places and times. And yet their engagement with ancient texts and the shifting provisionality of place itself disrupt rigid categorizations. Sermons are themselves sites of apocalypse, in both the world-altering and reality-disclosing understanding of that term. They weave together emplaced, present-day commitments, ancient narratives, and unknown futures as mats of welcome for the Word of God. In Fiji, mats are more than objects of comfort and beauty; they are spread to show honor, express grief, or claim a new relational reality. They open new possibilities between worlds and between communities.[44] A mat anticipates a guest. Sermons are also acts of anticipation. Through communal intimacy and expectation, they create places of innovation, with implications for both practical theology and the scientific community. Salmond notes that Indigenous leaders grappling with the climate crisis are "rebuilding their ships at sea—combining old understandings with those more recently arrived influences and ideas, including those of science . . . to generate distinctive existential possibilities."[45] To ignore this rebuilding work in the

[42] Rob Nixon, *Slow Violence and the Environmentalism of the Poor* (Cambridge, Mass.: Harvard University Press, 2011).

[43] Jerusha Matsen Neal, "The Edge of Water: Preaching Sovereignty in Rising Tides," *Interpretation: A Journal of Bible and Theology* 75, no. 2 (2021): 112–22 [121].

[44] The significance of mats as sites of encounter between worlds is beautifully described by Akanisi Sobusobu Tarabe in "Re-assessing Gender Roles: A Study of Indigenous Fijian Women's Role in Funerals" (master's thesis, School of Social Science, University of the South Pacific, Suva, September 2015).

[45] Salmond, "Re-building Ships at Sea," 221.

weekly sermons of Indigenous preachers is to ignore the creative contributions of communities that have the most to lose. For climate science, such a dismissal forecloses new insights in the field's fraught negotiation of nature and culture.[46] For practical theology, it distorts theologies of place in ways that serve the interests of a settled theological academy. And for preachers of the Christian Scriptures, to ignore such sermons is to discount the place of apocalypse in discerning the apocalypse of God's unveiling Word. For the Fijian climate-crisis sermons in this volume—and indeed, for the prophet Ezekiel—place is co-constitutive of human identity and permeable to catastrophe, making it a site of sovereign encounter, belonging, and loss. But in the displacement of exile, there is more than despair and denial. There is a holy call to covenantal commitment.

[46] Fikret Berkes critiques the bifurcation of "nature" and "culture" in conventional scientific approaches as "reductionistic" in *Sacred Ecology*, 3rd ed. (New York: Routledge, 2021), 261.

I

Whose Place and Experience?

Sovereignty, Belonging, and Loss

There was a hand-painted wooden sign that hung over the entrance to Davuilevu Theological College when I arrived in 2014: "Train the hands, train the head, train the heart." It was a motto central to the school's self-understanding—part boot camp and part monastic community. This oldest pastoral training school in the Fijian islands was built on theological affirmations that wove body to spirit, community to individual, and land to God. Indeed, the road that led to the school's summit was rarely traveled by Fijian laity unconnected to those training for ordained ministry. The grounds themselves were understood to carry *mana*,[1] imparting spiritual power to those who lived on them. A woman who bore a child at Davuilevu considered it a blessing to bury her child's umbilical cord beneath Davuilevu's palms, a sign of sustaining relationship between person and place.

At Davuilevu, the drumming of the *lali* wakes the community at 5:00 a.m. for morning prayer. By 6:00 a.m., students are working in the fields before the heat of the day, clearing grass with cane knives, preparing vegetable gardens, and planting *dalo*. By the time 8:00 a.m. worship begins, these men and women have done more physical labor than most of their Western theological counterparts do in a day. Ilaitia Sevati Tuwere's seminal theological text *Vanua: A Fijian Theology of Place* has been part of

[1] Ilaitia S. Tuwere describes *mana* as power or influence that shows itself in "physical force or excellence which a person possesses." He also notes the connection between *mana* and *tabu*—or the sacred. The world of *tabu* is not accessible to all, but only to those chosen. See his *Vanua: A Fijian Theology of Place* (Suva: Institute of Pacific Studies Press, 2001), 136. See also Matt Tomlinson and Sekove Bigitibau's discussion of *mana*'s connection to divine speech, the chiefly system, and Christian ethics in "Theologies of *Mana* and *Sau* in Fiji," in *New Mana: Transformations of a Classic Concept in Pacific Languages and Cultures*, ed. Matt Tomlinson and T. P. K. Tengan (Canberra: ANU Press, 2016), 239–57.

Davuilevu's required reading for decades. "It is . . . not accidental to trace the origin of the Fijian word for God—*Kalou*—to the yam garden," Tuwere explains, because "the God who is known as Savior through . . . acts in history is the same God who brings *Kalougata* and *sautu* [i.e., peace, fertility] to life on the land."[2] At Davuilevu, garden work is spiritual work, one of the many ways that faith is manifested in life. Holistic formation is what theological education is about, which means that ideas are inseparable from practices and practices from places.

Sigurd Bergmann describes the "spatial turn"[3] of Western theology in the closing years of the twentieth century as a response to postmodern philosophy's affirmation of place in the construction of identity—and indeed, the role of human agency in creating places themselves. He argues, however, that this is only part of the story. Behind this philosophical concern, a burgeoning environmental conversation drove this renewed interest in the significance of place for human well-being.[4] Globalization and industrial sprawl were severing Western communities from their "sense of place,"[5] and in response to an "unsettled age,"[6] new attention to rootedness, connection, and emplacement was required. This ephemeral hunger for a renewed relationship to the local would be named repeatedly in Western philosophies, ethnographies, and theologies of the late twentieth century,[7] but often with-

[2] Tuwere, *Vanua*, 147.

[3] Sigurd Bergmann, "Theology in Its Spatial Turn: Space, Place and Built Environments Challenging and Changing the Images of God," *Religion Compass* 1, no. 3 (2007): 353–79 [353].

[4] Bergmann, "Theology in Its Spatial Turn," 355–56.

[5] The phrase "sense of place" is used regularly to describe what has been lost in the industrial West—and alternately, what Indigenous cultures have retained. See Walter Brueggemann's description of a hunger for a "sense of place" in *The Land: Place as Gift, Promise, and Challenge in Biblical Faith* (Minneapolis: Fortress, 2002), 3–4; also Steven Feld and Keith Basso, eds., *Senses of Place* (Santa Fe, N.Mex.: School of American Research Press, 1996); Geoffrey Lilburne, *A Sense of Place: A Christian Theology of the Land* (Nashville: Abingdon, 1989).

[6] Keith Basso lists the "unsettled" quality of industrial society alongside the displacement of Indigenous people groups in the opening of his essay, a dated wordchoice given contemporary discourse on the collusion between industrialism and settler approaches to land: "Wisdom Sits in Places: Notes on a Western Apache Landscape," in Feld and Basso, *Senses of Place*, 53–90.

[7] See, for example, Edward Casey, *The Fate of Place: A Philosophical History* (Berkeley: University of California Press, 1998); Keith Basso, *Wisdom Sits in Places: Landscape and Language among the Western Apache* (Santa Fe: University of New Mexico Press, 1996); John Inge, *A Christian Theology of Place* (New York: Routledge, 2003). For a helpful overview of the ethnographic debates in the literature around place during this

out noting the painful irony at the conversation's heart. This unsettled disconnect between people and place was occurring precisely in communities deeply invested in the benefits, privileges, and economic structures of settler colonialism. The communities actually struggling with displacement, occupation, and forced relocation were asking different questions. They did not need theology to tell them how to recover a lost "sense" of place. They had lost land—and not on account of their own carelessness, lack of attention, or greed. They had lost land because of outsized and often violent geopolitical, economic forces over which they had little control.

There are clear resonances between the proliferating theories of place in late twentieth-century Western philosophy and Indigenous communities' affirmation of the bond between land and identity. Many thinkers in the West's spatial turn link their projects to the wisdom of Indigenous communities explicitly.[8] But despite these similarities, people who have lost the most from settler-colonial policies have been regularly ambivalent about the West's sudden "topophilia"[9]—and not only because of its late arrival to a very old party. Colonized, enslaved, displaced communities have a clear-eyed diagnosis of what took the West so long. A long-held trope in the West's own interpretation of its theological avoidance of place is the trauma of the Second World War and corresponding distrusts of sacralizations of land and culture in their various forms. Theological discussions of place were seen to be, at best, the products of a naïve liberalism and, at worst, the tools of nationalistic violence—tools useful to the Nazi regime.[10] A less-acknowledged, but more foundational, reality is highlighted by authors like Willie James Jennings in *The Christian Imagination: Theology and the Origins of Race*. Avoiding the theological implications of place was deeply convenient to Western sensibilities, and not only because of

---

time period, see Wolfgang Kempf, Toon van Meijl, and Elfriede Hermann, "Movement, Place-Making and Cultural Identification: Multiplicities of Belonging," in *Belonging in Oceania: Movement, Place-Making and Multiple Identifications*, ed. Elfriede Hermann, Wolfgang Kempf, and Toon van Meijl (New York: Berghahn Books, 2020), 1–24.

8 The three Western thinkers I discuss in this chapter are in this category.

9 Yi-Fu Tuan's influential *Topophilia: A Study of Environmental Perceptions, Attitudes, and Values* (New York: Columbia University Press, 1974) maps the significance of the displacement of the European medieval worldview by a scientific worldview for environmental engagement.

10 The opening pages of Bernard Anderson's *Rediscovering the Bible* (New York: Haddam House Books, 1951) summarize this view, linking a "world war, a world depression, the demonic furies of Hitler, the Spanish prelude to a second global war, [and] atomic explosions in Japan" with an eclipsing of a "biblical world view" by a worldview of "natural law" and Western culture's "secular spirit" (1–2).

the West's long commitment to philosophical idealism. Western economic powers had been in the business of commodifying and exploiting other places and persons over centuries of colonial extraction.[11] Colonial expansion in North America was premised on the objectification of land as a thing to be possessed and made productive. To understand place's intimate connection to human thriving brought the evils of the settler-colonial project into sharp relief and pressed a colonial reckoning. Minimizing place's theological importance was the easier path.

In this light, the West's renewed interest in place seems a hopeful step forward, not only for environmental preservation, but also for the work of decolonial justice. I sincerely hope this is the case. But I share the reservations of decolonial activists who see in "place-based" education perpetuations of "European universalism" on the one hand, and depictions of Indigenous peoples as "repositories of static cultural knowledge" on the other.[12] There are all kinds of ways to avoid a reckoning. The Western approaches to place described in this chapter reimagine place's significance in understanding human knowledge, subjectivity, and agency. Each draws on the thinking of Indigenous communities in crafting—or at least in authorizing—their proposal. I am sympathetic to their work, as the problems they address in the West's globalized, industrial context are real. But the questions they explore are not the paramount questions of the Indigenous communities they reference. They are questions of the Western academy, shaped by those with considerable power and privilege who are only recently feeling that privilege's weight. They are questions of cross-contextual relevance, effectiveness, and creative impact—questions which inform Western projects of globalization and control. Different questions rise from communities whose land is under threat: questions of *sovereignty*, *belonging*, and *loss*.

In 2016, when Cyclone Winston hit Viti Levu, the carefully painted affirmation of holistic education that stood proudly at Davuilevu's entrance snapped in two. "Train the hands, train the head, train the heart" became a piece of waterlogged scrapwood. It is possible to have a conversation about place that affirms the statement on the sign without taking seriously the sign's destruction. It is possible to have a conversation about "place and

[11] Jennings insists that the "story of place" is the "story of race," and an avoidance of the one leads to an avoidance of the other: *The Christian Imagination: Theology and the Origins of Race* (New Haven: Yale University Press, 2011), 289.

[12] Eve Tuck, Marcia McKenzie, and Kate McCoy, "Land Education: Indigenous, Post-colonial and Decolonizing Perspectives on Place and Environmental Education Research," *Environmental Education Research* 20, no. 1 (2014): 1–23 [1].

experience" that does not center the place and experience of the physically *dis*placed—overlooking critical questions about place raised by climate crisis, colonial history, and exiled communities themselves. In ignoring the questions of such communities, it is possible to leverage Indigenous insights about place in ways that do not serve the interests of Indigenous people. In short, one can speak about place and neglect righteousness.

It's a problem I know well because of my own identity and vocation. I am a white, mainline, North American preacher, and the homiletic communities that trained me have had their own twentieth-century conversations about place. Despite the dead ends and sharp turns in those conversations, they continue still. Preachers understand that place matters, even if we disagree on why and how. The significance of language, landscapes, and communities in shaping the worlds of listeners is a preacher's bread and butter. Practical theology is regularly stereotyped as a slow learner among the academic disciplines, following behind the philosophical trends of weightier guilds. But in this case, North American homiletics was at the forefront of theology's "spatial turn"—for better and for worse. Twentieth-century philosophers regularly cited in place-based conversations, thinkers like Martin Heidegger or Ludwig Wittgenstein, cast long shadows in the white, North American homiletic field.[13] My discipline has

13 The most obvious example of Heidegger's influence is seen through the impact of the "New Homiletic" in the guild's late twentieth-century scholarship and its contemporary reassessment of that movement. Growing out of the New Hermeneutic of Gerhard Ebeling and Ernst Fuchs, the New Homiletic attempts to connect the place of the biblical text with the places of contemporary listeners through a "word-event" that occurs in preaching. As Dawn Ottoni-Wilhelm summarizes in her historical mapping of the New Homiletic (and her critique), Heidegger's ontological understanding of language as disclosive of being, as well as being's "house," lies at the root of this conversation: "New Hermeneutic, New Homiletic and New Directions: A U.S.-North American Perspective," *Homiletic* 35, no. 1 (2010): 17–31. See Heidegger's discussion of language as the "house of being" in his 1947 "Letter on Humanism," in *Basic Writings*, ed. David Krell (New York: Harper & Row, 1977), 189–242 [239]. Wittgenstein's assertion of the intimate relation between a person's beliefs and attitudes and their locations in physical and social spaces continues to play out in postliberal and radical orthodox approaches to preaching that posit preaching's telos as the formation of a community in an ecclesial, narrative grammar; Charles L. Campbell's early writing on Hans Frei traces these connections in *Preaching Jesus: The New Directions for Homiletics in Hans Frei's Postliberal Theology* (Eugene, Ore.: Wipf and Stock, 1997). For a more contemporary collection of sermons influenced by this nonfoundational approach, see Alison Milbank, John Hughes, and Arabella Milbank, eds., *Preaching Radical and Orthodox* (London: SCM Press, 2017).

learned the costs of these shadows the hard way, for all their appreciation of human situatedness.

This chapter will outline three approaches to place championed by contemporary Western thinkers that are inadequate to the experiences of displaced preachers in Fiji and, I will argue, inadequate to an understanding of place in climate preaching more broadly. The point is not to tear down those who are doing significant restorative work in Western settings. The point is to name the insufficiency of their thought in relation to needs and questions of physically displaced communities. Their insights are valuable, but their experiences are particular and their agendas blinkered. The questions they skirt can create dangerous approaches to place in practice. At stake is my larger argument that the voices of displaced preachers are critically important in any constructive theological conversation about place, creation care, and decolonial justice. This is not only because displaced communities push the West's questions about place out of the shallows. It is also because "place and experience" are not neutral, universal categories. In both the biblical text and the contemporary world, place and experience are tied to communal histories, imperial politics, and divine covenants. We may stand on the same land or preach from the same text, but the implications of that text for our lives are different because of who we are. When sovereignty, belonging, and loss are treated as issues of significance, the questions of *who* and *whose* become central theological concerns, making the sermons of the displaced, in the biblical text and around the world, necessary conversation partners.

There is one more reason why this book on Fijian Methodist climate preaching begins with a critique of Western theories of place. I am a realist when it comes to the co-opting influence of this world's powers and principalities (Eph 6:12), and the sins of colonialism have far from vanished. They manifest themselves in global inequities of influence and resources—even within the environmental movement.[14] For those who wish to engage the work of climate justice, a healthy skepticism about one's limits and motives is appropriate. All too often, the very good we want to do, we do not do—and the very good we desire, we subvert (Rom 7:19). I would imagine that there are few, if any, readers of this book who explicitly support colonial constructions of place as historically defined:

[14] For example, Karin Louise Hermes names the "tokenism" she has experienced as a person of color in environmental circles and a lack of attentiveness to Indigenous concerns: "Why I Quit Being a Climate Activist," *Vice*, February 6, 2020, https://www.vice.com/en_us/article/g5x5ny/why-i-quit-being-a-climate-activist.

as an inanimate possession to be dominated, as empty space to be discovered, or as essentialized markers of ethnic identity.[15] The proposals of place described below deconstruct these understandings in significant (if partial) ways. And yet colonial ghosts of insularity masked as universality haunt these theories in ways that are not always perceived, especially when they are leveraged prescriptively across contexts. This makes their intersections with the environmental movement troubling. A powerful person who is convinced that she is "helping," "discovering," "fixing," or "teaching" can be a dangerous thing—which is, again, something that I know from the work of my life. I am a practical theologian, a preacher, and a former mission partner. I believe, in fact, that each of those roles is critical to the work of climate justice. Disengagement is no answer. But revelatory engagement with others—engagement that transforms participants in life-giving ways—requires reflexivity and no small amount of soul-searching. To receive environmental insights from exilic biblical texts and exiled preachers, an interruption of Western certainties and conversations is required.

For those new to philosophical conversations about place, I will signpost where this critique is headed. In spite of their many insights, I will argue that each of the thinkers named falls short of the generative complexity evidenced in displaced communities' relationships to place. Tim Ingold's model of "meshwork" emphasizes place's interconnections at the expense of particular identities. Jeff Malpas' more insular "topographical" understanding focuses on the belonging of the self to place rather than the welcoming and belonging of others. And Keith Basso's description of "place-making" overemphasizes human agency in place's creation, neglecting the implications of how place makes the human. The questions and experiences of displaced communities themselves, particularly those of the Fijian Indigenous community, disrupt these polarities. I will suggest that this disruption has implications for how the place of the biblical text relates to the place of the pulpit in sermons addressing climate justice.

[15] For an overview of these colonial constructions, see Willie James Jennings' discussion of how concepts of self-possession and productivity relate to the claiming of "empty" or "unproductive" lands, slavery, and colonial history: Kenan Institute for Ethics, "Facing the Anthropocene: A Conversation with Willie Jennings," March 11, 2021, https://www.youtube.com/watch?v=SGMXeE6p-tQ. Jennings also discusses how colonial essentialisms between place and identity are racialized in *Christian Imagination*, 29–31.

Tim Ingold, Wayfaring, and the Meshwork of the World

Late in Tim Ingold's beautifully written, iconoclastic *Being Alive: Essays on Movement, Knowledge, and Description*, there is a sentence that sings the contents of his heart: "Where studying *of* is a process of othering, studying *with* is a process of togethering."[16] Ingold's rumination on the work of drawing, the movement of lines, and the world as "meshwork"[17] is an interdisciplinary manifesto against understandings of place that minimize process and relationality, imaging a new metaphor for understanding how places and persons connect through their environments. Norman Wirzba lucidly describes how understandings of places as either "containers" that humans reside within or "globes" that humans stride upon have shaped the philosophies and economic systems of environmental crisis.[18] Ingold's meshwork does more than critique these models or imagine a new structure; it reconceives place as "lines of becoming,"[19] *events* of entanglement between humans themselves and with the nonhuman world, lived out through processes of communal knowledge creation.[20] As an anthropologist who began his work with Indigenous Sami communities in Finland, Ingold is interested in description that honors the unfolding of relational knowledge over time—a kind of "wayfaring"[21] that feels its way forward. The world is not something one holds in one's hand or observes with objectivity. Neither are human persons repositories of expertise, accessed like books in a library. They are lines of moving, lived experience, knotted to the world and to each other. Human experience is changed by those ties, even as it extends beyond them. Given the ways that anthropology has historically served the purposes of the cataloging, colonial gaze, Ingold's insistence on the possibility

[16] Tim Ingold, *Being Alive: Essays on Movement, Knowledge, and Description* (New York: Routledge, 2011), 226 (emphasis original).

[17] Tim Ingold, "On Human Correspondence," *Journal of the Royal Anthropological Institute* 23, no. 1 (2017): 9–27 [11].

[18] Norman Wirzba, *This Sacred Life: Humanity's Place in a Wounded World* (Cambridge: Cambridge University Press, 2021), 90–124. Wirzba's critique of the "globe" as a model for place draws on Tim Ingold's "Globes and Spheres: The Topology of Environmentalism," in Ingold's *The Perception of the Environment: Essays on Livelihood, Dwelling, and Skill* (London: Routledge, 2000), 214.

[19] Ingold builds on the language of "lines of becoming" as described by Gilles Deleuze and Félix Guattari in *A Thousand Plateaus: Capitalism and Schizophrenia*, trans. Brian Massumi (London: Continuum, 2004), 224–25.

[20] "Persons are not beings that move," Ingold states. "They *are* their movement" (*Being Alive*, 168, emphasis original).

[21] Ingold, *Being Alive*, 148–50.

of connection and co-creation in the work of anthropological research is a fight for the legitimacy of the discipline itself.

It is a vision that is deeply compelling to me, as "studying *with*" was the impetus behind my decision to move to Fiji. I didn't travel to Fiji as an ethnographer, hoping to study exotic grammars of Christian discourse or give thick descriptions of Indigenous rituals. I came at the request of Fijian church leaders who needed teachers to cover the classroom work of local scholars who were pursuing their doctoral degrees. I did not come to study global theological education; I came to be a part of it. And I was. Alongside my Fijian colleagues, I shook my head at opaque accreditation requirements; I chafed under the condescension of foreign visitors; I was baffled by how to craft meaningful, contextual theological formation using a library of English textbooks, a library where even the postcolonial resources were penned with Western audiences in mind. I prayed with students on my front porch in Fiji's late afternoon light and sang hymns alongside them in the still-dark dawn. The shared work and shared worship of the school revealed insights that I would have never gleaned scratching notes into a notepad.

Ingold's work also feels familiar to me as a white homiletician whose mid-1990s Introduction to Preaching course began with a sermon by Fred Craddock. Associated with the inductive, participatory rhetoric of the New Homiletic,[22] Craddock's 1979 *As One without Authority* also stressed the *eventfulness* of revelatory knowledge and the entangled, dialogical relationship between speaker and listener. Sermons are not anthropology, of course. But there was hope in Craddock's proposal for preaching *with* a community—a hope that fought for the legitimacy of preaching in light of postmodern assaults on objectivity and preacherly authority. I read Ingold, therefore, with a sense of déjà vu. The warmth and creativity of Craddock's proposal is present, but so are its weaknesses. In foregrounding the event of encounter between speaker and listener in the discovery of knowledge, the New Homiletic brought energy to a whole generation of preachers feeling their way forward in the pulpit. It was less

22 Ottoni-Wilhelm notes that "during his sabbatical at Tübingen and its Institute for Hermeneutics in 1968–69, Craddock was greatly influenced by the scholarship of Gerhard Ebeling and Ernst Fuchs. He adapted their new "theology of speaking" (i.e., the New Hermeneutic) to a new "theology of preaching," what would be known in North America as the New Homiletic": "New Hermeneutic," 19. One of the New Homiletic's major emphases was the participatory eventfulness of preaching. See Fred Craddock, *As One without Authority* (Nashville: Abingdon, 1979).

effective in paying attention to the borders and limits of experience and to the power differentials in the room. Inductive preaching has been accused of being as naively universal in its assumptions as its modernist predecessors and just as presumptuous in its centering the preacher's experience.[23]

Similar criticisms have been made of Ingold's "meshwork" model, even by those who appreciate the approach overall. Heather Giddens notes that meshwork thinking does not "in itself . . . identify socially recognized labels of identity or group affiliations. Also these relational models do not engage with relations of power or domination, . . . [or] social hierarchies."[24] While the approach describes processes of discovery and relationality, it makes few distinctions about the identities of those involved in the relation. And sometimes the question of *who one is* matters very much. Several years ago, when Ingold sat down for an interview with anthropological scholars in Brazil, he was asked about the difficulties of Indigenous anthropologists who were trying to study their own contexts with Western academic tools. His response downplayed the uniqueness of this situation and stressed the universality of scholarship's common life. "We are all indigenous people of the planet," he said.[25] It is a point that is technically correct, but critically misses what is at stake in his interlocutor's question: the sovereignty of Indigenous people to name the borders of their own knowledges and their own places for themselves.

This lack of attention to borders and sovereignty is not only seen in the expansions of Western power operationalized through globalized political economies. It is also seen in the history of the West's environmental movement. In their provocative essay "Land Education: Indigenous, Post-colonial, and Decolonizing Perspectives on Place and Environmental Education Research," Eve Tuck, Marcia McKenzie, and Kate McCoy note the ways that certain North American conversations about place serve

23 David Lose's "Whither Hence, New Homiletic?" (archived paper of the Academy of Homiletics' 35th Meeting, Perkins School of Theology, Southern Methodist University, Dallas, December 1, 2000) describes how the questions posed by New Homiletic sermons grow out of the preacher's experience and assume a hegemonic universality. Charles L. Campbell critiques the individualistic, experiential nature of Craddock specifically in *Preaching Jesus: The New Directions for Homiletics in Hans Frei's Postliberal Theology* (Eugene, Ore.: Wipf and Stock, 1997), 133–35.

24 Heather L. Giddens, "Neolithic Meshworks: A Multi-scalar Approach to Understanding Social Relations within the LBK" (PhD diss., School of History, Archaeology and Religion, Cardiff University, 2015).

25 Tim Ingold et al., "Wayfaring Thoughts: Life, Movement and Anthropology," *Ponto Urbe* 11 (2012): 1–15, https://doi.org/10.4000/pontourbe.341.

settler-colonial erasures of Native particularity, problematically "adopting a 'national goal' of 'becoming native . . . to the continent.'"[26] Elsewhere, Tuck has argued that this sort of "settler nativism" is a "move to innocence" that forestalls difficult conversations about local sovereignty, repatriation of land, and Indigenous history.[27] Fiji's colonial legacy is different from that of the United States. The British crown combined settler-colonial approaches to land with exploitative colonial models in complicated ways. But in the tendency of environmental advocates to assume Native authority in decisions affecting a common global home, settler nativism finds contemporary expression. Cresantia Frances Koya-Vaka'uta describes the impact of these entanglements in her poem "The Line and the Cord: A Response to Blood in the Kava Bowl." "We listen to reports that we are sinking or growing or something," she writes, "and the bottom line always is that the outsider knows what is best for us."[28]

One might posit that this is the opposite of Ingold's intention in his description of "wayfaring" as an anthropological process—but in its lack of attention to borders, identities, and power, it obscures the very conversations about place that have led to the crisis in which we find ourselves. It is not that there is no way forward, but the way forward must pay attention to the complexities of history—what Taniela Balenaikorodawa describes as "mud."[29] When Balenaikorodawa conceives a practical theological method grounded in Fijian understandings of place, he also foregrounds relationality and co-creation in the practical theological process. His model of *sosolevaki* (communal work) draws on an image of entangled lines as well—in this case, the communal work of the

26 Tuck, McKenzie, and McCoy, "Land Education," 16–17. Wes Jackson, for example, uses this "native" language in *Becoming Native to This Place* (New York: Counterpoint, 1996), 3. As an alternative, see Robin Kimmerer's more nuanced description of settlers who become "naturalized" to new places: *Braiding Sweetgrass: Indigenous Women, Scientific Knowledge, and the Teachings of Plants* (Minneapolis: Milkweed Editions, 2015), 214.

27 Eve Tuck and K. Wayne Yang. "Decolonization Is Not a Metaphor," *Decolonization: Indigeneity, Education and Society* 1, no. 1 (2012): 1–40 [10].

28 Cresantia Frances Koya-Vaka'uta, "The Line and the Cord: A Response to Blood in the Kava Bowl," in *Relational Hermeneutics: Decolonising the Mindset and the Pacific Itulagi*, ed. Upolu Lumā Vaai and Aisake Casimira (Suva: University of the South Pacific Press, 2017), 115–17 [117].

29 Taniela Balenaikorodawa, "Sosolevaki: A Collaborative Learning Model for the Methodist Church in Fiji, to Enable Transformational and Sustainable Christian Education among the Laity" (PhD diss., Pacific Theological College, Suva, forthcoming 2025).

weaving of a pandanas leaf mat. The similarities between Balenaikorodawa's and Ingold's models are striking, but the differences are as important. For Balenaikorodawa, mats are woven on the ground—touching the soil of local communities. They point toward the value of communal commitment, but they also create sites where conflict is mediated, power is negotiated, and the excluded are welcomed into conversation. In short, they are places where the identities of the weavers are not eclipsed, but revealed. In the mud of communal learning, those identities might even be interrogated and transformed.

Davuilevu Theological College was founded by Australian and British missionaries in 1907.[30] It was first a school for lay preachers, where Indigenous Fijians were taught by ordained foreigners—white men, with the paternalism characteristic of colonial missionary endeavors.[31] There were white Christians who recognized the strength of the Indigenous faith and fought for the full rights and inclusion of Fijian leaders.[32] But

30 The origins of the theological college are somewhat older. It had been operating in the Rewa Delta on land at Navuloa since 1873, after beginning in Kadavu in 1861; see Raeburn Lange, *Indigenous Ministers: Indigenous Leadership in Nineteenth Century Pacific Islands Christianity* (Christchurch: Pandanus Books, 2005), 143–45.

31 Although the Wesleyan Mission had more than 100 "native agents" by 1856 (8 "native assistant missionaries" [the designation equates "native" with a subordinate role] and 107 teachers), it was not until 1865 that "Joeli Bulu . . . participated, along with six white missionaries, in the ordination ceremony for two new native ministers" (Lange, *Indigenous Ministers*, 145). James Calvert predicts that the "time is very far distant—if indeed it should ever come—when [native ministers] . . . will be able to labour effectively without the direction and oversight of the missionary" (*Fiji and the Fijians: Mission History* [London: Alexander Heylin, 1858], 61). See also Andrew Thornley, "'Through a Glass Darkly': Ownership of Fijian Methodism, 1850–80," in *Vision and Reality in Pacific Religion*, ed. Phyllis Herda, Michael Reilly, and David Hilliard (Christchurch: Pandanus Books, 2005), 132–50. Jane Samson succinctly summarizes the situation: "Once again, a denomination originally born in dissent had become intimately linked with authority, in this case racial authority" ("Fijian and Tongan Methodism," in *The Oxford History of Protestant Dissenting Traditions*, vol. 4, *The Twentieth Century Traditions in a Global Context*, ed. Jehu Hanciles [Oxford: Oxford University Press, 2019], 409–32 [416]).

32 For example, nineteenth-century missionary John Hunt is remembered for, among other things, his careful translation and ministry collaboration with Noa Koroinavugona; see Andrew Thornley, *The Inheritance of Hope: John Hunt, Apostle to Fiji* (Suva: Institute of Pacific Studies, 2000). In the twentieth century, Allan Tippett helped draft the 1946 church constitution that paved the way for Indigenous ecclesial autonomy in 1964, supporting the Fijian majority over against the Board and European Synod. For a summary of Tippett's work, see Darrell Whiteman, "The Legacy of Allan R. Tippett," *International Bulletin of Missionary Research* 16, no. 4 (1992): 163–66.

more common were assumptions of Indigenous intellectual, spiritual, and administrative inadequacy and a thinly veiled racism.

Much had changed by the time my family and I arrived. The church had been autonomous from the Australian Methodist Church for fifty years, and my spouse and I were the only Western faculty members. But colonial legacies are slippery and strong. As in nearly all Fijian schools, English was still the language of instruction. The British had brought thousands of Indian indentured laborers to work their sugarcane plantations in the nineteenth century, and the descendants of these laborers now made up nearly 40 percent of the population. These Indo-Fijians spoke Hindi at home, while Indigenous Fijians spoke *iTaukei*. Davuilevu also trained a handful of Banaban and Rotuman students, who had distinct languages of their own. To avoid partiality, all students spoke in a language other than their mother tongues—as did every faculty person but myself. English was the language of instruction for other reasons, as well. Few theological resources exist in *iTaukei*, Rotuman, or I-Kiribati (a common Banaban dialect), and Hindi resources are rarely written with Oceania in mind. More to the point, doctoral programs require English facility, be they in Australia, New Zealand, the United Kingdom, or the United States. If students expect to thrive in these settings, years of training in academic English are required, alongside the study of biblical languages.[33] Fiji's system of theological education was shaped by these global realities—as were its four academic divisions, mirroring the guilds of the European university system that shaped the training of nineteenth-century missionaries. There were times when I wondered if my very presence on that *mana*-saturated ground drew the lines of Fijian pastors even further into the gravitational pull of Western conversations and categories.

It is not that Ingold is wrong about the relationality and eventfulness at the heart of knowledge creation. But to wrestle with the depth of the climate crisis, a more complex assessment of place is necessary. Such an assessment would give attention to the knots and shards of history that create particular agencies and vulnerabilities. It would also acknowledge that the entanglements of knowledge production are not necessarily benign. The lines of lived experience do not extend freely in theoretical space; they drag the painful flotsam of colonial history behind them. Those pollutants cannot be ignored in present-day climate conversations. The British ships

[33] A regular joke among Davuilevu faculty was the description of English as "the other sacred tongue."

that shadow Fiji's past and the foreign tankers that presently dock in Suva's harbor make ambivalent any conversation about "discovery." It matters *who* is doing the "wayfaring."

Jeff Malpas, "Romantic Materialism," and the Embeddedness of Experience

Jeff Malpas begins his article "Is There an Ethics of Place?" with a problem. Philosophical projects that emphasize place's significance for human life have fueled reactionary politics that exclude outsiders and spark violence. Commitments to place have exacerbated tensions in the Middle East and ethnonationalist sentiment in Japan. They were used to rationalize the ugliest of Nazism's "blood and soil" horrors.[34] But denials of place's significance have been similarly dangerous. Malpas notes the removal of Indigenous communities from their lands and the contemporary environmental crisis as just two examples of how anemic philosophies of place threaten human communities. How, Malpas wonders, does one hold onto the indispensability of place in conversations about communal regeneration, repatriation of land, and environmental sustainability while mitigating its historic dangers?

This is not a disinterested question. Malpas wants to set the record straight regarding the work of his life. Place has been a central theme in his writing for more than two decades, recentering place in the philosophical conversation and the broader academy through accessible, carefully argued texts. Over those years, there have also been recurring critiques of the political lacunae in his approach. At times, this critique has taken the form of respectful invitations to add a political-ethical dimension to his thinking on place and subjectivity.[35] At others, it has been expressed through scholarly impatience with Malpas' heavy reliance on Martin Heidegger and his bracketing of Heidegger's Nazi past from his celebration of Heideggerian approaches to "dwelling" and "home."[36] To

[34] Jeff Malpas, "Is There an Ethics of Place?" *Localities* 2 (2012): 7–31. Malpas notes that sites like Kosovo, Palestine, and even the Korean peninsula suggest that ideas of belonging to a place foster "an ethics and politics of violence and exclusion"; he lifts up the twentieth-century Nazism, alongside the Japanese nationalist thought of Tetsuro Watsuji, as further examples (9–12).

[35] John Cameron, "Some Implications of Malpas's *Place and Experience* for Place Ethics and Education," *Newsletter of the Society for Phenomenology and Architecture* 15, no. 2 (2004): 5–9.

[36] David Clarke, "Heidegger, Hermeneutics and History: Undermining Jeff Malpas's Philosophy of Place," *Philosophia* 42, no. 3 (2014): 571–91.

his credit, Malpas' recent companion volume to the influential *Place and Experience* takes these concerns seriously and deals with the ethical and normative implications of his work head-on.[37]

But even in his recent foregrounding of ethics, Malpas is clear that his politics of place proceed from the "topographical" relation between place and identity that informs his prior projects, a topography that shapes human life prior to sociocultural norms. "Place is not itself political," he states, differentiating his project from many other contemporary place-based discussions that begin with the political consequences of land, migration, and colonialism.[38] Place is "the very frame within which the political itself must be located."[39] Significantly, the very discussions he references suggest a false choice in Malpas' thinking about politics and place. The question is not, Is paying attention to place ethically dangerous or ethically indispensable? Few contemporary scholars would argue that place can be ignored as a prism for understanding ethics; it is not that Malpas wants to talk about place and others do not. Malpas wants to talk about place in a particular way, using a particular method and asking particular questions. Drawing heavily on English romantic poets and German philosophy, there is a "romantic materialism"[40] to his description of *what* place is and *how* it situates the human person. More importantly, these European thinkers heavily influence *why* place interests him in the first place. Malpas' key questions are about the agency of the self and its ability to know the places that shape it, in spite of place's complexity and the self's embeddedness in the world. In other words, if place shapes one's experience from the ground up, is one free to act, choose, and create? Similarly, if place can only be experienced through one's subjective point of view, does place exist as anything other than a projection of oneself?

Malpas uses the metaphor of topographical mapmaking to describe the ways that places are experienced and open themselves to individual

[37] Jeff Malpas, *The Ethics of Place* (London: Routledge, forthcoming 2025).

[38] Jeff Malpas, *Place and Experience: A Philosophical Topography*, 2nd ed. (New York: Routledge, 2018), 12.

[39] Malpas, *Place and Experience*, 215.

[40] Ingo Farin, "Place and World: On Jeff Malpas's Second Edition of *Place and Experience*," *International Journal of Philosophical Studies* 28, no. 2 (2020): 254–66 [255]. Malpas defines "romantic materialism" as "the idea that what is felt, thought, and remembered (the 'romantic' if you will) is given only in and through the placed materiality of the world—in earth and stone, in air and water, in flesh and bone, in light and sound—while at the same time that materiality is taken up into and imbued with the romantic" (*Place and Experience*, 192).

lives. A person cannot know a place without traversing it, mapping the ways that objects relate to the self and other objects as one moves through space and time. Malpas uses this metaphor to break down the dichotomy between subject and object in conceptions of place, noting how the objectivity of place is revealed through our subjective relationship to it, and also how our subjectivity finds expression within place's bounds. There is an "interdependence" between the two.[41] It is not just that we are free in spite of place, able "to think, to feel, to grasp, [and] to act"; we are free *because* of place.[42] For economies that have treated places as inert commodities, this is a wake-up call. Malpas argues that the very things that Western cultures have prioritized most—the freedom to choose, the power to act, and the ability to discover—are "inextricably bound to place,"[43] both made possible and redefined by the same.

It is worth pausing for a moment to consider how differently this inquiry might have looked if it had been shaped by different questions and different experiences. In the introduction to *Place and Experience*, Malpas quotes Māori leader and scholar Te Rangi Hīroa to show the profound importance of place in Indigenous communities. Rangi Hīroa states, "Captives in distant lands have begged for a pebble, a bunch of leaves, or a handful of earth from the homeland that they might weep over a symbol of home, . . . it is the ceaseless crooning of the waves against a lone shore that perpetuates the sound of voices that are still."[44] Place is clearly central in Rangi Hīroa's understanding of himself. But as his words suggest, the problem of place is not finally about its relationship to the self's freedom, power, and discovery. It is about the self's ability to belong in the world when a particular place has been lost.

The loss of place is not a major conversation topic in Malpas' work, in part because of his complicated relationship to place's particularity. On the one hand, place can only be known in the particular. There is no general theory of place that can substitute for a lived engagement with place's opening of itself through space and time (i.e., through "topographical" transversal). On the other hand, Malpas resists normative hierarchies of place, that is, particular places that matter more than others for the shaping of human identity. He is interested in a general theory of how

41 Farin, "Place and World," 259.

42 Malpas, *Place and Experience*, 13.

43 Malpas, *Place and Experience*, 12.

44 Te Rangi Hīroa, *The Coming of the Maori* (Wellington: Maori Purposes Fund Board, 1949), 381, quoted in Malpas, *Place and Experience*, 4–5.

particular places shape human life, be they "derelict houses . . . the mining site . . . [or a] national park."[45] What this frame of reference excludes are communal histories that tie people and places together. "Topographical" mapping will not tell a person about the umbilical cords buried in the ground—or the buried bones. Neither will it reveal the far-off political decisions that have impacted the cleanliness of the water or how particular places haunt those who have been driven from their homelands. While movement *within* a particular place is key to Malpas' thought, movement *between* places is underdeveloped. If Ingold stresses place's global interconnections, Malpas' topographical model tends toward more insular metaphors and descriptions.[46] It is no accident that Malpas repeatedly returns to novels and poetry that explore the inner workings of the self in relation to place, rather than the well-being of communities.[47] For critics like Ingo Farin, this leads to one of *Place and Experience*'s greatest shortcomings: its lack of attention to immigration policy, refugees, and asylum seekers.[48] For Farin, these are not simply "political" addendums. They are core to understanding place at large. For if place is meant to be "humanized and humanizing," a phrase from Seamus Heaney that is particularly meaningful for Malpas,[49] it must "certainly be characterized by hospitality."[50] For Farin, overlooking the "welcome of the foreigner coming from outside" as part of *what* place is and *how* and *why* it matters opens Malpas' project to the very "topographical foundationalism"[51] that he is intent to

45 Malpas, *Place and Experience*, 201. Malpas does speak of place's "fragility," but describes its relation to individual memory more than political community. Place's "fragility" is also disconnected from conversations about hospitality toward others who have been displaced: "The Remembrance of Place," in *Exploring the Work of Edward S. Casey: Giving Voice to Place, Memory, and Imagination*, ed. Azucena Cruz-Pierre and Donald Landes (New York: Bloomsbury Academic, 2013), 63–72.

46 It is telling how differently each responds to the work of Deleuze and Guattari, *Thousand Plateaus*. Ingold borrows and builds on Deleuze and Guattari's concept of "lines of becoming" in creating his meshwork model (*Thousand Plateaus*, 224–25). Malpas intentionally distances himself from their thinking on "nomadism" as overly "strategic" and "political" (*Place and Experience*, 8).

47 Malpas' repeated reflection on Marcel Proust's *In Search of Lost Time* or William Wordworth's "Michael: A Pastoral Poem" shows his prioritization of the individual in theorizing place, rather than a community (*Place and Experience*, 1–15).

48 Farin, "Place and World," 261.

49 Seamus Heaney, "The Sense of Place," in *Preoccupations: Selected Prose, 1968–1978* (London: Faber & Faber, 1984), 145.

50 Farin, "Place and World," 261.

51 Farin, "Place and World," 260.

avoid. In his postponing of the ethical question, Malpas allows places to become the ground of identity in ways that can lead to the dangers that worry him most. Farin summarizes his critique this way: a philosophy of welcome "is essential to place, for one cannot inhabit a place as one's own without acknowledging its interconnections with other places."[52] The question is not only *how* a human belongs to a place, a question central to Malpas' work. The question is also *who* belongs to a place.

Indigenous *iTaukei* have been grappling with this question through their relationships to their Indo-Fijian neighbors since the British first brought *girmitiya* indentured laborers to Fijian shores to work the colonial sugar plantations in the late nineteenth century. The British promoted distrust between these people groups, going so far as to restrict Indo-Fijian visits to *iTaukei* villages, ostensibly to protect the cultural integrity of the native population.[53] Such essentialisms of ethnicity and culture served colonial interests, corresponding with white-supremacist obsessions with difference, genealogy, and racial purity. Practically speaking, it was easier to control a divided population. Post-independence, as the Indo-Fijian population increased, Indigenous fears about loss of land led to a series of ethnonationalist coups. The *iTaukei* Methodist military colonel Sitiveni Rabuka used the language of the biblical exodus and exile to justify his takeover of a newly elected, multiethnic Fiji Labour Party coalition, promising to free the land from foreign influence.[54] Critically, Rabuka was not describing freedom from the British Commonwealth with this rhetoric; he was describing freedom from the political influence of Hindu and Muslim immigrants who had been living and working Fijian land for nearly a hundred years.[55]

[52] Farin, "Place and World," 260.

[53] Sashi Kiran, "For Peace Which Is True," *Fiji Times*, September 29, 2016, http://www.fijitimes.com/story.aspx?id=372628. See also Martha Kaplan's description of how Britain used Indo-Fijian claims for political rights as a "counterpoint . . . to unite [*iTaukei*] Fijians to colonial projects," in "Promised Lands: From Colonial Lawgiving to Postcolonial Takeovers in Fiji," in *Law and Empire in the Pacific: Fiji and Hawai'i*, ed. Sally Engle Merry and Donald Brenneis (Santa Fe, N.Mex.: School of American Research Press, 2003), 153–86 [162].

[54] Lynda Newland traces the cultural (and missional) lineage of this rhetoric in "The Lost Tribes of Israel—and the Genesis of Christianity in Fiji: Missionary Notions of Fijian Origin from 1835 to Cession and Beyond," *Oceania* 85, no. 3 (2015): 256–70.

[55] Rabuka's political trajectory continues to shift. Replacing Voreqe Bainimarama as prime minister in the 2022 election, Rabuka was part of a multiethnic coalition, publicly apologizing for ethnonational sentiments expressed during the 1987 coup. The implications of that apology are still unfolding. I will discuss the Fijian Methodist Church's complicated relationship to Rabuka in chapter 4. Natasha Frost, "In Fiji's

When one reads Ilaitia S. Tuwere's *Vanua: A Fijian Theology of Place*, published in 2001, one sees Malpas' ethical quandary about place engaged twenty years prior to his framing of the question.[56] The problem for both is similar: how does one honor the co-constitutive nature of place and identity while at the same time making room for "the other"? But there are key differences. Tuwere's approach to the place of Fijian land and culture (*vanua*)[57] integrates the theological, contextual, and practical implications of place for his community because the *sautu* (peace, fertility) of his people is at stake. His approach to place is not only about individual belonging. It addresses the ethical decisions that ground communal life, deconstructing the false dichotomy between colonial essentialism and cultural extinction. Tuwere's work argues for a hospitality of place that does not end in dissolution of community and culture.

When we arrived in Fiji in 2014, the military dictator Voreqe Bainimarama was in power. Bainimarama had taken control of the government through yet another coup in 2006, and because of the Methodist Church's prior political influence, it was seen as a threat. The Methodist Annual Conference was suspended, as was its national, annual choir festival. Methodist Church leaders were jailed, and the government monitored what was preached in Methodist pulpits.[58] Following these tumultuous years, sociologist Lynda Newland describes a shift in the language and metaphors of the Methodist Church leadership from that of land to sea.[59] The significance and character of the *vanua* was increasingly defined through images of movement and migration—specifically, through the image of the *drua*. The *drua* is an Indigenous ocean-crossing vessel that played a significant migratory role in Pacific history and was guided by traditional

---

Election, One Former Coup Leader Replaces Another," *New York Times*, December 20, 2022, https://www.nytimes.com/2022/12/20/world/asia/fiji-election.html.

56 Malpas' certainty that a "place-based ethics and politics is not currently to be found in any adequate or developed form in the existing literature" rings somewhat hollow given the extent of Fiji's engagement with this theme theologically and politically ("Is There an Ethics of Place?" 24).

57 *Vanua* is a culturally rich description of not only Fijian land but all that Fijian land sustains, including communities, flora, fauna, and traditions. It does not recognize a nature/culture dichotomy. "Place" is an English approximation, but only if place is understood as an active agent in a living, communal relationship.

58 Sam Sekek, "Police Again Detain Fiji Church Leaders," ABC News, July 22, 2009, http://www.abc.net.au/news/2009-07-22/police-again-detain-fiji-church-leaders/1363198.

59 Lynda Newland, "From the Land to the Sea: Christianity, Community and State in Fiji—and the 2014 Elections," in *The People Have Spoken: The 2014 Elections in Fiji*, ed. Steven Ratuva and Stephanie Lawson (Canberra: ANU Press, 2016), 109–34.

navigational practices. It was large enough to carry an entire community to new places. Newland describes this shift as holding together *iTaukei* particularity with an openness to a multicultural future,[60] honoring rootedness while charting a course through change. This shift in language was part of the Methodist Church's "New Exodus," and while Newland describes its political and social impact, she misses its sermonic, biblical context. The biblical and ecclesial significance of Fijian Methodism's "New Exodus" and its *drua* is the subject of chapter 4.

This ability to honor the distinctiveness of place while extending hospitality to the stranger is, of course, more than a Fijian concern. It is also the concern of any preacher who has struggled with the exclusionary history that marks the place of the pulpit.[61] During the 2020–21 academic year, Duke University funded a research project that examined sermons that had been preached at Duke University Chapel either on the subject of immigration or preached by first-generation immigrants themselves.[62] In the chapel's digital repository of sermons recorded between the years of 1946–2002, researchers struggled to find immigrant voices represented. They named all sorts of ways that the place of the chapel shaped the rhetoric of its preachers. Like Malpas, they recognized how preacherly witness was "interdependent" with ecclesial space and time, finding its identity in relation to the chapel's stone and stained-glass horizons. The acoustics, aesthetics, memories, and biblical images that surrounded the pulpit impacted the choices that preachers made and the subjects to which they gave voice. Certainly, the witnesses of preachers shaped the place of the chapel as well, expanding congregational expectations as to how sermons might creatively engage a biblical text or inviting vocal responses from the

[60] Newland, "From the Land to the Sea," 125. Maebh Long agrees, noting that—even if the political implications are still in flux—traditional descriptions of *vanua* hold a potential for "fluidity belied by its groundedness," in "Vanua in the Anthropocene: Relationality and Sea Level Rise in Fiji," *symploke* 26, nos. 1–2 (2018): 51–70 [64].

[61] John McClure's *Otherwise Preaching: A Postmodern Ethic for Homiletics* (St. Louis: Chalice Press, 2001) addresses multiple facets of this exclusionary history, constructing his other-centered proposal for preaching around Emmanuel Levinas' critique of Heidegger.

[62] Peace Pyunghwa Lee and Miriam Cho, project managers, "When I Was a Stranger: Immigration, Preaching, and Religious Imagination (2020–2021)," Bass Connections program at Duke University, Summer 2020 through Spring 2021, https://bassconnections.duke.edu/project-teams/when-i-was-stranger-immigration-preaching-and-religious-imagination-2020-2021.

pews. But a larger question was framed by students like Crystal Chiu.[63] Did the academically rigorous, liturgically conservative place of that high-steeple pulpit overdetermine *who belonged* in the space? Had a desire to preserve the pulpit's distinctive character created a lack of hospitality to other narratives, memories, and persons, preserving a stunted, regulated place, rather than a place of welcome?[64]

This ability to remain rooted in place, even as one remains hospitable to the foreigner, is a key challenge for present-day responses to the climate crisis. Rev. James Bhagwan, general secretary of the Pacific Council of Churches, points out that Fiji's courage in facing this issue in relation to Indo-Fijians is just the first of many courageous hospitalities that will be required. Climate change will require the relocation of entire communities. Up until this point, those relocating communities have been able to move to higher ground within their own territories. But the day is coming when a more "radical hospitality" will be needed.[65] Communities will need invitations to live on land that is not their own. In spite of Fiji's historic generosity to climate refugees,[66] Bhagwan is honest about the conflict that land reallocation will require. "We haven't gotten there yet," he states. When place is interpreted through the lens of migration and loss rather than individual agency, who will decide who belongs?

Keith Basso, Place-Making, and the Transmuting of Indigenous Practices

Keith Basso spent decades of his life listening to and living among Western Apache communities in the southwestern United States. His ethnographic study *Wisdom Sits in Places: Landscape and Language among the Western Apache* has had a profound impact on both ethnographic method

63 Crystal Chiu, "Reimagining the Sermon: Stories of Immigration," directed by Miriam Cho, July 2, 2021, https://www.youtube.com/watch?v=isnn2ke_HJY&t=44s.

64 Peace Pyunghwa Lee and David Stark address this question in light of the legacy of Duke Chapel's segregated history in "The Towering Sermon: Duke Chapel as Monument to White Supremacy," in *Unmasking White Preaching: Racial Hegemony, Resistance, and Possibilities in Homiletics*, ed. Lis Valle-Ruiz and Andrew Wymer (London: Lexington Books, 2022), 45–54.

65 Rev. James Bhagwan (general secretary of the Pacific Council of Churches), personal online interview, September 28, 2021.

66 John Campbell describes the 1945 Banaban relocation to the Fijian island of Rabi following Banaba's destruction by colonial phosphate mining; see "Climate Induced Community Relocation in the Pacific: The Meaning and Importance of Land," in *Climate Change and Displacement: Multidisciplinary Perspectives*, ed. Jane McAdam (New York: Bloomsbury, 2010), 57–79.

and the field's attentiveness to Native American relationships to land. Basso's work is grounded, specific, and slow, and it is clear that he wants to honor the voices of his interlocutors at each step. Of particular interest to Basso are the Western Apache approaches to naming places and how these practices differ from settler-colonial approaches. "Place-names" grow out of Apache oral histories, tied to communal engagement with places over time, ancestral memories, and landmarks that connect landscapes with stories.[67] Critically, Basso stresses the ways that places have healing properties for communities, especially when their names are learned and integrated into one's own life. Places are where "wisdom sits."[68]

Near the start of his volume, Basso coins a term that has taken on a life of its own over the last twenty-five years: *place-making*. For Basso, the term refers to "a universal tool of the historical imagination," a kind of "retrospective world-building" that constructs history through stories of the past that "interpret their significance and imagine them anew."[69] In the final chapter of his book, Basso makes clear that this retrospective revisioning has present-tense implications. Through the contemplation of place inspired by and necessary for the "narrative art"[70] of making places, human persons and their communities are changed. There is a "reciprocal relationship" between place and person, "a relationship in which individuals invest themselves in the landscape while incorporating its meanings into their own most fundamental experiences."[71] Basso mediates this reciprocal relationship through human language and agency. "Places may seem to speak," he states. But the voices are "not their own. . . . They are animated by the thoughts and feelings of persons who attend to them."[72] One understands Basso's reticence in giving land its own agency in his analysis; he is, after all, writing for the Western academy, and there is a careful veering away from interpretations of his work that might seem unscientific, unsupported by continental theory, or overly mystical.[73]

Basso notes, however, that his analysis differs from ways the agency of land is described by his Western Apache conversation partners. The

67 Basso, *Wisdom Sits in Places.*

68 Basso, *Wisdom Sits in Places*, 121.

69 Basso, *Wisdom Sits in Places*, 5–6.

70 Basso, *Wisdom Sits in Places*, 33.

71 Basso, *Wisdom Sits in Places*, 102.

72 Basso, *Wisdom Sits in Places*, 108.

73 In his final chapter, for example, when describing the reflexivity of place-making, Basso is careful to frame and authorize his findings through references to Martin Heidegger, Edmund Husserl, and Jean-Paul Sartre (*Wisdom Sits in Places*, 106–8).

words of community elder Annie Peaches are particularly striking. Land is always "stalking people," she states. "The land makes people live right. The land looks after us."[74] *Place-making* seems a strange term for the relationship that Peaches describes. There is a dual agency and reciprocity in her description akin to Tuwere's well-known claim that, in *iTaukei* understanding, one "does not own land, the land owns him [*sic*]."[75] In *iTaukei* contexts, land is "seen to act *upon* people, . . . [it is] an agent which can affect humans."[76] The land, it seems, is also doing the "making"—which means that Basso's term for this Western Apache practice is one-sided in its privileging of human agency and determinative power.

If the term were concretely tied to and defined by Western Apache practices and norms, this might not be a problem, but over the decades, place-making has become a common theme in discussions of architecture,[77] art,[78] environmental science,[79] community development,[80]

74 Basso, *Wisdom Sits in Places*, 38.

75 Tuwere, *Vanua*, 49. In comparison with Basso, Jeff Malpas is bolder in speaking of place as having its own agency, rather than simply reflecting the agency of humans. He echoes Tuwere's sentiment: "To have a sense of place is not to own but to be owned by the places we inhabit" ("Place and Human Being," *Environmental and Architectural Phenomenology Newsletter* 20, no. 3 [2009]: 19–23 [23]).

76 Matt Tomlinson, "Sacred Soil in Kadavu, Fiji," *Oceania* 72, no. 4 (2002): 237–57 [247].

77 Amir Sasan Hadian, "Toward a Placemaking Approach to Building Architecture and Structures," *International Journal of Architectonic, Spatial, and Environmental Design* 13, no. 4 (2019): 41–56; for a critique of "place-making" as a conservative force for tradition and stasis in architecture, see Sean Griffiths, "In the Name of Placemaking, Architects Are Often Complicit in Social Cleansing," *dezeen*, January 25, 2018, https://www.dezeen.com/2018/01/25/sean-griffiths-opinion-place-making-social-cleansing/. Griffiths draws on Ingold's attention to daily practices, rather than "essences," for architectural inspiration.

78 Jennifer Craft, *Placemaking and the Arts: Cultivating the Christian Life* (Downers Grove, Ill.: IVP Academic, 2018).

79 Authors like Stuart Kirsch look at the impacts of ecological degradation on the place-making of local populations; see "Changing Views of Place and Time along the Ok Tedi," in *Mining and Indigenous Lifeworlds in Australia and Papua New Guinea*, ed. Alan Rumsey and James Weiner (Wantage, UK: Sean Kingston, 2004), 182–207. See also Amrita Sen and Harini Nagendra, "The Role of Environmental Placemaking in Shaping Contemporary Environmentalism and Understanding Land Change," *Journal of Land Use Science* 14, nos. 4–6 (2019): 410–24.

80 See, for example, the University of Delaware's community-planning toolbox, highlighting "place-making" as a way to "reinvent public spaces" ("Placemaking," Delaware Complete Communities Planning Toolbox, accessed October 24, 2021, https://www.completecommunitiesde.org/getting-started/).

contemplative spirituality,[81] and theology.[82] These uses are far removed from Basso's original conversations with Western Apache communities. Too often, rather than providing critiques of colonial approaches to land, the language of place-making has been used to justify the gentrifications of settler communities, making places that accommodate the stories and values of those with resources and power. Again, the key question is which communities have agency in "making places" and which do not. If ethnographer Wolfgang Kempf is right, place-making is more than a narrative art. It includes the "historical processes, political projects and social practices that contribute to the ongoing constitution of place, . . . in the context of local, regional, and global entwinements."[83] This means that place-making is not, in and of itself, restorative. Nor is it only the work of Indigenous communities. Empires have been in the business of making places since the earliest civilizations. A focus of Kempf's work, for example, is how Oceanic communities respond when places are made unlivable through the unsustainable development of corporations.[84] In certain ways, the language of place-making mirrors colonialism's larger project of making the world into the conqueror's home.

The connotation would be quite different if Basso had chosen a term like "place-relating" or even "place-covenanting" to name the Apache practices he describes.[85] But such language requires a more radical

[81] Douglas Christie, "Place-Making as Contemplative Practice," *Anglican Theological Review* 91, no. 3 (2009): 347–72; also Christie Purifoy, *Placemaker: Cultivating Places of Comfort, Beauty, and Peace* (Grand Rapids: Zondervan, 2019).

[82] Craig Bartholomew, *Where Mortals Dwell: A Christian View of Place for Today* (Grand Rapids: Baker Academic, 2011). Bartholomew's final chapter deals with place-making practices specifically, intersecting John Inge's discussion of Christian shrines in *Christian Theology of Place.*

[83] Wolfgang Kempf, Toon van Meijl, and Elfriede Hermann, "Movement, Place-Making and Cultural Identification: Multiplicities of Belonging," in *Belonging in Oceania: Movement, Place-Making and Multiple Identifications*, ed. Elfriede Hermann, Wolfgang Kempf, and Toon van Meijl (New York: Berghahn Books, 2014), 1–24 [7]. The authors follow Setha M. Low and Denise Lawrence-Zuniga's "Locating Culture," in *The Anthropology of Space and Place: Locating Culture*, ed. Setha M. Low and Denise Lawrence-Zuniga (Malden, Mass.: Blackwell, 2003), 1–47, in their argument.

[84] Wolfgang Kempf and Elfriede Hermann, "Reconfigurations of Place and Ethnicity: Positionings, Performances, and Politics of Relocated Banabans in Fiji," *Oceania* 75, no. 4 (2005): 368–86.

[85] At one point, Basso does draw on the covenantal language of marriage to describe the ways that human consciousness and place are "wedded" to each other; see *Wisdom Sits in Places*, 107.

decentering of the human than he thought to embrace. It is a step that has proved similarly difficult for Western environmental science. In 2005, Australian anthropologist Michael Jackson gave a public lecture to a group of North American theologians and scientists working on ethical responses to climate catastrophe. He suggested that, given the severity and complexity of the climate crisis, the action of Western science could not save the planet any more than Indigenous practices of lighting torches could stop a thunderstorm. Both sets of practices were significant in how they transformed the practitioner, holding the potential to serve the purposes of hope, communal connection, and a restorative relationality with the natural world. But neither could predictably control the weather.[86] The response to his paper, from a scientist speaking with theological cadence, was withering. Jackson, it was argued, must not believe in "sin" or "wisdom" or the possibility of "good work."[87] Note the connection here between "good work" and control—or at least, efficacy. Fifteen years later, Jackson's paper seems much less radical. It sounds more like Norman Wirzba's honest redefinition of hope in light of humanity's growing realization of its powerlessness to stop the disasters on the way. It is "not that the future doesn't matter," Wirzba states, "or that we should dismiss scientific forecasts about rising sea levels or cataclysmic weather events. Rather, the cultivation of hope depends on . . . practices that position people to live *now* in ways that affirm, nurture, and heal life, ways that draw people more sympathetically into relationships with others and their shared places."[88] Wirzba summarizes his point through the words of Wendell Berry: "Hope lives in the means, not the ends."[89]

86 Michael Jackson, "A Walk on the Wild Side: The Idea of Nature Revisited," in *Diversity and Dominion: Dialogues in Ecology, Ethics, and Theology*, ed. Kyle Van Houtan and Michael Northcott (Eugene, Ore.: Cascade, 2010), 46–63. Jackson's original remarks were part of a jointly sponsored lecture series by the Nicholas School of the Environment and Earth Sciences and Duke Divinity School.

87 Kyle Van Houtan, "Response: Good Work," in Van Houtan and Northcott, *Diversity and Dominion*, 64–68.

88 Norman Wirzba's original lecture "Can the Land Forgive? Searching for Hope in a Wounded World" is summarized in his presentation "The Song of the Meadow Lark: A Conversation with Norman Wirzba," by Mari Joerstad, *Facing the Anthropocene* webinar series, Kenan Institute for Ethics, April 15, 2021, https://kenan.ethics.duke.edu/the-song-of-the-meadow-lark/.

89 Wendell Berry, "Discipline and Hope," in *A Continuous Harmony: Essays Cultural and Agricultural* (Berkeley, Calif.: Counterpoint, 2021), 83–164.

In the same year that Jackson's essay was published, Tongan scholar Jione Havea made a similar observation about the dangers of a "savior-complex" in the political agendas of the climate-change movement. It's not that "saving" the land is unimportant; Havea sees the losses and griefs that are on the horizon. But given the very likely fact that it is already too late to save the island nations being washed away, Havea wonders if the politics of climate change shift "attention away from the disadvantaged and victims themselves, from people and their cultures, toward the assumption (or illusion) that their land could be saved" by the power brokers who know how to make places whole again.[90] The interests of these power brokers too often overshadow the interests and voices of those whose lands are actually being destroyed. Over a decade later, Samoan-born theologian Faafetai Aiava echoes Havea's concern in his paper "Talking God in a Divided House: Rebuilding Spirituality in God's Pacific Household." In trusting the "hands of humans" to fix the earth, an "all-too-familiar savior narrative" is redeployed. In Aiava's view, the world is trying to "solve a problem with the same level of consciousness that created it."[91]

What is being advocated by these Oceanic theologians is not passivity, but relationality. They recognize that one's ability to "make" a world—through cultural grammars,[92] science, or any other technology of control—is both limited and made possible by the agency of others and by the world itself. Preaching has had to face its own limits and shortcomings in this regard. The power of preaching to "evoke a world,"[93] be it the

[90] Jione Havea, "The Politics of Climate Change: A *Talanoa* from Oceania," *International Journal of Public Theology* 4 (2010): 345–55 [354].

[91] Faafetai Aiava, repurposing a quote from Albert Einstein, in "Talking God in a Divided House: Rebuilding Spirituality in God's Pacific Household," paper presented at Iliff School of Theology, Denver, Colo., October 22, 2021.

[92] Kyle Whyte argues that "settler grammars" disrupt Indigenous "resilience, adaptive capacity, and sustainability" ("Settler Colonialism, Ecology, and Environmental Injustice," *Environment and Society: Advances in Research* 9, no. 1 [2018]: 125–44 [140]). The term "settler grammars" is used by Mishuana Goeman to describe the sedimentation of language and thought patterns which undermine Indigenous sovereignty; see "Disrupting a Settler-Colonial Grammar of Place: The Visual Memoir of Hulleah Tsinhnahjinnie," in *Theorizing Native Studies*, ed. Audra Simpson and Andrea Smith (Durham, N.C.: Duke University Press, 2014), 235–65.

[93] Sally A. Brown uses this language, which draws on Paul Ricoeur's description of the "world opened by the referential claims of the text," in *Interpretation Theory: Discourse and the Surplus of Meaning* (Fort Worth: Texas Christian University Press, 1976), 37. Critically, Brown nuances its use. She argues that while "all biblical narratives evoke a world of assumptions and possibilities," it is the preacher's job to not only allow the text to "draw us into [its] world . . . [but] return us to our own," in

world of the biblical text or some idealized eschatological future, can do damage. One danger is seen in sermons that silence the laments and apocalypses of marginalized voices in the present. In this respect, the place of Duke Chapel is again revealing. On September 22, 1963, the Sunday after the 16th Street Baptist Church bombing in Birmingham, the sermon preached from the chapel pulpit makes no mention of the deaths of the four girls that lost their lives in that act of terrorism. Elizabeth Grantland was part of a research team examining how the preaching of Duke Chapel engaged key moments in the Civil Rights Movement, and she notes how often the place of Duke's pulpit seemed far removed from what was happening outside the sanctuary.[94] Relating the place of the text to the place of the congregation is fraught with similar negotiations of sovereignty, belonging, and power as are described in this chapter, and Grantland's observation presses the question: *Whose community* is centered in the places a sermon makes and in the world the text evokes?

But there is another danger, as well. Ignoring the relational limits of place-making can also give preachers—in their role as narrative place-makers—a false sense of control.[95] Charles L. Campbell describes how narrative sermonic structures press preachers to proclaim a "cheap hope" in relation to the climate crisis, resolving the unresolvable in tidy twenty-minute arcs.[96] This approach is not only untenable in its pretense; it is unethical in its proclaiming of peace where "there is no peace" (Jer 6:14). Climate change reminds preachers that the pulpit is not only fraught with power; it is fraught with powerlessness. Language of "making" or "evoking" aside, the displacements of environmental, colonial,

---

Sunday's Sermon for Monday's World: Preaching to Shape Daring Witness (Grand Rapids: Eerdmans, 2020), 142. For Brown, the world "made" by the biblical text (and faithful preaching) exists in vulnerable relation to the world of the listener—and the world outside the sanctuary walls.

94 Elizabeth Grantland shares her reflections in "The Sunday After," an unpublished reflection on "Preaching and Protest: Sermons from Duke Chapel during the Civil Rights Era," a project funded by the John Hope Franklin Humanities Institute at Duke University, 2018; Luke Powery, dean of Duke University Chapel, sponsored the project, and Peace Pyunghwa Lee served as graduate mentor (https://www.youtube.com/watch?v=kQYA5Oxflus).

95 In contrast to this view, Jennifer Allen Craft describes how certain crafting traditions approach processes of "making" as a releasing of artistic control rather than its reification; see *Placemaking and the Arts*. But preaching's monological form and emphasis on persuasive power regularly overshadow this insight.

96 Charles L. Campbell, *The Scandal of the Gospel: Preaching and the Grotesque* (Louisville: Westminster John Knox, 2021), 68.

white supremacist violence are more than linguistic experiences. They are losses. When one is standing in the burnt rubble of a bombed sanctuary, or when one sees a community's coral reefs bleached like bone,[97] words will not turn back time. Place-making has its limits.

When Fijian biblical scholar and preacher Tui Nuku Smith describes the work of biblical interpretation for his local community, he describes it through a very particular role in a traditional Fijian village: the *daunivucu.*[98] A Fijian *daunivucu* is responsible for teaching a particular *meke* (dance) to a particular village at a particular time. Spiritual discernment, as well as artistic skill, is required. In pre-Christian Fiji, a *daunivucu* was more like a prophet than a historian or artist, using trance-like states to communicate with the spirit world. After learning from the spirits of the *vanua* (Fijian land and culture) the communal dances meant for a particular community, the *daunivucu* would teach the people new dances of old stories that would shape the village's *vanua* practices into the future. Critically, the community owned those dances. They could not be performed in communities other than those to whom the dances had been given. In comparing this role to the role of biblical interpreter—and therefore, to preacher—Smith lingers on both the transcendent revelation assumed in interpretive work and interpretation's communal grounding.[99] For Smith, biblical interpretation is the spiritually vulnerable work of receiving new dances that tell old stories for the thriving of local communities. In that work, the *daunivucu* is both an actor and one who is acted upon. The community does not simply make the *vanua.* The community is also made by the *vanua.* More than this, the *vanua*'s practices and narratives are revealed (and at times changed) through the *daunivucu*'s transcendent encounter.[100]

Smith did not arrive at this interpretive metaphor in a vacuum. It is no accident that the subject of his biblical studies thesis is the significance of the biblical exile for modern-day Fiji, focusing on the witness of the prophet Jeremiah to his exiled community (Jer 29:1–23). In his own reflec-

[97] The connection between bleached coral and Ezek 37's valley of dry bones is fleshed out by Rev. James Bhagwan in his Bible study for the 15th Uniting Church of Australia (UCA) Triennial Assembly, July 2018.

[98] Tui Nuku Smith, "Reading Jeremiah 29:1–23 in Light of the *Daunivucu ni Vanua* in the Fijian Methodist Context" (unpublished bachelor of theology thesis, Davuilevu Theological College, Fiji, 2017).

[99] Smith, "Reading Jeremiah," 58–60.

[100] Smith, "Reading Jeremiah," 7–9.

tion on Fiji's colonial history, its recent ethnonationalist coups, and the Indigenous community's fears of cultural and ecological atrophy, Smith describes the exilic losses of Fiji's present day in relation to Jeremiah's tears (cf. Jer 9:1). In such a season, more than naïve trust in a human capacity for place-making is required. More than even a trust in the land itself is necessary, for it is precisely the land that is threatened. A greater Maker of places is needed—as well as a prophet who is willing to listen and teach the community new dance steps, reenacting old stories on displaced ground.

Climate Justice Preaching and the Place of the Biblical Text

In orienting the reader to contemporary conversations about place in the Western academy and in introducing the broad contours of the Fijian context, I have highlighted the insights of displaced communities regarding place and climate justice. Ingold, Malpas, and Basso all make significant contributions to the West's rethinking of place, critiquing Enlightenment bifurcations of nature and culture in necessary ways. But the questions of sovereignty, belonging, and loss that grow out of even a cursory reading of Fiji's history add complexity and cautionary nuance to each proposal. *iTaukei* practices of place have a tensiveness that hold together relation with difference, continuity with change, and agency with dependence. These tensions grow out of lived experiences of loving land, recognizing that land's claim on one's identity, and seeing that land eroded by the rising tides of globalized power—whether in the nineteenth century or the twenty-first. Climate scientists are increasingly acknowledging the importance of such contributions in reframing the questions they foreground, their definition of the crisis at hand, and their criteria for assessing sustainability.[101] Western theology's "spatial turn" would benefit from a similar attentiveness to displaced wisdom.[102]

This chapter has also illumined the importance of place for the work of preaching, and particularly preaching on climate change. Preaching

[101] Peter Rudiak-Gould notes how broad, culturally based definitions of "climate change" in the Marshall Islands are helping climate scientists think about holistically about sustainability: "Promiscuous Corroboration and Climate Change Translation: A Case Study from the Marshall Islands," *Global Environmental Change* 22 (2012): 46–54.

[102] Mary Emily Briehl Duba's doctoral work is an example of the constructive possibilities of looking at theologies of place through the lens of displacement: "God is Here: A Theology of Place and Displacement" (PhD diss., University of Chicago Divinity School, 2018).

has long recognized the importance of place in interpreting the "means of persuasion" effective in particular communities.[103] But place has also framed the extended conversations around biblical interpretation and preaching that occupied white, North American homiletics for decades of the twentieth century. These conversations championed particular relationships between the world of the text, the world behind the text, and the world in front of the text, not to mention the world of the biblical canon and ecclesial tradition.[104] It is no surprise that there are resemblances between the approaches to place described in this chapter and any number of homiletical proposals growing out of these debates.[105] These projects flow from common philosophical tributaries. It is also no surprise that they share blind spots. Like Ingold, Malpas, and Basso, homiletic conversations around place can mute the questions of displaced communities in the contemporary world and in the biblical text

[103] Aristotle's means of persuasion, *ethos*, *pathos*, and *logos*, require contextual knowledge of the audience's place (i.e., their values, expectations, and relation to the argument being made) to be successfully deployed. Aristotle also drew on *topoi* (places) as a metaphor for regularly used tropes in rhetorical arguments. Places, in this sense, were recognizable paths through which an argument moved. See George Kennedy's discussion in his translation of Aristotle's *On Rhetoric: A Theory of Civic Discourse* (Oxford: Oxford University Press, 1991), 45.

[104] The "three worlds of a text" is language generally associated with Paul Ricoeur (see, for example, *The Conflict of Interpretation* [Evanston, Ill.: Northwestern University Press, 1974]), though the terms have become ubiquitous. Heidegger's emphasis on the embeddedness of hermeneutics in the world as well as Hans-Georg Gadamer's description of texts as places with "horizons" have also contributed to the trope of texts as places: Hans-Georg Gadamer, *Truth and Method*, ed. Garrett Barden and John Cumming (New York: Seabury Press, 1975). For a slightly different relation of place to biblical interpretation, see Brevard Childs, *Biblical Theology in Crisis* (Philadelphia: Westminster, 1970). Childs repeatedly turns to the context of the text within the larger biblical canon and its function as Christian scripture as the locus of interpretation.

[105] For example, New Homiletic approaches that emphasize the speech-event of the sermon, inductive models that stress a listener's participatory "journey," narrative preaching that uses form to create movement and structure, nonfoundational models that transport listeners into the story of Scripture, or liberation models of preaching that maintain the value of eisegesis when leveraged for positive social change. Cleophus LaRue describes the dislocation he felt as a black homiletic scholar during this period, as so little of this conversation engaged the preaching of the African American church; see "Black Preaching and White Homiletics," in *I Believe I'll Testify* (Louisville: Westminster John Knox, 2011), 19–36. For a redirection of the discussion of place for preaching in a postcolonial frame, see HyeRan Kim-Cragg, "Place: Grappling with Colonial Realities and Locating One's Own Social Location," in *Postcolonial Preaching: Creating a Ripple Effect* (Lanham, Md.: Lexington Books, 2021), 47–66.

by foregrounding questions of relevance and efficacy (i.e., universality and control) over questions of sovereignty, belonging, and loss. This focus on transcending, preserving, or fixing places rather than engaging them in attentive relation is a stubborn hallmark of Western practical theological method and has significant detriments for preaching and biblical interpretation that engage the climate crisis.[106] It has not only impacted how the Bible is leveraged in sermons to advance particular agendas; it has also impacted how the Bible is read and understood, as my next chapter will show. For preachers, this means that climate advocacy requires a careful, reflexive, biblical turn.

Climate-justice preaching requires more than increased numbers of displaced persons engaging in climate conversations. It requires a relational posture of biblical interpretation that draws on the insights and challenges of displaced communities who identify with land that has been lost or is under threat. It requires, in other words, a biblical hermeneutic that attends to place's bordered, relational complexity. It is possible to analyze exiled prophets in the biblical text, prophets like Ezekiel, through assumptions and questions normed by Western approaches to emplacement. One could evaluate displaced preachers in the Pacific through the same shuttered lens. Such analyses might take a variety of guises. A preacher could dissolve differences of context in ways that ignore the particularities of identity and power.[107] She could

106 Again, Courtney Goto's *Taking on Practical Theology* (Leiden: Brill, 2018) is an excellent overview of this tendency in the North American guild, but see also Elaine Graham's discussion of sources and norms in "The State of the Art: Practical Theology Yesterday, Today, and Tomorrow," *Theology* 120, no. 3 (2017): 172–80. For an example of how descriptive ethnographies can also assume a naïve confidence in their knowledge "about" the world, see Christian Scharen, "Theological Ethnography and World Christianity: A Response," *Journal of World Christianity* 10, no. 1 (2020): 109–19.

107 Given the interstices of this project, it is intriguing that the dissolving of particular contexts and identities is seen repeatedly in ecological readings of Scripture that attempt to protect and speak for land. For example, Norman Habel describes Ezekiel as an "urbanite," rather than as an exiled refugee, a two-dimensional characterization that impacts Habel's reading of Ezekiel's temple vision, in "The Silence of the Lands: The Ecojustice Implications of Ezekiel's Judgement Oracles," in *Ezekiel's Hierarchical World: Wrestling with a Tiered Reality*, ed. Stephen L. Cook and Corrine L. Patton (Atlanta: SBL, 2004), 127–40 [139]. See also Rebecca Copeland's translation of traditional ecological knowledge (TEK)—and Ezekiel's theocentric ecology—through the lens of "rational responsibility" and Western science, in "'Their Leaves Shall Be for Healing': Ecological Trauma and Recovery in Ezekiel 47:1–12," *Biblical Theology Bulletin* 49, no. 4 (2019): 214–22 [221].

overemphasize place's insularity and disregard a shifting politic,[108] or she could make the places of Scripture and pulpit familiar and tame.[109] Each of these approaches would sidestep the hard-fought wisdom of displaced communities: a relation that attends to difference, a continuity that leaves room for change, and an agency chastened by dependence. And as I have argued, such sidestepping is inadequate to address the colonial roots of our present need.

The most challenging questions of environmental preaching are not the questions of efficacy or how to winsomely translate scientific data. The hardest questions to answer are those that address the relationships and responsibilities between the places of others and the places that preachers and their hearers occupy, the limits of our understanding and our power, and the boundaries of our belonging. They are the questions of displaced communities, but they are also the questions of biblical interpretation when a preacher stands in muddy[110] solidarity with the same. From that ground, a more difficult question finds voice. It is the question of hope—what it means and where it is found, and how communities continue to dance the steps they have been given when it is unseen.

[108] This sort of insulation can dismiss the text as "bizarre" and irrelevant to contemporary conversations, a reaction that Ezekiel scholar Daniel I. Block laments in his encounters with nonspecialists; see "In Search of Theological Meaning: Ezekiel Scholarship at the Turn of the Millennium," in Cook and Patton, *Ezekiel's Hierarchical World: Wrestling with a Tiered Reality*, 227–40 [227]. But such insulation can also ignore the ways that the social locations of various communities illumine the text differently. Mark Brett describes, for example, how experiences of colonization shape textual reception and illumine strands of meaning in Scripture: *Political Trauma and Healing: Biblical Ethics for a Postcolonial World* (Grand Rapids: Eerdmans, 2016).

[109] Preachers create the textual worlds they wish to see in a variety of ways, such as ignoring textual contradictions, spiritualizing expressions of violence, and allowing intertextual allegory to overdetermine meaning. John F. A. Sawyer, for example, discusses how the "closed door" in Ezek 44:2 was used as a proof text for the virgin birth of Christ in Christian tradition; see "Ezekiel in the History of Christianity," in *After Ezekiel: Essays on the Reception of a Difficult Prophet*, ed. Paul Joyce and Andrew Mein (New York: T&T Clark, 2011), 1–10 [7–8].

[110] I draw here on Balenaikorodawa's description (in "Sosolevaki") of "mud" as the grassroot engagement necessary for contextually grounded, historically attentive theology.

2

Place in the Apocalypse of Empire

Sovereignty, Belonging, and Loss in Ezekiel

When Apenisa Waqalerua first came to his pastoral post on Totoya Island, he was disturbed by how close the pastor's home stood to the sea. There were only twelve paces between his front door and high tide. It hadn't always been this way. The tide had been steadily eroding the shoreline in recent years. Palm trees were dying, and the graves of ancestors were being washed away. In response, Waqalerua began a village-wide prayer and fasting program on the week following the village's monthly communion service. He called for a recommitment to spiritual practices and holy living. He poured a cup of blessed, sacramental juice along the shoreline, praying that the blood of Jesus would protect the community from the rising waves. He scoured Scripture for biblical passages about God's power over the sea, and he preached those passages with conviction. Nearly five years later, when Waqalerua was a student at Davuilevu Theological College, I heard his proclamation about this early season in his ministry.[1] He was convinced, then and now, that this spiritual intervention halted rising tides.

I was less convinced. I was skeptical of uncorroborated, anecdotal evidence, yes. But more significantly, I was worried about the theology implied by Waqalerua's approach. His insistence that spiritual vigor could push back waves implied that a spiritual lethargy existed in communities that had been displaced. It left open the possibility that rising tides were a divine punishment for lack of purity and resolve in spiritual practice—adding guilt to grief. Scientists have worried that such "promiscuous"[2] approaches to climate analysis mix apples and oranges. When nature and culture are

1 Apenisa Waqalerua, "The Testimony of Dravuwalu Village," sermon preached at Davuilevu Theological College, Nausori, Fiji, September 2016. My thanks to Rev. Waqalerua for the use of his sermon transcript.

2 Peter Rudiak-Gould complicates this critique in "Promiscuous Corroboration and Climate Change Translation: A Case Study from the Marshall Islands," *Global Environmental Change* 22 (2012): 46–54.

collapsed, they argue, communities become distracted by spiritual practices that have no material impact on their situation when urgent attention is needed elsewhere. Decolonial theorists insist that underscoring local responses to climate crises leaves the sins of geopolitical, economic forces across the sea unnamed, a form of blaming the victim.[3] And then there is the basic theological quandary of divine transcendence. What is God's role in climate catastrophe and human displacement? The question is, in part, a question of causation (i.e., did God create the crisis?), but it is also a question of how restoration happens. In numerous white, evangelical spaces in the United States, claims of God's sovereignty over weather patterns and climatic events have been used to rationalize human passivity and diminish the importance of environmental science. If God has ultimate power over the wind and the waves, such theologies argue, then scientific hand-wringing about the global temperature is wasted energy.[4] For a white, Western theological educator who cares about the environment, Waqalerua's testimony raises all sorts of specters.

These specters are familiar to those who grapple with the book of Ezekiel. Turning to Ezekiel to discuss contemporary conversations around environmental justice can seem a singularly strange choice. The book is not a favorite of environmental advocates for many of the reasons that made me uncomfortable with Waqalerua's sermon. Ezekiel's ferocious emphasis on divine punishment blames the victims of Babylonian violence—the Judean people and the land itself—for their own desolation.[5] Ezekiel's focus on ritual purity can spiritualize a catastrophe better explained through historical, economic, and sociological analysis.[6] Additionally,

3 John Cox et al., "Disaster, Divine Judgement, and Original Sin: Christian Interpretations of Tropical Cyclone Winston and Climate Change in Fiji," *Contemporary Pacific* 30, no. 2 (2018): 380–411.

4 Wylie Carr, Michael Patterson, Laurie Yung, and Daniel Spencer describe a spectrum of perspectives on climate change within the U.S. evangelical community in "The Faithful Skeptics: Evangelical Religious Beliefs and Perceptions of Climate Change," *Journal for the Study of Religion, Nature and Culture* 6, no. 3 (2012): 276–99.

5 Norman Habel critiques Ezekiel's "blaming the victim" in "The Silence of the Lands: The Ecojustice Implications of Ezekiel's Judgement Oracles," in *Ezekiel's Hierarchical World: Wrestling with a Tiered Reality*, ed. Stephen L. Cook and Corrine L. Patton (Atlanta: SBL, 2004), 127–40 [136–39].

6 Rather than ritual contamination of the temple as the cause of its destruction, for example, Bob Becking notes the role of global warming and drought in instigating the deportation of the Judaean exiles: "Global Warming and the Babylonian Exile," in *By the Irrigation Canals of Babylon: Approaches to the Study of the Exile*, ed. John Ahn and Jill Middlemas (London: T&T Clark, 2012), 49–62.

Ezekiel's envisioning of the land's restoration as the miraculous act of a transcendent God creates a passive, almost magical approach to restorative justice—resonating with conservative Western theologies that use arguments about God's power to wash human hands of creation's travail.[7] Such rationalizations seem increasingly dangerous and self-serving in a season of climate catastrophe.

Environmental advocates have dismissed Ezekiel's theology for any number of reasons: he was an urbanite with no appreciation for wildness,[8] controlling and hierarchical,[9] or xenophobic.[10] One of scholars' most effective tools of marginalization is the use of trauma studies—not to engage Ezekiel's witness—but to pathologize and dismiss it. When trauma-based approaches to interpretation are used in this way, Ezekiel's witness becomes little more than a cautionary description of the ways that traumatic displacement distorts the theologies of suffering communities. Ezekiel's witness becomes useful as diagnostic data, but not as generative theology.[11] Contemporary Indigenous advocates have signaled their frustration with the deployment of trauma-based analyses in present-day contexts that marginalize rather than center the questions and concerns of displaced communities.[12]

But if the witnesses of displaced communities are significant for theologies of place that address the climate crisis, then Ezekiel's exhortation deserves careful attention—as does Waqalerua's sermon. These are preachers

7 In the words of Keith Carley, "Hope is so focused on divine initiative that it obscures the issue of human responsibility for Earth and the environment": "From Harshness to Hope: The Implications for Earth of Hierarchy in Ezekiel," in Cook and Patton, *Ezekiel's Hierarchical World*, 109–26 [120].

8 Julie Galambush notes how "wildness" in Ezekiel—whether in people, flora, or fauna—is associated with "opposition to YHWH": "God's Land and Mine: Creation as Property in the Book of Ezekiel," in Cook and Patton, *Ezekiel's Hierarchical World*, 91–108 [104].

9 Carley critiques the relationship between God and the Earth as a "dominator hierarchy": "From Harshness to Hope," 110.

10 Nathan MacDonald interprets Ezek 44:7, 9 as a ban of ethnic "foreigners" from YHWH's temple and a rebuke of Isaiah 56's inclusive welcome of the same: *Priestly Rule: Polemic and Biblical Interpretation in Ezekiel 44* (Berlin: De Gruyter, 2015), 26.

11 Daniel Smith-Christopher, a scholar who has championed more nuanced use of trauma studies in his analysis of exilic texts, acknowledges ways in which psychologically diagnosing a biblical author can be a method of "avoidance": *A Biblical Theology of Exile* (Minneapolis: Augsburg Fortress, 2002), 89.

12 Natalie Clark, "Shock and Awe: Trauma as the New Colonial Frontier," *Humanities* 5, no. 14 (2016): 1–16.

and prophets who are working out their theology of place from displaced ground. The circumstances of their prophetic witnesses are distinct; they are separated by millennia, oceans, and particular histories of struggle. Both, however, grapple with what it means to be faithful to God's call when one's place of ministry is actively disappearing—or, in the case of Ezekiel, when one's place of ministry has been systematically destroyed.

Contemporary scholars have any number of theories as to why the book of Ezekiel begins with reference to "the thirtieth year" (Ezek 1:1),[13] but Origen's ancient suggestion is particularly compelling. Origen thinks that Ezekiel is talking about his age.[14] A Zadokite priest's thirtieth year is no small thing. After seasons of training, a thirty-year-old priest becomes eligible to offer temple sacrifices for the first time.[15] It is a year of fulfillment, a year for stepping into the most holy aspects of the priestly call. But these acts of worship, intercession, and encounter are mediated through access to a particular place: YHWH's temple. And in Ezekiel's thirtieth year, that temple lies hundreds of miles west. Ezekiel has been exiled to foreign territory, and in that displacement, he has also been exiled from the telos of his vocation.[16]

It is a detail that throbs at the heart of the book's witness, though it might be easy to overlook. The pain Ezekiel describes is graphic. There are piled corpses (11:7), horrific depictions of sexual violence (chs. 16; 23),[17] and traumatic silences (3:26–27; 24:15–25). His visions are haunted by toppled regimes, humiliations, and severed families. These losses are more

[13] Moshe Greenberg attends to the possibilities of interpretation (e.g., thirty years from a particular historical event, a later fragmentary addition) and the inadequacies of each proposal: *Ezekiel 1–20* (New York: Doubleday, 1983), 39–40.

[14] Origen, *Homiliae in Ezechielem* 1.4, cited by Margaret Odell, *Ezekiel* (Macon, Ga.: Smyth and Helwys, 2005), 16.

[15] Odell notes the significance of the thirtieth year in the vocation of a Levite (Num 4:3, 23, 30) and notes that, given this timeline, Ezekiel's final vision would take place at fifty, "the age at which Levites retire from service" (*Ezekiel*, 16).

[16] Despite his skepticism that Ezek 1:1 is a reference to Ezekiel's age, Baruch Schwartz sees Ezekiel's vocational crisis, resulting from his displacement from the temple, as a central theme of the book. It is not possible, Schwartz argues, for Ezekiel to perform the role of a priest without the "place" of God's sanctuary: "A Priest Out of Place: Reconsidering Ezekiel's Role in the History of the Israelite Priesthood," in Cook and Patton, *Ezekiel's Hierarchical World*, 61–72.

[17] Daniel Smith-Christopher presents these horrific texts as more than allegory, noting parallels between the sexual violence depicted and Babylonian war practices. See "Ezekiel in Abu Ghraib: Rereading Ezekiel 16:37–39 in the Context of Imperial Conquest," in Cook and Patton, *Ezekiel's Hierarchical World*, 141–58.

than metaphors. Ezekiel's careful stamping of place and time, his chronologies of events and attention to locations (e.g., 1:1–2; 3:15; 8:1, 3; 11:1, 13, 24; 30:20; 31:1; 32:17; 33:2), make clear something concrete is at stake. Ezekiel has lost land, and there are political, economic, and psychological consequences to this material loss. But his first words speak to a different category of consequence—a consequence that the prophet will trace through the rest of the book. His displacement has resulted in spiritual loss and vocational crisis. For Ezekiel, the loss of his land is a theological gash with personal and communal urgency, and his questions spiral. Who is he without his place? Who are his people? And who, more importantly, is God?

This vocational framing makes the book a particularly compelling text for preachers who are asking the deep theological questions that accompany the climate crisis. Ezekiel gives no easy answers to these questions, but in wrestling with them, the contours of his theology of place emerge. It is a theology of place that contrasts with the imperial place-making of the Babylonian empire, foregrounding different commitments and different aims. It is also a theology of place that illumines the complexity of what is at stake in Ezekiel's emphases on divine punishment, ritual purity, and human impotence. Proclamation is place specific. It does different work on different soil. The very words that raise specters for Western readers, readers used to objectifying and commodifying creation, can have restorative implications for those whose relationship to creation and place is grounded in divine encounter, particularly when those communities have lost places central to their identities.

Proclamation, of course, can also be co-opted by the powers that be. In her critique of the horrific misogyny and fetishistic visions of rape in Ezekiel 16 and 23, Carolyn Sharp suggests a far more ominous possibility. "Texts produced in cultural contexts of colonization do not simply represent exile and oppression," she notes. "They can collude in the colonizer's ongoing exercise of power."[18] She raises a question at the heart of this project: "What kind of sacred language can withstand the dehumanizing threat under which the colonial subject lives?"[19] When the apocalypse of empire leads to an apocalypse of place, can liberative preaching survive? I have argued that those who are displaced have a privileged location in

[18] Carolyn J. Sharp, "Sites of Conflict: Textual Engagements of Dislocation and Diaspora in the Hebrew Bible," in *Interpreting Exile: Displacement and Deportation in Biblical and Modern Contexts*, ed. Brad E. Kelle, Frank Ames, and Jacob Wright (Atlanta: SBL, 2011), 365–76 [365].

[19] Sharp, "Sites of Conflict," 365.

articulating faithful theologies of land in the context of the climate crisis. But one could make the opposite claim. Particularly when caught in the crosshairs of neocolonial power, displaced communities can parrot the ideology of the oppressor and "project their anxiety and rage onto a safer target: fellow citizens in the subaltern state."[20] Anyone familiar with the ethnic tensions that continue to mark the relationship between Fiji's primarily Indigenous Christian population and the nation's Hindu-Muslim Indo-Fijian minority will recognize the relevance of Sharp's assertion for the Fijian context. The complicity of the Methodist Church in Fiji (MCIF) in the ethnonationalist coups of the late 1980s and early 2000s has been clearly documented.[21] The New Exodus described in chapter 4 is, in part, the MCIF's answer to Sharp's concern, reflexively addressed from Fijian pulpits and from Fijian leaders themselves.

This chapter, however, has a humbler goal. I will not try to justify or explain away Ezekiel's emphases on punishment, purity, or human passivity. Each of these emphases raises real challenges for discussions of creation justice, particularly when co-opted by economically secure, settled communities. I will, however, reflect on what these emphases reveal about the driving concerns shaping Ezekiel's theology of place and the significant differences they connote between his approach and the imperial approach taken by the Neo-Babylonian empire. These emphases also suggest foundational assumptions about place that are distinct from those of many Western scholars and environmental activists. Tellingly, Ezekiel prioritizes questions about place and displacement that are similar to those of Fijian thinkers introduced in the last chapter. Specters are not the only things these communities share. They share similar investments in questions around *sovereignty*, *belonging*, and *loss*. Ironically, though perhaps unsurprisingly, it is often in Ezekiel's negotiations of punishment, purity, and passivity—those emphases that cause Western-trained, environmentally minded interpreters like myself discomfort—that these investments find voice. It is worth wondering what that irony reveals about the insularities

20 Sharp, "Sites of Conflict," 366.

21 See, for example, Lynda Newland's mapping of this history in "Religion and Politics: The Christian Churches and the 2006 Coup in Fiji," in *The 2006 Military Takeover in Fiji: A Coup to End All Coups?* ed. Jon Fraenkel, Stewart Firth, and Brij V. Lal (Canberra: ANU Publications, 2007), 187–207, or Matt Tomlinson's succinct account, "The Generation of the Now: Denominational Politics in Fijian Christianity," in *Christian Politics in Oceania*, ed. Matt Tomlinson and Debra McDougall (New York: Berghahn Books, 2013), 78–102 [80–84].

and assumptions of Ezekiel's readers, from their various social locations. "Reflexivity" is not solely the responsibility of the displaced community.[22] Sharp's fears of imperial collusion apply to contemporary biblical interpreters as readily as they apply to the texts being interpreted. That collusion needs careful excavation, particularly when trying to discern faithful, communal responses to such proclamation.

Finally, it is faithful, communal response that separates trauma-informed readings marked by "professional distancing and avoidance"[23] from those that "attend to violence and its effects in the world we inhabit"[24] with courage and care. For Natalie Clark, who facilitates trauma-informed groups for Indigenous girls, such faithful responses name the intersecting, ongoing colonial oppressions of racism, violence to Indigenous lands, and European hegemony. But they also center the "ongoing resilience, survivance, and positive resistance" of Indigenous communities themselves.[25] This chapter holds open the possibility that the specters in Ezekiel's theology of place are more than desperate accommodations to the trauma of displacement. They also resist imperial ideologies that displace. As such, they have constructive value in reframing colonial conceptions of place and land. For all of the book's complexity, there is more going on in this problematic prophet than pathology. Ezekiel foreshadows

22 Daniel Smith-Christopher notes the importance of "reflexivity" in cross-contextual, cross-temporal trauma studies and in the work of biblical interpretation, calling the biblical guild to a "serious debate about the role of social reflexivity in the biblical analysis of the events of the exile": "Reading Exile Then: Reconsidering the Methodological Debates for Biblical Analysis in Dialogue with Sociological and Literary Analysis," in Ahn and Middlemas, *By the Irrigation Canals of Babylon*, 139–57 [157].

23 Sarah Jobe's creative reframing of trauma-informed care as a methodological, rather than a diagnostic, tool prioritizes restorative relationship as the telos of analysis. She notes that therapeutic diagnoses are not meant to be ends in themselves. They are meant to illumine "what sorts of interventions and what modes of listening might be most helpful in the continued relationship": "A Hermeneutics of Trauma: Trauma-Informed Care as Method for Biblical Interpretation," in *A Hermeneutics of Trauma* (Sheffield: Sheffield Phoenix Press, forthcoming 2025).

24 Anathea Portier-Young argues that responsive solidarity in relation to the complex horrors of violence in the contemporary world is correlated with the biblical task of "gnawing on the shards" of violence in our Scriptures. She explicitly notes the generative relationship between biblical studies and practical theology in this work. See "Drinking the Cup of Horror and Gnawing on Its Shards: Biblical Theology through Biblical Violence, Not around It," in *Beyond Biblical Theologies*, ed. Heinrich Assel, Stefan Beyerle, and Christfried Böttrich (Tübingen: Mohr Siebeck, 2012), 387–408.

25 Clark, "Shock and Awe," 2.

searing questions around sovereignty, belonging, and loss from displaced ground, centering the concerns of communities that have lost land. Such centering should move critics to slower, more circumspect engagement. Spectral or no, Ezekiel's questions crucially disrupt Western conceptions of place that are simplistically universal, insular, or constructivist in ways that are dangerous for theological educators or preachers to dismiss. These disruptions are also the soil from which his critique of imperial, Neo-Babylonian approaches to place and place-making emerge.

Place-Making in the Babylonian Empire

Daniel Smith-Christopher has argued that "we do not 'hear' Ezekiel clearly unless we also 'hear' the imperial voice of Nebuchadnezzar the conqueror."[26] A champion of social-scientific analysis in biblical interpretation, Smith-Christopher has worked to take seriously the "exile as a real event where human beings deeply suffered."[27] More is at stake in Smith-Christopher's argument than the biblical guild's dismissal of Ezekiel as a "neurotic psychopath" or stylized poet.[28] He has repeatedly named the proclivity of scholars to diminish the impact and import of displacement on exiled communities. The horror of the biblical exile included violent deportation, loss of security and status, and separation from family.[29] It almost assuredly included forms of sexual humiliation and rape.[30] There were also more

26 Smith-Christopher, *Biblical Theology of Exile*, 104. Hendrik Bosman sketches this insight's significant history in biblical studies, beginning with Hermann Gunkel's 1895 observations about the similarities between Babylonian myths and "myths of YHWH." See "Myth, Metaphor, or Memory? Allusions to Creation and Exodus in Isaiah 51:9–11 as a Theological Response to Suffering during the Exile," in *Exile and Suffering: A Selection of Papers Read at the 50th Anniversary Meeting of the Old Testament Society of South Africa, Pretoria, August 2007*, ed. Bob Becking and Dirk Human (Leiden: Brill, 2009), 71–82.

27 Bosman, "Myth," 89.

28 Bosman, "Myth," 104. Smith-Christopher names David J. Halperin's *Seeking Ezekiel: Text and Psychology* (University Park: Pennsylvania State University Press, 1993) and Walther Zimmerli's classic *Ezekiel*, trans. Ronald E. Clements, 2 vols. (Philadelphia: Fortress, 1979) as indicative of these respective approaches.

29 Brad E. Kelle notes that, in contrast to prior consensus that "Judean life in exile was a relatively benign existence . . . interpreters in the last three decades . . . view the Judean experience of exile as a severe and traumatic personal, social, and psychological crisis that entailed suffering and domination and . . . destabilizing recalibrations of the social and theological identity": "An Interdisciplinary Approach to Exile," in Kelle, Ames, and Wright, *Interpreting Exile*, 5–40 [13–14].

30 Smith-Christopher, "Ezekiel in Abu Ghraib," 148–54; Yagail Yadin, *The Art of Warfare in Biblical Lands in the Light of Archaeological Study* (London: Weidenfeld and Nicolson, 1963), 396–97.

subtle assaults to language, identity, and faith commitments. Exile required a reframing of beliefs about core theological claims at the same time that one "resist[ed] . . . a vast empire with an advanced cultural tradition in the form of literature, architecture, military power, economic development and religious symbols."[31] The fact that Ezekiel's prophecies make muted reference to Babylon does not negate its influence. After Babylon's decisive defeat of Assyria at Harran in 609 BCE and its defeat of the Egyptians at the Battle of Carchemish four years later, the empire's supremacy was assured. By 601 BCE, "all the nations of Syria-Israel had become colonies of the Neo-Babylonian empire."[32] The imperial voice of Nebuchadnezzar II thundered across the region, loud and clear.

Babylon's faith in its superior "pan-Mesopotamian" self-consciousness[33] was based on more than military might. The empire's "advanced knowledge of irrigation agriculture, urban comforts, and justly famous literary tradition"[34] amplified its conviction of cultural primacy. Its complex Akkadian cuneiform, as well as alphabetic Aramaic, had become the "*lingua franca* of the Ancient Near East."[35] Babylon boasted in its hanging gardens, metal works, and pottery. But more foundational to its understanding of its relationship to the world was the "language of supremacy [included] in a variety of historical, mythical and literary texts."[36] The official titles ascribed to the Babylonian emperor in court documents are one example. One text describes Nebuchadnezzar II accepting a command from Marduk, the Babylonian deity, to "rule the totality of all peoples," wielding a "just scepter to lead all populated regions aright and make humanity thrive."[37] The ruler's divinely appointed task, thus stated, was to govern less civilized societies and impose justice. But "justice" is a term which can mean different things to conqueror and conquered. Pressing back on arguments

31 Leo Perdue and Warren Carter, *Israel and Empire: A Postcolonial History of Israel and Early Judaism* (London: Bloomsbury, 2015), 81.

32 Perdue and Carter, *Israel and Empire*, 69.

33 J. Richard Middleton, *The Liberating Image: The* Imago Dei *in Genesis 1* (Grand Rapids: Brazos, 2005), 197. See Peter Machinist, "On Self-Consciousness in Mesopotamia," in *The Origins and Diversity of Axial Age Civilizations*, ed. Shmuel N. Eisenstadt (Albany, N.Y.: SUNY Press, 1986), 183–202.

34 Middleton, *Liberating Image*, 197–98.

35 Perdue and Carter, *Israel and Empire*, 72.

36 Perdue and Carter, *Israel and Empire*, 72.

37 VAB 4 112 (Nbk 12=Zyl III, 5) I 13–17, quoted in Perdue and Carter, *Israel and Empire*, 73.

that Neo-Babylonian imperial rule was somehow more civilized and less cruel than the empire of the Assyrians, Smith-Christopher notes that even Babylon's academic apologists cannot deny "the brute facts of subjection . . . [in a society] focused on the domination and exploitation of non-Babylonian populations for the benefit of a ruling elite."[38] Tribute, human labor, and wealth flowed from conquered nations back to Babylon's ruling classes and temples. The empire may have claimed a vocation of universal benevolence, but Babylon was clearly the beneficiary of that universe. More particularly, Babylon's king was the beneficiary. The palace owned or controlled the majority of land within the state.[39]

One of Babylon's most powerful tools of domination was the mass deportation of populations and the strategic exile of leaders who resisted its demands or were perceived as threats. When Nebuchadnezzar II's armies threatened to destroy Jerusalem in 598 BCE following Jehoiakim's unsuccessful attempt at rebellion (2 Kgs 24:1), a large group of Judahite leaders—including Jehoiachin, Jehoiakim's successor—were taken into exile. Ezekiel was one of their number. His prophecies recount his grief and fury at watching Zedekiah, Nebuchadnezzar II's choice for Jehoiachin's successor, swear loyalty to Babylon in 594 BCE while simultaneously working behind the empire's back to garner Egyptian support for a second rebellion.[40] Following Zedekiah's refusal to pay tribute, Nebuchadnezzar II destroyed Jerusalem and the temple in 587 BCE.

These sieges and deportations were accompanied by military tactics that polluted and destroyed natural resources. Brad E. Kelle has argued that the preponderance of language about nature's devastation and rejuvenation in Ezekiel is likely to have "an identifiable connection to actual practices of warfare known throughout the Assyrian and Babylonian periods."[41] Siege warfare often included intentionally destroying trees, vineyards, and agricultural networks, sowing the soil with salt, and stripping

38 Smith-Christopher, *Biblical Theology of Exile*, 77. To make his point, Smith-Christopher quotes David Stephen Vanderhooft, *The Neo-Babylonian Empire and Babylon in the Latter Prophets*, Harvard Semitic Monographs 59 (Atlanta: Scholars Press, 2000), 6.

39 Perdue and Carter, *Israel and Empire*, 75.

40 Ezekiel 17 presents this history allegorically.

41 Brad E. Kelle, "Dealing with the Trauma of Defeat: The Rhetoric of the Devastation and Rejuvenation of Nature in Ezekiel," *Journal of Biblical Literature* 128, no. 3 (2009): 469–90 [485]. For examples of Ezekiel's description of nature's destruction at God's hand, see Ezek 6:1–14, 7:1–4, and 20:45–49.

the land to make it susceptible to insect infestation.[42] Unlike Assyria, which regularly rebuilt the cities it destroyed, Babylon had a "devasted earth policy" when it came to conquered territories, leaving them demoralized and unable to subsist on the land.[43] Oded Lipschits suggests that in Israel's case, this destruction was meant to serve as a kind of natural "dead zone" between Babylon and Egypt, providing a human-made buffer of uninhabitable land between the Ancient Near Eastern superpowers.[44] In its relation to the natural world, its building of temples and cityscapes, and the expansion of its borders, Babylon's imperial policies reflect a society that believed it held the power of life and death over creation.

J. Richard Middleton points out the ways this ideology is reified in one of Babylon's most popular cosmologies: the *Enuma Elish*. In Middleton's estimation, this creation story explains a particular social system in which humanity is "created to serve gods and relieve them of burden."[45] Humanity does not have intrinsic worth, nor does the earth. The goal of both is to provide food for the gods, who do not wish to engage in the time-consuming work of mass-scale agriculture themselves.[46] The earth's utilitarian value is underscored by the *Enuma Elish*'s violent depiction of its creation. The earth has a decidedly disrespected, rebellious antecedent: it is carved from the dismembered body of Marduk's defeated foe and grandmother, Tiamat. After killing this matriarchal symbol of chaos, Marduk flies into a "celebratory rage or victory fever, tramples [her] corpse, crushes her skull, and scatters her blood upon the winds."[47]

Given the fact that the Babylonian king was reaffirmed as Marduk's "image" and representative in annual ceremonies like the Akitu Festival,

42 Kelle points readers to Jacob L. Wright, "Warfare and Wanton Destruction: A Reexamination of Deuteronomy 20:19–20 in Relation to Ancient Siegecraft," *Journal of Biblical Literature* 127 (2008): 423–58, and Michael G. Hasel, *Military Practice and Polemic: Israel's Laws of Warfare in Near Eastern Perspective* (Berrien Springs, Mich.: Andrews University Press, 2005) for fuller descriptions of Babylonian siege practices.

43 For further discussion of sixth-century Babylonian imperial policies, see Brad E. Kelle, *Ancient Israel at War 853–586 BC*, Essential Histories 67 (Oxford: Osprey, 2007), 57–75, and Oded Lipschits, *The Fall and Rise of Jerusalem: Judah under Babylonian Rule* (Winona Lake, Ind.: Eisenbrauns, 2005), 36–65.

44 Lipschits, *Fall and Rise of Jerusalem*, 68.

45 Middleton, *Liberating Image*, 149.

46 Perdue and Carter describe how Babylonian gods "were dressed in expensive clothing, dwelt in luxurious temples and received the daily care of priests who had overseen their crafting and engaged in rites of 'opening the mouth' to bring them to life" (*Israel and Empire*, 72).

47 Middleton, *Liberating Image*, 163.

this cosmology became enfleshed in the king's "unique, unparalleled cultic status, including a priestly role as mediator of social harmony and cosmic fertility from heaven to earth."[48] Resistance to the king was resistance to Marduk's "sovereignty over creation."[49] In the context of this rigid, centralized sociopolitical order, Middleton argues that the creation story in Genesis 1 serves as an ideological critique of its bloody Babylonian counterpart—a story where creation is named "very good" (Gen 1:31) and where all humanity, men and women alike, share the *imago dei* (Gen 1:26).[50]

I appreciate Middleton's hopeful contrast, but I highlight a different point: the striking similarities between Neo-Babylonian conceptions of place/place-making and the conceptions that grounded the European colonial project. H. Daniel Zacharias notes the ways that the success of the European conquest was rooted in and "has contributed to . . . [a] landless anthropology and universalizing theology of land."[51] This hegemonic disregard of borders, combined with the assumption that human persons do not have a "deep and essential"[52] connection to the land in which they live, lies at the root of empires that expand and displace. One of the ugly contradictions of such a worldview is how an empire's assumed universality creates basic categories of value that are unchanging across contexts, which in turn make unending change a painfully necessary survival strategy for all who do not reside at the universe's gravitational center. In his exegesis of European colonialism's deportations, genocides, and slave trades, Willie James Jennings traces the violent implications of this denial of place's specificity. "Everything—from peoples and their bodies to plants and animals, from the ground and the sky—was subject to change, subjects for change, subjected to change. The significance of this transformation cannot be overstated," he states. "The earth itself was barred from being a constant signifier of identity."[53] And yet, alongside this transmuting of the world—this making and remaking of places and localities to serve

48 Middleton, *Liberating Image*, 172.

49 Perdue and Carter, *Israel and Empire*, 72–73.

50 Middleton, *Liberating Image*, 185–212.

51 H. Daniel Zacharias, "The Land Takes Care of Us: Recovering Creator's Relational Design," in *Theologies of Land: Contested Land, Spatial Justice, and Identity*, ed. K. K. Yeo and Gene L. Green (Eugene, Ore.: Cascade, 2021), 69–97 [93].

52 Zacharias, "Land Takes Care of Us," 93.

53 Willie James Jennings, *The Christian Imagination: Theology and the Origins of Race* (New Haven: Yale University Press, 2011), 43.

one's interests—a new "constant signifier" emerges: the assertion of imperial superiority. Babylon may have put great faith in the transformational power of its Mesopotamian ways, but its hierarchies were "immutable."[54]

Middleton is not alone in suggesting that Genesis 1 is an ideological critique of this Babylonian worldview. Wanda Deifelt agrees that the first chapter of the Hebrew Bible was "meant to elevate an enslaved and exiled people."[55] But if these scholars are correct, the similarities between the Neo-Babylonian and Western colonial conceptions of place and place-making should give preachers and practical theologians pause. Texts that were meant for liberation in one context, codified and recorded by displaced populations, can be interpreted for domination in another. They can even be used to serve imperial projects similar to those they were written to subvert.[56] "Landless" textual interpretation shares the dangers of "landless anthropology." Texts have localities, and they require interpreters who see what is at stake, not only in the places they were written, but in the places they are given voice.

Ezekiel's Unlikely Places of Resistance

My description of Waqalerua's sermon at the opening of this chapter leaves out some key information. In response to rising tides, his village had applied to purchase a seawall that would cost millions of dollars, putting it in significant debt to the military government of the time. Waqalerua's distrust of this expensive line of action is clear, as is his distrust of "lobbying bigger nations" for help. His wariness is rooted in the dangers of outside political influences on village life.[57] But it is also rooted in his clear

54 Middleton, *Liberating Image*, 199.

55 Wanda Deifelt, "And G*d Saw That It Was Good—*Imago Dei* and Its Challenge to Climate Justice," in *Planetary Solidarity: Global Women's Voices on Christian Doctrine and Climate Justice*, ed. Grace Ji-Sun Kim and Hilda P. Koster (Minneapolis: Fortress, 2017), 119–34 [128–29].

56 See, for example, Tony Ballantyne, "Genesis 1:28 and the Languages of Colonial Improvement in Victorian New Zealand," *Victorian Review* 37, no. 2 (2011): 9–13.

57 Voreqe Bainimarama's government, responsible for the nation's 2006 coup, had not yet been democratically elected when Waqalerua's sermon was preached, and its relationship with the Methodist Church in Fiji was antagonistic. See, for example, David Crampton, "Fiji's Government Again Cancels Methodist Conference," *Christian Century*, August 25, 2011, https://www.christiancentury.org/article/2011-08/fiji-government-again-cancels-methodist-conference. Fiji's relationship to "bigger nations" is marked by colonial history and neocolonial globalization. Fiji obtained independence from British rule in 1970.

vision about the desperateness of the situation. A rising ocean will not be stopped by a seawall. He has seen numerous seawalls fall. Waqalerua's call to prayer and fasting may resonate with certain Western fundamentalist approaches to climate change that relinquish human responsibility, but his sermon has very different goals. He underlines the reality of the crisis, and he calls his church to place its relationship to the ocean and land at the center of its spiritual practices. Rather than embracing resignation, his insistence on divine sovereignty grounds his argument that human engagement is necessary and significant. More importantly, he resists abdicating the village's economic and decision-making sovereignty to governmental or foreign control.

In her influential study of apocalyptic literature during the Hellenistic rule in Judea, Anathea Portier-Young argues that a literary genre that has often been characterized as escapist and otherworldly in its orientation can be a "resistant counterdiscourse"[58] against imperial oppression. An apocalyptic worldview, she argues, can redefine "what [is] possible and . . . real" through a "radical redefinition of power."[59] Portier-Young draws on the work of James Scott[60] to describe a spectrum of resistance strategies employed by such communities, including "answering myth with myth," offering "alternative cosmologies," appealing to history to reveal "the contingency of present realities," or destabilizing the "logic and coherence of a social reality" believed to be governed by empire.[61] Portier-Young's work is careful to acknowledge not only the diversity of resistance strategies employed but also the diverse responses of threatened communities themselves. "Resistance" is not their only alternative. As Sharp has stressed, oppressed communities can co-opt and be co-opted by colonial ideologies of power. Indeed, co-optation and resistance are often muddled together in the necessities of survival. Simple narratives are rare and often flawed. But Portier-Young leaves her readers with an insight that stretches beyond the tight frames of Hellenistic Judaism and the biblical apocalyptic genre: colonial resistance can show up in unlikely places.

[58] Anathea Portier-Young, *Apocalypse against Empire: Theologies of Resistance in Early Judaism* (Grand Rapids: Eerdmans, 2011), xxii.

[59] Portier-Young, *Apocalypse against Empire*, 4.

[60] James Scott's *Domination and the Arts of Resistance: Hidden Transcripts* (New Haven: Yale University Press, 1990) is particularly significant in Portier-Young's analysis, as is his *Weapons of the Weak: Everyday Forms of Peasant Resistance* (New Haven: Yale University Press, 1985).

[61] Portier-Young, *Apocalypse against Empire*, 13.

Waqalerua's sermon points to the diverse ways that spiritual practices function in different communities; what can appear escapist in one context can be "resistant counterdiscourse" in another. Minimally, his sermon assumes a very different understanding of relationality between God, human, and nonhuman creation than the "landless," "universal" conceptions of place critiqued by Jennings and Zacharias. I will argue that the same is true for Ezekiel's approach to place. As a preacher, I've learned to pay attention to texts and situations that disrupt my comfort and my categories. Ezekiel's insistence on God's punishment of the land, his framing of land in terms of purity and impurity, and his passive depiction of ecological restoration disrupt both. I will argue that in these complicated responses to land loss, Ezekiel destabilizes the "logic and coherence"[62] of Neo-Babylonian approaches to place, resisting its hegemony and refusing its terms.

YHWH's Punishment of the Land and the Question of Sovereignty

For many scholars who specialize in creation-centered interpretations of Ezekiel, YHWH's punishment of Judah's land causes outrage. It is not only that earth is described as "a passive object of horrifying maltreatment,"[63] it is also the apparent injustice of the land's suffering. In Norman Habel's words, "Why should the trees . . . suffer . . . at God's hands? The trees have not sinned and yet they have experienced the wrath of God."[64] Julie Galambush points out the uniqueness of Ezekiel's approach. Throughout the rest of Scripture, the land is portrayed as a victim of human sin, but in Ezekiel, it is "a party to its inhabitants' actions."[65] In its "metonymous" representation of Judah's people, the land serves as a "site of injury" for no other purpose than to "display YHWH's power."[66] Noting the land's inability to defend itself against this injustice, Kalinda Rose Stevenson imagines a courtroom where the earth charges God with "assault and battery, transference of guilt, and discrimination."[67] The judgment against the mountains in Ezekiel 6 serves

62 To use Portier-Young's language.

63 Carley, "From Harshness to Hope," 109.

64 Habel, "Silence of the Lands," 134. He speaks specifically here of the forest of Negeb (Ezek 20:45–48).

65 Galambush, "God's Land and Mine," 100.

66 Galambush, "God's Land and Mine," 101.

67 Kalinda Rose Stevenson, "If Earth Could Speak: The Case of the Mountains against YHWH in Ezekiel 6:35–36," in *The Earth Story in the Psalms and the Prophets*, ed. Norman Habel and Shirley Wurst (Cleveland, Ohio: Pilgrim Press, 2001), 160.

as an example. The mountains will have a "sword" brought against them (v. 3). They will be defiled by corpses (v. 4), and the cities that shelter in the mountains will be "waste" and "ruined" (v. 6). There is particular focus on the sin of idolatry, with the "high places" of the mountains and its idolatrous altars being "broken and destroyed" (v. 6). Using a verb that recalls the judgment of the Flood, Ezekiel prophesies that works of the mountains will be "wiped out" (v. 6; cf. Gen 6:7; 7:4, 23).

Kelle frames the land's punishment in Ezekiel as an assertion of sovereignty, by which he means divine sovereignty. In his creative weaving together of recent threads in Ezekiel scholarship, he brings analyses of Ezekiel's priestly focus on purity, ecotheological readings, and trauma studies into conversation. For Kelle, Ezekiel's central issue is the crisis of YHWH's apparent defeat by Babylon and YHWH's lack of sovereignty over the cosmos. Ezekiel's emphasis on the land's impurity puts YHWH back in control, causally linking the devastation of the land to YHWH's will. Ezekiel, Kelle states, "allows the Babylonians no role."[68] Through the lens of trauma, Kelle explains this vindictive portrayal of the deity as an "attempt to cope with unstable social conditions."[69] He echoes the thought of Ruth Poser, who has described Ezekiel's focus on God's punishment as a way to maintain human agency in abusive conditions. "Although there is no *objective* reason for [the exiles of Judah] to blame themselves," she suggests, ". . . [the] assumption of guilt after trauma is often connected to the fact 'that it can be easier to maintain one's mental balance if one was guilty rather than completely helpless.'"[70] Kelle and Poser both describe Ezekiel's depiction of YHWH's punishment of the land as the attempt of a traumatized prophet to give coherence to a horror that has befallen his community. The theme of God's punishment not only "takes power away from the enemy and gives it to God,"[71] it also "reshapes Israel's identity from defeated warrior, to survivors who continue to live under the . . . sovereignty" of God.[72]

68 Kelle, "Dealing with the Trauma of Defeat," 489.

69 Kelle, "Dealing with the Trauma of Defeat," 480.

70 Ruth Poser, "No Words: The Book of Ezekiel as Trauma Literature and a Response to Exile," in *Bible through the Lens of Trauma*, ed. Elizabeth Boase and Christopher Frechette (Atlanta: SBL, 2016), 27–48 [41]. Poser quotes Angela Kühner, *Kollektiv Traumata—Annahmen, Argumente, Konzepte: Eine Bestandsawfnahme nach dem 11. September*, Berghof Report 9 (Berlin: Berghof Forschungszentrum für Konstrucktive Konflicktbearbeitung, 2002), 32.

71 Kelle, "Dealing with the Trauma of Defeat," 489.

72 Kelle, "Dealing with the Trauma of Defeat," 490.

There are certainly ways that this analysis of YHWH's punishment of the land in Ezekiel resists Neo-Babylonian approaches to place. Most basically, it denies Marduk supreme authority over the world. It stands on the banks of an irrigation ditch and, against all evidence to the contrary, claims YHWH as sovereign Lord. C. A. Strine sees a similar move in Ezekiel 38–39, interpreting Gog, of the land of Magog (Ezek 38:2), as Marduk and YHWH as victoriously engaging in "*Chaoskampf* against empire."[73] There are also ways, of course, that such readings reinforce the categories and assumptions of imperial logic, even as they invert its hierarchies. Regardless of the restoration following the land's devastation in Ezekiel, YHWH's physical onslaught against Judah's land, forests, and mountains brings to mind Marduk's own bloody abuse of creation. Ezekiel's theology of place can be viewed as a re-performance of Babylon's *Chaoskampf* script with different actors in different roles. For the exiled community, asserting YHWH's power in the face of empire is significant theological work, but it does not necessarily subvert an imperial orientation toward place and creation.

There is, however, a more radical protest against imperial "universality" and "landless anthropology" embedded in Ezekiel's discussion of the land's punishment. This disruption also has to do with the question of "sovereignty," but with an eye toward the word's broader semantic range. Ezekiel asserts *the land's sovereignty*, defined within the particular context of covenantal relation to YHWH and God's people. In contrast to scholarly descriptions of the land as object, when Ezekiel holds the land accountable for sin, he presumes the land's "autonomy, independence, and responsibility."[74] He presumes it is a subject. In this difficult theme of punishment, there is a radical deconstruction of approaches to place that treat creation as a dead, transferable thing at the disposal of whichever human sovereign is most powerful. The particularity, agency, and relationality of land fundamentally inform Ezekiel's theology of place and destabilize imperial definitions of what embodied, creaturely sovereignty entails.

[73] C. A. Strine, "Chaoskampf against Empire: YHWH's Battle against Gog (Ezekiel 38–39) as Resistance Literature," in *Divination, Politics, and Ancient Near East Empires*, ed. Alan Lenzi and Jonathan Stökl (Atlanta: SBL, 2014), 87–108.

[74] Wojciech Pikor, "The Subjecthood of the Land of Israel in the Book of Ezekiel," *Biblical Annals* 3, no. 1 (2013): 29–46 [37].

Taking seriously the land as an accountable subject that participates in God's covenant is, perhaps, a new idea for many Western preachers. But it is not a new idea to biblical scholars.[75]

It is, in fact, a recurring theme in Scripture. Wojciech Pikor begins his study of the land in Ezekiel with the startling claim that the uniqueness of the prophet's view of land is not its passive objectification, but rather its "subjectification."[76] Building on Galambush's insight about Ezekiel's holding the land accountable for sin but taking her argument in a different direction, Pikor claims that "far from being a passive object, the land is presented as the protagonist of Israel's unfolding history."[77] He begins by deconstructing Galambush's claim that the land is a "metonymy" for the people in Ezekiel's text, noting how the book never allegorizes the land in the manner of Jerusalem and how the punishments that the land undergoes (such as depopulation; see 6:6, 7, 14) assume that the land and the people remain separate entities.[78] Ezekiel 22:23–27 underscores the separation between the land and the leaders of Israel in ways more familiar to Scripture as a whole: the leaders have failed to protect the land and are "within her" (22:25) like a roaring lion devouring prey.

The implications of taking the land's subjectivity seriously are provocative. The land is addressed by God through prophetic oracles, and though the land is not anthropomorphized, it is treated "as a moral subject . . . aware of good and evil, . . . free and responsible for [its] behavior."[79] The land is understood as belonging to God ("my land," 36:5) just as God's people are described as belonging to God ("my people," 36:8–12). But, similar to Fijian understandings of *vanua*, the people are also described as belonging to the land. Ezekiel 36:13–15 repeatedly assures the land that "your people" ("your nation," [גּוֹיַיִךְ] NRSV) will no longer be bereaved.

In Ezekiel, the subjectivity of land is inseparable from relational solidarity with God's people. The connection is less of a metonymy—and more of a "kinship" bond.[80] When the people defile the land through sins

75 See, for example, Christopher Wright's much referenced language of an "ecological triangle" between God, human, and nonhuman creation in *Old Testament Ethics for the People of God* (Leicester: InterVarsity Press, 2004).

76 Wojciech Pikor, *The Land of Israel in the Book of Ezekiel* (New York: T&T Clark, 2018), 38.

77 Pikor, *Land of Israel*, xxi.

78 Pikor, *Land of Israel*, 35–36.

79 Pikor, *Land of Israel*, 38.

80 Mari Joerstad, *The Hebrew Bible and Environmental Ethics: Humans, Nonhumans, and the Living Landscape* (Cambridge: Cambridge University Press, 2019), 139.

of injustice (Ezek 22:6–13) or idolatry (Ezek 6), the land is held "responsible for its relationships."[81] This may seem unfair to modern sensibilities, but Mari Joerstad agrees that Ezekiel describes a world where people, animals, and land "are all part of a differentiated whole."[82] She notes that "the corollary of a personalistic, relational view of nonanimal nature is that nonanimal nature can fail in much the same ways that humans fail."[83] In wrestling with present-day implications of this idea, William Briggs turns to the impact of toxic industrial chemicals on the town of Anniston, Alabama, dumped by a company that knew it was poisoning the land but "continued its operations anyway."[84] Briggs describes the further "devastation" that has been required to remove the toxic pollution—nothing less than a stripping of the earth. "In a created, relational world, human crimes of polluting land result in the land polluting humanity, requiring that both be cleansed in the end."[85] In Ellen Davis' words, "People and land are bound together in their well-being—for better, for worse."[86]

This relationality of place and person implies at least two things about creaturely sovereignty. First, if humans were made to be in relation to each other and to land, the boundaries of emplaced communities are to be respected. Borders matter to Ezekiel. Locality matters. The subjectivity and relationality of land speak to his material concern with place's particularity. Land is not interchangeable, nor is it an artifact to be crafted and sold. Land has sovereignty as a relational subject created by YHWH. Concretely, this is seen in Ezekiel's fury at Edom's occupation of Judean land following the deportation. When Edom claims that Judah's land will now belong to them to be used for food since the mountains are "desolate" (35:12), YHWH comes to the mountain's defense. In the biblical text, land relationships are not universal.

Second, in the world Ezekiel describes, creaturely sovereignty is defined by relational accountability. Sovereignty is not the freedom to act unencumbered. It is the freedom to be responsible for one's place

81 Pikor, *Land of Israel*, 50.

82 Joerstad, *Hebrew Bible and Environmental Ethics*, 139.

83 Joerstad, *Hebrew Bible and Environmental Ethics*, 137.

84 William Briggs, "Idols and Land Grabs, Ancient and Modern: Creation and Ecotheology in Ezekiel 6, 35:1–36:15," *Horizons in Biblical Theology* 40, no. 1 (2018): 41–64 [51].

85 Briggs, "Idols and Land Grabs," 52.

86 Ellen Davis, "The Pain of Seeing Clearly: Prophetic Views on the Created Order," in *Biblical Prophecy: Perspectives for Christian Theology, Discipleship, and Ministry* (Louisville: Westminster John Knox, 2014), 83–110 [88].

in the world. One of the challenges in contemporary Indigenous conversations about "sovereignty" has been how to assert the autonomy and agency of Indigenous people without simply adopting the categories and definitions of colonial rhetoric. "Sovereignty!" writes Laura Harjo. "The discourse surrounding the term . . . transforms it into a strange and meaningless word."[87] Patricia Monture-Angus insists that, to truly resist imperial logic, sovereignty needs to be both asserted and redefined. "Sovereignty," she states, "when defined as [the] right to be responsible, . . . requires a relationship with territory (and not a relationship based on control of that territory); . . . the Aboriginal request to have our sovereignty respected is really a request to be responsible."[88] In Ezekiel's vision of radical relation between people and place, something like Monture-Angus' definition emerges. There is both an assertion of borders and an assumption of co-responsibility. Questions remain about how this understanding of sovereignty shapes Ezekiel's community and its relationship to YHWH, and the objections scholars raise to YHWH's violence are unresolved. But these questions and objections are undergirded by a crucial insight: a conception of place distinct from Neo-Babylonian constructions. Ezekiel's theology of place is not a theology of empire.

The Purity of the Land and the Question of Belonging

When compared to other prophetic texts, scholars regularly note that Ezekiel's diagnosis of Jerusalem's destruction emphasizes failures of cultic purity and idolatry over "social justice concerns" (such as treatment of the poor and vulnerable).[89] There are denunciations of judicial and economic abuses described in the text (e.g., Ezek 13:18–19; 18:5–9; 22:6–12), but these are mostly assessed as marginal. Whether resulting from his vocational training as a priest[90] or his status as an elite member

[87] Cited in Stephanie Nohelani Teves, Andrea Smith, and Michelle H. Raheja, *Native Studies Keywords* (Tucson: University of Arizona Press, 2015), vii.

[88] Patricia Monture-Angus, *Journeying Forward* (Halifax: Fernwood, 1999), 36.

[89] Andrew Mein, *Ezekiel and the Ethics of Exile* (Oxford: Oxford University Press, 2001), 3. As I describe, I find more references to economic and judicial justice in Ezekiel than are regularly acknowledged; at minimum, there is an interdependence between cultic and "social justice" concerns.

[90] Baruch Schwartz notes that Ezekiel "explains what went wrong, depicts the results of what went wrong, and predicts the eventual rectification of everything that went wrong, from a thoroughly priestly standpoint": "Priest Out of Place," 62.

of Judean society,[91] Ezekiel's attentiveness to questions of pollution and purity—and his mapping of these categories onto Judah's destruction and restoration—are recurring touchstones in the book's interpretation.[92] The theme has not endeared Ezekiel to the modern academy.

A century ago, critiques of cultic purity could take on an ugly, antisemitic tone. Julius Wellhausen describes the Priestly tradition's cultic concerns as "a systematic decline into heathenism."[93] In recent scholarship, Ezekiel's focus on purity and pollution has been judged for encouraging "shame and self-loathing"[94] or creating unwieldy hierarchies. Stephen Cook and Corrine Patton note that the prophet's interest in preserving purity is connected to Ezekiel's desire to create a space where a holy God can dwell. "The stress on spatial order, separation, and gradation, which borders on hyperbole, unfolds a new 'failsafe' reality in Israel that ensures the permanent dwelling of God's glory (Ezek 43:7, 44:2)."[95] Cook and Patton are agnostically evenhanded about whether such hierarchies work for society's health or harm, but other scholars are more skeptical. Particularly among those who prioritize environmental concerns, Ezekiel's attention to structure and hierarchy is interpreted as an embrace of urban, or even explicitly Babylonian, norms. Galambush notes that "Ezekiel's extreme preference for the ordered world—to the point of wild nature's exclusion . . . reflect[s] the sociopolitical tensions of the exile."[96] She reads his interest in purity and pollution, order and hierarchy (8:10; 16:30–32; 22:9–10; 39:12, 14, 24; 44:4–27) as a way to deal with the exiles' "profound land anxiety"[97] through structures of control. Norman Habel sees Ezekiel's cultic interests, and particularly his focus on the temple, as

91 Mein, *Ezekiel and the Ethics of Exile*, sees Ezekiel's concerns as characteristic of a member of a "privileged class dealing with change" (75).

92 Pikor provides a counterpoint to this approach, taking the time to delineate Ezekiel's critique of "social sins" against the powerless and muddying the dichotomy between cultic and justice concerns. He notes, for example, the relationship between idolatry, bloodshed, and child sacrifice (*Land of Israel*, 51–57).

93 Julius Wellhausen, *Prolegomena to the History of Ancient Israel* (New York: Meridian Books, 1957), 297, quoted in Daniel Smith-Christopher, *The Religion of the Landless: The Social Context of the Babylonian Exile* (Eugene, Ore.: Wipf and Stock, 2015), 139.

94 Carley, "From Harshness to Hope," 114.

95 Stephen L. Cook and Corrine L. Patton, "Hierarchical Thinking and Theology in Ezekiel's Book," introduction to Cook and Patton, *Ezekiel's Hierarchical World*, 1–26 [13].

96 Galambush, "God's Land and Mine," 106.

97 Galambush, "God's Land and Mine," 108.

indicating an antipathy toward "rural orientation[s]"[98] and a rejection of the earth's innate goodness.

Embedded in this scholarly discomfort is a different ideology of purity at work. Creation is understood to be most liberated when unsullied by human hands. In one way, this affirmation protests the settler-colonial conflation of land and productivity. Such conflations also fueled the Neo-Babylonian propaganda that undergirded the imperial economy: humans and creation were only valuable for their ability to produce food for the gods and the gods' representatives. Habel is correct that creation has innate value; its creation in Genesis precedes that of humans, and it has more than a utilitarian function.[99] But when Galambush worries that fertile land is still "worked" in Ezekiel's visions of restoration[100] or Habel balks at Ezekiel's vision of a world "that apparently begins and ends with the city,"[101] there are echoes of a different colonizing danger. Western environmentalism has a history of "fetishizing pristine nature" and "disconnecting nature from natives."[102] Steed Vernyl Davidson describes the contemporary tourism industry as an attempt to find a lost paradise, "a replica of Eden that remains unspoiled by human sin."[103] Noel Salazar critiques romantic projections of the natural world grounded in "the myth of the unchanged, the myth of the unrestrained and the myth of the uncivilized."[104]

98 Habel, "Silence of the Lands," 133.

99 Habel notes that the intrinsic value of the earth is a fundamental principle of the Earth Bible project ("Silence of the Lands," 127–28).

100 Galambush, "God's Land and Mine," 102.

101 Habel, "Silence of the Lands," 130. There appears to be a contradiction between Habel's stated principles and his analysis. In justifying a hermeneutic that "see[s] things the way Earth would view them"—an admittedly "precarious" task—Habel insists that "we . . . are an integral part of the Earth," rejecting a "false dualism" between human and nonhuman creation (128). His critique of Ezekiel's "urban cosmogeny," however, portrays this interrelation as suspect (129).

102 S. Lily Mendoza and George Zachariah, introduction to *Decolonizing Ecotheology: Indigenous and Subaltern Challenges*, ed. S. Lily Mendoza and George Zachariah (Eugene, Ore.: Wipf and Stock, 2022), 1–16 [4]. For example, J. M. Bacon outlines the historical connections between the U.S. conservation movement and discourses of "national expansion" and Indigenous displacement in "Settler Colonialism as Eco-social Structure and the Production of Colonial Ecological Violence," *Environmental Sociology* 5, no. 1 (2019): 59–69 [61–62].

103 Steed Vernyl Davidson, "Lost Paradises: Tracing the Imperial Contours of Modern Tourism on Land and Peoples," in *People and Land: Decolonizing Theologies*, ed. Jione Havea (Minneapolis: Fortress Academic, 2019), 15–34 [22].

104 Noel B. Salazar, *Envisioning Eden: Mobilizing Imaginaries in Tourism and Beyond* (New York: Berghahn Books, 2010), 43.

Mari Joerstad's reading of prophetic approaches to creation brings a helpful clarity. "The prophets do not ask themselves how to preserve remote nature—but how to live well with others!"[105] Neither do cultic laws and the natural world work against each other. Joerstad notes how, in the Priestly tradition, significant festivals are tied temporally to a celestial calendar. Human obedience is related to an "obedient cosmos."[106] Her point refocuses Ezekiel's concerns with purity on right relationality—right relation with God, nonhuman creation, and other human persons.

The connection between Ezekiel's interest in purity and his commitment to a relationship with YHWH may be the most obvious of these three. Building on Zimmerli's suggestion that the land has a sacral character in Ezekiel,[107] Pikor asserts that "the land is the space in which Israel's communion with God will be fully realized."[108] This is what makes the purity of the land and the community's right relationship to the land so significant. Without such a relationship, "[the land] ceases to function as a space of contact between God and humanity."[109] What is less clear is how Ezekiel's concerns with purity ground his understanding of the land's ability to create life-affirming contact and connection "between humans themselves."[110] Purity laws seem designed to separate, categorize, and distinguish. In Ezekiel, however, these laws serve an additional purpose for human relationships. They also give shape and permeability to the category of belonging. Belonging is a crucial and disputed signifier in a world where land and people are bound together and where displacement from land is an ongoing reality. Part of what is at stake is the definition of social identity: how does one define a "people" when it has been dispersed? But there is another, often-overlooked aspect to the question of belonging highlighted by Ezekiel's attention to purity practices: the land's hospitality to others. "Who," in Pikor's words, "has the right to . . . inhabit the land?"[111]

As discussed in chapter 1, there is a dangerous shadow to theologies that emphasize the integral relationship between people and place. Such

105 Joerstad, *Hebrew Bible and Environmental Ethics*, 155.

106 Joerstad, *Hebrew Bible and Environmental Ethics*, 169. Joerstad engages the work of Jeffrey L. Cooley, "Psalm 19: A Sabbath Song," *Vetus Testamentum* 64, no. 2 (2014): 177–95, to flesh out this claim. She also extends his argument, noting that "sacrifice, purity laws, and agricultural regulations are all tied to calendric concerns."

107 Zimmerli, *Ezekiel*, 1:261.

108 Pikor, *Land of Israel*, xxi.

109 Pikor, *Land of Israel*, 43.

110 Pikor, *Land of Israel*, 43.

111 Pikor, *Land of Israel*, xix.

approaches can become rigidly exclusive, defining "belonging" through blood ties and sedimentations of ethnicity. Ezekiel's emphasis on purity does something different. It defines the relationship to land in performative rather than ontological terms. Pikor notes Ezekiel's criticism of "the belief that entitlement to the land stems from the mere fact of the land's inhabitancy" (Ezek 11:15). This would foreclose the exiles' land claim through the simple fact of their absence. But Ezekiel also criticizes those who codify the relationship between people and land through "blood ties to Abraham, to whom (and to whose offspring) God gave the land (Ezek 33:24)."[112] In the words of Daniel Block, YHWH's gift of land is not mechanistically tied to "Darwinian materialism."[113] Belonging is legitimated through fidelity to YHWH and actions of "righteousness and justice" (Gen 18:19).

There is a tension, then, created by Ezekiel's emphasis on purity—a tension revealed in two distinct portraits of the prophet in scholarly work. Sometimes Ezekiel is understood to be foreshadowing Ezra and Nehemiah, delineating and maintaining social boundaries as a mechanism of survival. Smith-Christopher argues that laws governing cultic purity are a potent method for "preserving and maintaining" a distinct minority identity in the face of displacement.[114] Scholars like Mark Brett paint a starkly different picture, emphasizing Ezekiel's distinction from Ezra and Nehemiah's concerns. He notes, for example, how the land provisions Ezekiel makes for foreigners (*gerim*) in his final vision (Ezek 47:22) do not mirror Ezra's wariness about "mixing . . . with the peoples of the lands" (Ezra 9:1–2).[115] Brett connects Ezekiel, instead, with the pragmatic, Priestly approach to displacement and difference addressed through Leviticus' Holiness Code. The Code's attention to justice for the stranger (e.g., Lev 19:33–34; 24:22)[116] honors the complexities and vulnerabilities of displaced people, underscored through the experience of exile. It speaks, as well, to the challenges that displacement and difference raise for relational understandings of land. Brett argues that Ezekiel's cultic concerns negotiate these

112 Pikor, *Land of Israel*, 3.

113 Daniel I. Block, *The Book of Ezekiel: Chapters 25–48* (Grand Rapids: Eerdmans, 1998), 260.

114 Smith-Christopher, *Religion of the Landless*, 149.

115 I will return to the complexities surrounding Ezekiel's provision of land to the *gerim* in chapter 5.

116 Mark Brett, *Political Trauma and Healing: Biblical Ethics for a Postcolonial World* (Grand Rapids: Eerdmans, 2016), 91–109. The Holiness Code is found in Lev 17–26.

questions in similar ways, insisting that strangers share a responsibility for maintaining the land's sanctity,[117] but allowing for a world in which "permeable social boundaries need not undermine the core identity and vocation of Israel."[118]

This negotiation of belonging is visible in many displaced Indigenous communities today. Terikata Teikabua is a Banaban Methodist pastor who has lived his entire life on Fijian soil. The history of Banaba is a textbook example of colonial disregard for colonized land and peoples, and the relationship between them. After the Second World War, when industrial phosphate mining left the Banaban homeland uninhabitable, the British government relocated the Banaban people a thousand miles away to Rabi, an island in the Fijian archipelago.[119] When Teikabua preached to Davuilevu Theological College in 2016 on practices of "righteousness" in relation to the earth, he spoke as an insider, with deep knowledge of Fijian customs and norms, and also as an outsider displaced from the land of his ancestors.[120] In describing righteous ecological practice, Teikabua notes the importance of Banaban *tabu* rituals that alternate prohibitions between deep-sea fishing and reef fishing in order to sustain fish populations.[121] He also notes the importance of Fijian Christian rituals surrounding *sevu*, the "first fruits" offering brought to the church after harvest. This *sevu*, he argues, ties people to place. While his sermon has all sorts of generative possibilities for ecotheology, it also raises compelling possibilities for the thorny political and cultural questions around "who belongs"—possibilities that surface precisely because of his social location as a displaced, Indigenous person. Teikabua does two things simultaneously. He asserts the importance of

117 Ezekiel is not, of course, envisioning a modern, secular democracy. There are regulations that "bind all who live in the land (Ezek. 20:2–5)" (Brett, *Political Trauma*, 109).

118 Brett, *Political Trauma*, 109.

119 Wolfgang Kempf's "Climate Change, Migration, and Christianity in Oceania" provides a succinct history of the Banaban migration to Rabi, Fiji, in *Climate Change and Human Mobility*, ed. Kirsten Hastrup and Karen Fog Olwig (Cambridge: Cambridge University Press, 2012), 243–47.

120 My thanks to Rev. Terikata Teikabua for the use of his sermon manuscript in this analysis: "The True and Righteous Faith: Unwrapping the Oceanic Word of 'Tabu/Tapu,' Romans 4:13–25," sermon preached at Davuilevu Theological College, Nausori, Fiji, March 2015.

121 The practice, known as *katabuan marawa na vakatatabu ni waitui loa*, builds on the word *tabu* and its delineation of various locations of ocean as sacred and "off limits" in different seasons.

cultural distinctives (Banaban *tabu* practices) for communities separated from ancestral land, while he also argues for legitimate belonging in a new land based on his participation in the ritual practices (Fijian *sevus*) of a new community. In Teikabua's sermon, easy distinctions between cultic purity and social justice unravel. Both relate to the political, cultural, and ethical issues of belonging in displaced communities, as well as the belonging *of* displaced communities in the land where they now reside.[122] In this case, Teikabua's description of ecological righteousness preserves the Banaban community's cultural particularity, even as it asserts the legitimacy of their Fijian land claim.

Teikabua's example does not resolve the questions raised by Ezekiel's emphasis on cultic purity. Ezekiel's historical context is different, and there is much about the Judean experience of displacement that is unknown.[123] But Teikabua's sermon does underline how ritual practices of purity (e.g., practices of *tabu* and *sevu*) help negotiate belonging in displaced communities that affirm the interdependence of place and persons. Indeed, it demonstrates how Ezekiel's distinct caricatures in academic discourse (i.e., guardian of social identity or defender of the stranger) can coexist with coherence. Cultic rituals create modes of belonging within a community whose land has been stripped away, but they also clarify the performative basis on which a community belongs to land in the first place. In Ezekiel, the community's belonging to land, to YHWH, and to each other is more than an ethnic or political signifier. The answer to "who can inhabit the land" is not answered by blood relation or military might. It is answered through performances of responsible, relational fidelity

[122] Wolfgang Kempf and Elfriede Hermann focus on the challenges of "belonging" in light of Oceanic Indigenous displacement in "Uncertain Futures of Belonging: Consequences of Climate Change and Sea-Level Rise in Oceania," in *Belonging in Oceania: Movement, Place-Making and Multiple Identifications*, ed. Elfriede Hermann, Wolfgang Kempf, and Toon van Meijl (New York: Berghahn Books, 2014), 189–213.

[123] One significant point of contemporary conversation, for example, is Ezekiel's status relative to "the people of the land" and the impact of exiles' return on the unexiled population. "Empty land" theories have been persuasively challenged (Hans Barstad, *The Myth of the Empty Land: A Study in the History and Archaeology of Judah during the "Exilic" Period* [Oslo: Scandinavian University Press, 1996]), but the material conditions of the unexiled population shifts in various readings. Certain scholars describe a new "upper class" on friendly terms with Babylon (Smith, *Religion of the Landless*, 195), while others maintain that, even in exile, Ezekiel maintained "privileged class" status (Mein, *Ezekiel and the Ethics of Exile*, 75). See also Stephen L. Cook, *Prophecy and Apocalypticism: The Postexilic Social Setting* (Minneapolis: Fortress, 1995).

to YHWH's covenant—performances of faithfulness that differ depending on the particularities of one's identity and one's connection to the covenant community.[124] This is no simple solution, but it is Ezekiel's solution. It is a solution that allows Ezekiel to negotiate the grief-soaked questions of belonging experienced by displaced people. More subversively, it destabilizes imperial models of belonging grounded in conquest and fixed ethnonational distinctions.[125]

For readers of a different era, it is worth noting how Ezekiel's solution also destabilizes models of belonging more familiar to contemporary expressions of settler colonialism. Modern assertions of belonging between people and place may not be codified by royal land grants engraved on Babylonian stone stelae,[126] but they are asserted through other means: bills of sale, national citizenship, free-floating spiritualities, and fluent performances of globalized culture.[127] Ezekiel's emphasis on the covenantal commitment between people and place, in the context of covenantal commitment to YHWH, creates a model for belonging that is more challenging. He asserts a belonging that is both particular and permeable—with no guarantees.[128] There are all sorts of reasons why Ezekiel's emphasis on purity would make modern readers uncomfortable, not least of which is the prophetic insistence that places make claims on those who benefit from them.

124 Brett notes that the *gerim* were required to abide by certain regulations that bound all those who lived in the land (e.g., Lev 20:2–5), but they were not compelled to participate in Passover celebrations. If they wished to do so, male circumcision was required. "The *ger* had choices in relation to the cult that the Israelites did not enjoy" (*Political Trauma*, 109).

125 Perdue and Carter clarify that, in Babylon, "adopting the major culture features of the ruling metropole" did not create permeable categories of belonging. "Exiles [were] never accepted as equals by the Babylonian" regime (*Israel and Empire*, 81).

126 J. A. Brinkman, "Babylonian Royal Land Grants, Memorials of Financial Interest, and Invocation of the Divine," *Journal of the Economic and Social History of the Orient* 49, no. 1 (2006): 1–47, http://www.jstor.org/stable/25165127.

127 Eve Tuck and K. Wayne Yang describe various "moves to innocence" (3) enacted by settler communities that assert the legitimacy of settler land claims and their belonging to the land. These can involve the appropriation of Native designations or decontextualized spiritual practices that erase differences rather than engage them. They can also presume a global citizenship, where one is at home everywhere: "Decolonization Is Not a Metaphor," *Decolonization: Indigeneity, Education and Society* 1, no. 1 (2012): 1–40.

128 Pikor notes that Ezekiel's understanding of YHWH's promises of belonging between people and land does not constitute an "unconditional obligation" on YHWH's part (*Land of Israel*, 105).

Human Passivity in the Land's Restoration and the Question of Loss

In the last two sections, I have noted how Ezekiel's theology of place disrupts the universality of Babylonian imperial claims and ethnonational models of belonging from unlikely sites of resistance. Ezekiel's emphases on divine punishment and cultic purity have been scrupulously interrogated by modern interpreters, and rightly so. But I have argued that these problematic themes are vantage points from which Ezekiel's distinct theology of place comes into focus. His is a social imaginary that retains a "necessary ambiguity [and] polyvalence,"[129] holding together the tensive needs of displaced communities. Place is bounded and hospitable, responsible and responsive. Place is promised, even as place is provisional. This last section brings another such tension to the fore. It places Ezekiel's call for communal performances of faithfulness alongside Ezekiel's passive witness to the land's restoration. Ezekiel insists that human action will not be the catalyst of restoration for places that have been destroyed, a seemingly counterproductive assertion for those mobilizing communities to protect the planet. For many present-day environmental activists, passivity of any sort seems profoundly out of touch with the urgency demanded by the crisis.[130] I will argue again that this difficult vantage point becomes a site of resistance for displaced theologies of land, refusing to diminish the very real crisis of place's loss.

Ezekiel's focus on purity is remarkable for a reason unmentioned in the prior section. For all of the theme's centrality in his writings, Ezekiel is skeptical about its attainment. The prophet has a dim view of his community's capacity for performances of responsible, relational fidelity, which leads to one of the book's great challenges for environmental activists trained in Western spaces. Ezekiel insists that YHWH is the sole catalyst for the land's restoration and the people's return. Whether speaking of the exiles' inability to stop the disasters of the book's first half or their marginality in restoring creation in the book's final chapters,

[129] Oscar García-Johnson, "Faith Seeking for Land: A Theology of the Landless," in Yeo and Green, *Theologies of Land* (Eugene, Ore.: Cascade, 2021), 38–68 [62].

[130] Greta Thunberg's charge to "panic" in her 2019 speech to world leaders and economists in Davos, Switzerland, exemplifies this urgency. The transcript of her speech can be found at "'Our House Is on Fire': Greta Thunberg, 16, Urges Leaders to Act on Climate," *Guardian*, January 25, 2019, https://www.theguardian.com/environment/2019/jan/25/our-house-is-on-fire-greta-thunberg16-urges-leaders-to-act-on-climate.

Ezekiel's prophesies are marked by a "radical theocentricity."[131] YHWH is the central character in Ezekiel's story, and the people are passively dependent on YHWH's action.

It is not just that YHWH is powerful and humans are not. The actions of the people do not awaken divine power or change the divine heart. Human action is not the crux on which Ezekiel's plot turns. Patton notes that the "repeated pattern in the book is not sin-punishment-repentance-restoration, but sin-punishment-restoration-repentance."[132] Humans do participate in YHWH's work. The returned exiles assist in the land's purification (Ezek 39:12–16), Ezekiel prophesies when told to do so (37:4, 9), and his final temple vision has concrete implications for the people's economic and political life.[133] But none of this participation is possible without YHWH's prior initiative. For much of the book, Ezekiel is not even able to speak unless YHWH has given him something to say (3:26–27). This causes Block to contrast Ezekiel's description of divine abandonment and return with Babylonian motifs of the same. In Babylon, God returns after repentance, while "Ezekiel emphasizes that by a divine act the subjects' hearts will be changed (11:18–21)."[134]

Ezekiel's relentless prioritization of divine agency has led to descriptions of the prophet as an exilic "millennialist," a sociological term used to describe groups designated as apocalyptic by biblical scholars.[135] In

[131] Paul Joyce, *Divine Initiative and Human Response in Ezekiel* (Sheffield: JSOT Press, 1989), 89.

[132] Corrine L. Patton, "Priest, Prophet, and Exile: Ezekiel as a Literary Construct," in Cook and Patton, *Ezekiel's Hierarchical World*, 73–90 [83].

[133] Ellen Davis notes the "practical concern for human welfare" in Ezek 47 and the "intricate systems of interaction" between land and people and nonhuman creatures created by God's healing waters: *Biblical Prophecy*, 126.

[134] Daniel I. Block, "Divine Abandonment: Ezekiel's Adaptation of an Ancient Near Eastern Motif," in *The Book of Ezekiel: Theological and Anthropological Perspectives*, ed. Margaret S. Odell and John T. Strong (Atlanta: SBL, 2000), 15–42 [39].

[135] See, for example, Mein, *Ethics of Exile*, 240. For discussion of the term *millennial*, see Sylvia L. Thrupp, "Millennial Dreams in Action: A Report on the Conference Discussion," in vol. 2 of *Millennial Dreams in Action*, ed. Sylvia L. Thrupp (The Hague: Moutnon, 1962), 11–27 [11–12]. Scholarly distinctions between prophetic, proto-apocalyptic, and apocalyptic eschatologies are slippery. A key question in this designation is whether Ezek 40–48 indicates a reality meant to be realized in history or if it depicts an eschatological, otherworldly hope. Along with others, Barbara Rossing's work on Revelation sheds light on the overlaps and complexities in this dichotomy. Apocalyptic literature can engage history and eschaton simultaneously. See "Lost Land: Visualizing Deforestation and Eschatology in the Apocalypse of John and the Column of Trajan in Rome," in Havea, *People and Land*, 159–74.

these depictions, the combination of Ezekiel's "great distress" and "political helplessness" fuels his belief in YHWH's cataclysmic intervention in the social order. Andrew Mein argues, "Decisions and actions that might realistically bring about a national restoration no longer form part of [the exiles'] moral world."[136] Apocalyptic or no, Ezekiel's rejection of human solutions as "false hope"[137] is clear. Judah's salvation does not lie in the "international politics of Zedekiah."[138] Ezekiel disdains Zedekiah's attempt to circumvent displacement through foreign alliances (17:11–21). Ezekiel's final vision of the land's restoration (chs. 40–48) may have concrete political and economic implications, but it is not presented as a "blueprint" for restoration—at least not a restoration made possible by human means. For Mein, it serves more as a "public testimony to the nature of God."[139] Not even correct cultic behavior will save the day. Despite his fury about contamination of the temple by the preexilic priests (22:26), Ezekiel does not put the responsibility for the land's restoration in priestly hands. In Schwartz's reading, "It is too late for priestly teaching to rectify the problem—YHWH alone, not the Israelites, will perform the purification of the future."[140] The people remain, in Mein's words, "passive recipients"[141] of YHWH's promise.[142]

In her engagement of ecological trauma in Ezekiel 47:1–12, Rebecca Copeland worries about this denial of human agency and natural, causal processes. She finds Ezekiel's role as a "passive observer"[143] of YHWH's restorative work ecologically dangerous. Such depictions justify theologies that relegate the earth's restoration to divine acts of "miraculous healing . . . [and] supernatural powers"[144] when a more "rational" human response is necessary.[145] One does not address ecological

136 Mein, *Ethics of Exile*, 256.

137 Pikor, *Land of Israel*, 7.

138 Patton, "Priest, Prophet, and Exile," 89.

139 Mein, *Ethics of Exile*, 254.

140 Schwartz, "Priest Out of Place," 70.

141 Mein, *Ethics of Exile*, 255.

142 Jacqueline Lapsley teases out various theologies of moral agency in Ezekiel, but argues that the book's final move stresses God's gift of moral identity as revelatory knowledge—a move away from anthropocentric, action-focused models of the "virtuous moral self": *Can These Bones Live? The Problem of the Moral Self in the Book of Ezekiel* (New York: De Gruyter, 2000), 6–8.

143 Copeland, "Their Leaves Shall Be for Healing," 219.

144 Copeland, "Their Leaves Shall Be for Healing," 219.

145 Copeland, "Their Leaves Shall Be for Healing," 221.

traumas through "magical" means; they are addressed through clear-eyed scientific analyses.[146] There is a "can-do" optimism to the problems of ecological devastation in Copeland's essay. She postulates that, when interpreted correctly, Ezekiel's text holds traditional ecological knowledge (TEK)—ecological wisdom embedded in the ancient memories of Indigenous communities—that can provide the keys necessary for the "engineering and biological remediation" of salinized soil.[147] Reading Ezekiel 47 as if it were both an Indigenous excavation site and a scientific field journal, Copeland imagines how the river that flows from the temple would impact the water table of the region and notes the importance of shoreline tree planting for the land's thriving. Her argument is creative and interdisciplinary. She draws on scientific research from the Department of Primary Industries and Regional Development of Western Australia to link British colonial and Babylonian practices of irrigation with declines in the long-term health of ecosystems.[148] Troublingly, however, Copeland grounds her thesis in simplistic dichotomies between her "rational" approach and the "irrational"[149] approaches of traumatized communities that look for spiritual solutions to material problems. While she justifies her proposal as an example of TEK, she does not quote any traditional Indigenous community in her analysis,[150] nor does she speak to anyone displaced by the ecological trauma she describes—which gives me pause. The "idealization . . . of TEK" has been critiqued by certain Indigenous researchers as a way environmental education appropriates Indigenous knowledge.[151] But Copeland does more than idealize TEK; she constructs it. She is worried enough about Indigenous passivity that she removes Indigenous voices altogether. In her desire to fix, solve, and demythologize knowledge systems, the

[146] Copeland, "Their Leaves Shall Be for Healing," 220.

[147] Copeland, "Their Leaves Shall Be for Healing," 220.

[148] Specifically, Copeland draws on Katrina Proust's critique of British colonial irrigation practices that led to widespread salinization and infertility of soil ("Their Leaves Shall Be for Healing," 219). See Proust, "Salinity in Colonial Irrigation: British India and South-Eastern Australia," *Australian Geographer* 39, no. 2 (2008): 131–47.

[149] Copeland, "Their Leaves Shall Be for Healing," 221.

[150] Copeland does not mention if Proust has any relational or genealogical connection with the Indigenous communities impacted by her analysis.

[151] Eve Tuck, Marcia McKenzie, and Kate McCoy, "Land Education: Indigenous, Post-colonial, and Decolonizing Perspectives on Place and Environmental Education Research," *Environmental Education Research* 20, no. 1 (2014): 1–23 [10].

"vulnerability and uncertainty"[152] that marks generative engagements between Western science and Indigenous knowledge is muted.

Perhaps Ezekiel's focus on passivity is clear-eyed in ways that are harder for Western science to hear—ways that illumine brutal realities of power and the contours of a world that cannot be controlled. I would argue that there is a provocative consequence to Ezekiel's marginalization of human agency in relation to the action of YHWH. Passive dependence provides a precise counterpoint to the hubris of the Neo-Babylonian place-making project. Babylonian power was memorialized in monuments and temples. Gerdien Jonker describes Babylon as an ongoing construction zone, with "places of worship . . . continually being restored and expanded, . . . an ever-present witness to a glorious past."[153] Various scholars have speculated on the significance of these monuments for Ezekiel's final temple vision.[154] They note stylistic resonances, but also "subversive transformation[s],"[155] similar to the ways Mesopotamian ideology was challenged in the creation account of Genesis 1. The greatest contrast, however, lies not in the temple's structure, but in Ezekiel's role as a "passive observer." YHWH's temple and the miraculous stream that flows from it have been created by something more than human effort and advanced technology. In the prophet's passive observation of divine power, he resists Neo-Babylonian mechanisms of historical and ecological control. Ezekiel's vision proclaims that the empire's tools have not been able to dismantle YHWH's house with any finality—and neither have the empire's tools been used to rebuild it.

Ezekiel's passivity resists Neo-Babylonian understandings of place in a more vulnerable way, as well. It tells the truth about the loss his people have experienced and does not minimize the costs of displacement. Land can be lost in ways that no human technology, irrigation canal, or royal edict can restore. Devastation can be so severe and valleys of death so broad (Ezek 37:1–3) that only an act of YHWH can bring back life. Ezekiel's passivity

[152] Amiria Salmond, "Re-building Ships at Sea," in *Environmental Transformations and Cultural Responses: Ontologies, Discourses, and Practices in Oceania*, edited by Eveline Dürr and Arno Pascht (New York: Palgrave Macmillan, 2017), 215–26 [223].

[153] Gerdien Jonker, *The Topography of Remembrance* (Leiden: Brill, 1995), 36–37, cited in Adriane Leveen, "Returning the Body to Its Place: Ezekiel's Tour of the Temple," *Harvard Theological Review* 105, no. 4 (2012): 385–401 [389].

[154] See, for example, Tova Ganzel, *Ezekiel's Visionary Temple in Babylonian Context* (Berlin: De Gruyter, 2021). Margaret Odell also notes the parallels between Ezek 40–48 and building inscriptions describing King Esarhaddon's rebuilding of Babylon (*Ezekiel*, 483–84).

[155] Middleton, *Liberating Image*, 196.

is more than a psychological coping mechanism. It testifies to the severity of the exile's toll and argues that no human attempt at place-making can make it right. There is surely lament in this acknowledgment—but there is also anger. Passivity can be a protest.

The former prime minister of Tuvalu, Enele Sapoaga, has made it clear that there is no plan B for his country's survival. He has vehemently rejected proposals for relocation, even as Tuvalu is slowly being submerged in water. Sapoaga is choosing passivity in the face of rising tides. To do otherwise, he says, would hide the loss that forced migration would entail. "Where would we go," he asks, "and what would happen to our sovereignty?"[156] Forced migration would shift responsibility away from the nations most responsible for the crisis. It would solve the problem, not by addressing the roots of climate change, but through one more severing of people and land. Relocation would "only serve to let the culpable . . . countries off the hook."[157] Sapoaga's passivity refuses to minimize the crisis. He is done with human solutions that negate the suffering of his people. In refusing to act, he deliberately withholds false hope from a world that wants absolution for its complicity in displacement's pain. Sapoaga would rather they felt it.

Elizabeth Phillips argues that "what is central to the apocalyptic—indeed, what makes something an apocalypse—is not cataclysm, but unveiling, 'an imaginative vision of the future that functions to jolt us out of our status-quo perceptions and see the current social reality for what it truly is in relation to God.'"[158] Ezekiel's passivity has that quality of "unveiling." It protests false hopes of minimization, insisting that the depth of the problem be seen (Ezek 8:7–18). It protests materialistic solutions that dismiss the spiritual crisis of displacement (36:26–27). It refuses to construct a solution that absolves those most responsible (22:6–12). Most significantly, it refuses to pretend that all is well in the face of unimaginable loss. It is significant how much of Ezekiel's witness involves silence and

[156] Anthony Stewart, "Tuvalu PM Slams Kevin Rudd's Proposal to Offer Australian Citizenship for Pacific Resources as Neo-colonialism," ABC News, February 17, 2019, https://www.abc.net.au/news/2019-02-18/tuvalu-pm-slams-kevin-rudd-suggestion-as-neo-colonialism/10820176.

[157] Seforosa Carroll, "Displacing Liturgy: A Pacific Exploration," *Liturgy* 36, no. 2 (2021): 26–35 [27].

[158] Elizabeth Phillips, "Narrating Catastrophe, Cultivating Hope: Apocalyptic Practices and Theological Virtue," *Studies in Christian Ethics* 31, no. 1 (2018): 17–33. Phillips references her work in *Political Theology: A Guide for the Perplexed* (London: T&T Clark, 2012), 148.

stillness. He sits speechless (3:12–15, 26); he lies on his side (4:4–8). He withholds cries of mourning (24:15–17). Passivity becomes a way to name what is beyond the scope of words; it becomes a way to "unveil" the pretense of a world where human capacity and control are the final sources of salvation.

When Cyclone Winston demolished the northeastern corner of Viti Levu in 2016, Davuilevu students were deployed to villages across the country to provide trauma-informed pastoral care. They returned with stories of village elders, silent and still, sitting on destroyed porches or under makeshift tarps. The Davuilevu students did not preach sermons or organize building campaigns, though those would come. They sat alongside the grieving and listened to their stories. They heard halting accounts of flattened harvests, demolished schools, and departed family members. Places had been lost, and there was no clear way forward. False hopes—of quick recovery, foreign aid, or the praying away of catastrophic grief—were insufficient to the need. Real hope was veiled.

When YHWH asks Ezekiel if bones can live again, the prophet is equivocal, answering, "O Lord God, you know" (Ezek 37:3). Hope is not in Ezekiel's control.

The Specters in the Mirror and the Questions Not Asked

Specters are funny things. The ones we see are not necessarily the ones we need to worry about most. I've organized this chapter around Ezekiel's specters in Western biblical conversations: his chilling descriptions of punishment, his obsessions with purity, and his passive grief. For many scholars who prioritize ecological readings of Scripture, these specters cause consternation and anger. Ezekiel's descriptions of YHWH's destruction, his focus on hierarchy and cultic purity, and his dogged insistence that only YHWH can save seem profoundly out of touch when the world is burning. I have argued, however, that precisely in these difficult, contested aspects of Ezekiel's witness, he does vital work. He disrupts Neo-Babylonian conceptions of place, unveiling the dangerous universality, insularity, and hubris undergirding the apocalypse of imperial expansion. He points to different specters: place defined as object, place defined by blood, and place constructed by empire. The point is not to absolve Ezekiel of his violent depictions of the deity or suggest his solutions as master plans for contemporary practice. There are profound questions about how Ezekiel's theology of place would play out in a multireligious democracy or intersect the necessary work of mobilizing global action for climate justice. The point is not to collapse Ezekiel's world and our own, nor is it to exorcize his ghosts. For preachers and scholars trained in Western spaces, the point is to pay attention to ours.

The scholarly critiques of Ezekiel's witness lifted up in this chapter—particularly those made by those who prioritize the earth—suggest two things. First, many conversations about climate justice in the academic guild are not asking the questions Ezekiel asks. Ezekiel foregrounds questions of sovereignty, belonging, and loss in unexpected and tensive ways. In so doing, he holds seemingly contradictory truths side by side. Borders matter to Ezekiel, but so does the belonging of strangers. Purity matters to Ezekiel, even as its definition shifts. Hope matters to Ezekiel, but not a hope one can control. Ezekiel's attention to questions of sovereignty, belonging, and loss bring to mind the Fijian disruptions of the Western philosophies of place described in the last chapter. They are questions that emerge again in the Fijian Methodist sermons collected and analyzed in chapter 3. When a community deeply connected to land faces displacement, questions of *who is responsible*, *who belongs*, and *who can save* are fierce. This does not mean that these questions are resolved in the same way. Different histories lead to different negotiations of relation and difference, continuity and change, and agency and dependence. But displacement requires that a community negotiate each. It is significant that displaced communities are asking questions about place's import in ways that resonate. It is also significant that these questions are not those that many Western theologies of place prioritize.

This brings me to my second observation. The Western academy has its own specters—specters hovering just over our shoulders and out of our ranges of vision. A mirror may be necessary to see them. Most scholars no longer categorize place as an object or define it by blood, but biblical scholarship has its own universal approaches to place, its own romantic projections, and its own constructivist tendencies. My work owes much to the thought of Daniel Smith-Christopher, but when he equates the experiences of exiled refugees to the "exilic" experiences of settled faith communities in the United States in his early work, the particularities of his categories falter.[159] Decades later, he voices his concern about "level[ing] out the experiences of all peoples into a kind of generic 'trauma experience' that denies

[159] Smith-Christopher, *Religion of the Landless*, 206–7. Speaking of the U.S. church, he states, "We live in Babylon." The "theological state of the modern Christian is exile."

unique histories,"[160] clearly wrestling with the universalizing dangers of the cross-contextual projects that have formed the work of his life—and the work of mine. When Norman Habel "reads from the perspective of the Earth," desiring to "see things as Earth would view them," one wonders if place's particularities have collapsed altogether.[161] I am grateful for Habel's desire to bring voice to the land in his interpretative work, justifying the attempt through his insistence that human voices are part of the earth's voice.[162] We are not disconnected. But the dangers of projection are real. Speaking for the earth—like speaking for displaced communities—can replace the voices of others with those voices we would rather hear or, more dangerously, with one's own. Ethnographic scholars have grappled with the dangers of speaking for others for decades.[163] Finally, there are the biblical scholars who center the epistemologies of scientific categorization and control in ways that silence the witnesses of displaced communities themselves—whether in the text or in the world—through attempts to fix, diagnose, and educate. Ezekiel holds up a mirror to all of us.

During my years of teaching at Davuilevu, the sermons of Fijian preachers did the same.

When I heard Waqalerua's sermon in 2016, I was not a congregation member or a visitor. I was his instructor. Waqalerua's senior thesis, a thesis my husband was supervising, focused on the very theme that caused me discomfort: the power of prayer in relation to climate change. My husband and I were stumped. Should we let him make this sort of constructive theological claim—that prayer and fasting stopped rising tides—given all of the specters it raised? Or did we ask him to revise his thesis so that it sounded more scientifically and theologically respectable to our Western ears? Notice that neither of these options engaged Waqalerua and his community. I confess that it never occurred to my husband and me

160 Daniel Smith-Christopher, "Reading War and Trauma: Suggestions Toward a Social-Psychological Exegesis of Exile and War in Biblical Texts," in Kelle, Ames, and Wright, *Interpreting Exile*, 269.

161 Habel, "Silence of the Lands," 127.

162 Norman Habel describes the importance of the "mediation of . . . non-human voices to our consciousness by sensitive humans": "Guiding Ecojustice Principles," *Spiritan Horizons* 11, no. 11 (2016): 92–109 [101].

163 Drawing on the work of Glen Hinson, Luke Eric Lassiter gives a concise summary of this history in *The Chicago Guide to Collaborative Ethnography* (Chicago: University of Chicago Press, 2005). See also Elaine Lawless, "'I Was Afraid Someone Like You . . . an Outsider . . . Would Misunderstand': Negotiating Interpretive Differences between Ethnographers and Subjects," *Journal of American Folklore* 105, no. 417 (1992): 302–14.

that, given the seriousness of the crisis at hand, we should be fasting and praying too. Maybe the problem was not that Waqalerua's village was fasting in response to climate change; maybe the problem was that the world's biggest polluters were not.

This is not just a question about piety. It is a question about practical theological method—and about preaching. I have suggested that the most difficult homiletic issues raised by the climate crisis are not those of data transmission or persuasion. They are the unresolved questions of relationship and responsibility between the places of others and ourselves. Such questions interrogate our limits, our power, and the boundaries of our belonging. If this is true, then climate-justice preaching requires more than correct content. It requires courageous, vulnerable postures of relating to the world, the pulpit, and the biblical text—postures that are attentive to the particularities of where we stand and the specific ways we might be called to change.

3

Place in the Apocalypse of Rising Tides

Sovereignty, Belonging, and Loss in Fijian Pulpits

Marika Bale and his family were packed into a rented car over the Christmas summer holiday, heading through back roads in the coastal region of Ra on the island of Viti Levu. They were just hours from Fiji's capital, but they were a world away from Pacific Theological College in Suva, where Bale had recently graduated with honors from his master's program. The roads they drove down were unpaved, and a number of the villages they passed were still marked by the ravages of Cyclone Winston. Winston was the Southern Hemisphere's strongest cyclone on record when it hit Fiji in 2016, knocking down forests like matchsticks and leaving entire swaths of the interior without roofs or power. It was five years later, and there were still washed-out roads and communities relying on tents for shelter. COVID-19 had not eased the challenges of rebuilding. The country's pandemic travel ban had been in effect for six months, particularly isolating for regions inaccessible by Internet. Phone service was spotty because electricity was spotty; one needed a generator to keep a phone operating. So when the travel ban lifted in November, Bale decided that an in-person visit to the region was the only way he could complete his task: to gather hard-copy outlines of sermons that local Methodist ministers had been preaching on climate change.

After so many months in lockdown, Bale wasn't leaving his family behind at Christmas, and so the trip became a family affair. His three children watched attentively from the back seat as he drove through the Nakorotubu circuit in search of Rev. Viliame Donu. Rev. Donu had preached a sermon in February of 2021 on the rising tides in his village, and he had been unreachable by phone or Internet for months. The car crawled along the rutted road until it came to a muddy hill that could not be climbed. As the car wheels spun in the earth, a voice piped up from the back seat. Bale's young son had begun to pray. When his father described the moment to me several weeks later over an unstable Internet connection, he said that the prayer was a request for Jesus to help the car move up the hill "so we can tell people about climate change."

I was listening to Bale tell this story through a computer screen, sitting on the other side of the globe, humbled by his child's stake in the project and my dependence on that commitment. My wealthy university's resources couldn't get Bale's car up that muddy hill. They couldn't even get me into the country during that season of the pandemic. I had received grant funding for an in-person trip to Fiji to collect these sermons myself, but COVID-19 had disrupted that work over multiple years. Initially, working with an on-site research assistant had seemed a difficult compromise. But as I listened to Bale tell the story of his son, I realized again the gift of my absence. The subject of climate change was holy ground to this young family—and to this young boy. It was a subject worthy of prayer. More than that, the sermons of preachers facing the crisis, carefully recorded in outline form in sermon diaries, were holy ground. Despite my disappointment in not being able to return to communities I had lived in for three years, something about my distance felt appropriate and even generative. It was a concrete methodological reminder of my limits as a researcher and the necessary participation of communally accountable, Indigenous voices in conversations about Indigenous spirituality and land.[1] Distance can be a helpful thing. It can illumine the significance of local leadership in cross-contextual research and become a precursor to more genuine relation.

The first chapter of this book describes how the questions many displaced Indigenous communities are asking about place interrupt normative conventions in Western categories of the self, place, and human experience. These interruptions have particular importance for practical theology's prescriptive arc. Questions of *sovereignty*, *belonging*, and *loss* complicate scientific and theological interventions that advocate for concrete change across contexts. Just as critically, these questions necessitate such attempts. Climate change is no theoretical abstraction. Ethical responsiveness requires a dual awareness of how place limits our understanding, even as place connects and implicates us in relation to others.

[1] Amiria Salmond discusses her skepticism of ethnographic projects that seek to "seamlessly channel other people, ontologies, or life-worlds," in "Re-building Ships at Sea: Ontological Innovation in Action," in *Environmental Transformations and Cultural Responses: Ontologies, Discourses, and Practices in Oceania*, ed. Eveline Dürr and Arno Pascht (New York: Palgrave Macmillan, 2017), 215–26 [224].

Questions of sovereignty, belonging, and loss also complicate the work of preaching and biblical interpretation. Chapter 2 fleshes out a specific example: Ezekiel's anguished assertions about place's theological valance center questions of *who is responsible*, *who belongs*, and *who saves* when speaking of places that have been lost as sites of divine encounter. His answers can seem inadequate to many ecologically minded interpreters. But careful listening reveals Ezekiel's stalwart resistance to imperial projects that objectify, commodify, or construct place for the empire's benefit.

Together, these two chapters invite readers to acknowledge "where they stand"[2] in relation to places and communities impacted by climate catastrophe. The chapters intentionally distance settler theologies of place from the theologies of displaced communities with long-standing ties to land, even as they assert the latter's relevance for the former. To draw on a familiar biblical metaphor, these chapters invite Western readers to "remove the sandals from [their] feet" (Exod 3:5) and assume stances of vulnerable humility.[3] One can assess and diagnose too quickly, assuming an unearned familiarity. One can also dismiss apocalyptic loss as irrelevant to the contemporary conversations of insulated populations. The climate crisis forecloses both of these practical theological shortcuts. Places are not interchangeable; neither do they exist in isolation.

For all of their distance, the sermons described in this chapter offer reorientations to Western conversations about climate justice in ways that are uncomfortably close to the bone. They stretch secular, liberal assumptions about identity and human responsibility in foundational ways. In a manner resonant with the biblical witness of Ezekiel, these Indigenous Fijian Methodist preachers foreground questions of sovereignty, belonging, and loss, and in so doing, resist the colonial, consumerist conceptions of place that have seeded climate catastrophe.

A Relational Methodology

Soliciting the fifteen sermons gathered in this collection was no simple matter. Permission to request the participation of local Methodist pastors in this study was obtained through written requests to the leadership of

[2] To reference the language of this book's introduction.

[3] In contemporary Fijian life, removing one's sandals is also a customary practice upon entering a village Methodist church. It is what one does when one approaches "holy ground" (Exod 3:5). Exodus 3 describes God's call to Moses to lead the Israelite people out of slavery in Egypt, speaking through a burning bush that is not consumed.

the Methodist Church in Fiji (MCIF) over multiple years.[4] These written requests, as is appropriate in the Fijian context, were followed up by an in-person request from a person who could vouch for the legitimacy and relevance of the project. In this case, Rev. Dr. Upolu Lumā Vaai, principal of Pacific Theological College, was instrumental in speaking personally to the project's import and in requesting the MCIF's formal recognition of Rev. Bale as a trusted, communally authorized research partner. Because of Bale's academic record, his thesis in recent Fijian history,[5] and his former pastoral appointment in the remote islands of Lau, Vaai argued for Bale's unique expertise in moving the project forward, and the church concurred.

Indeed, Bale's knowledge of regions that were undergoing climate catastrophe in its multiple valences was critical to the sermon solicitation process. I had knowledge of displaced villages like Vunidogoloa because its experience had been described in published accounts.[6] I had visited the island of Ovalau and sections of Ra following Cyclone Winston in 2016 and knew the severity of that storm's devastation in numerous local communities. I had heard the testimony about the climate crisis from students while I lived in Fiji. But I knew less about the deadly impact of the more recent storms, Cyclone Yasa and Cyclone Ana, that hit Vanua Levu in January 2021. I had little contact with pastors in the remote eastern islands of Lau, where sea-level rise was undeniable. And I knew little about the climate challenges facing villages in the nation's interior. Increasingly fierce storms, dying coral reefs, and sea-level rise were only part of the story. Water pollution, floods, and landslides had also been linked to local and international development projects. Bale was able to direct our sermon solicitations to pastors serving in a variety of village ecosystems across the country, and fifteen pastors responded.

4 I am grateful for the blessing of Rev. Dr. Epineri Vakadewavosa and Rev. Ili Vunisuwai, the qase levu vakacegu and current qase levu of the Methodist Church. I am also grateful for the support of Rev. Iliesa Naivalu, the late MCIF general secretary, and Rev. Dr. Jolame Lasawa Sedra, the former principal of Davuilevu Theological College and current MCIF general secretary.

5 Marika Baleidelabaci (Bale) Tamanikaigavo, "Racial Integration in Transition: A *Magimagi* Historical Assessment of the Methodist Church in Fiji's Response to Identity and Race Issues in Fiji" (unpublished master of theology thesis, Pacific Theological College, September 2021).

6 Justin Worland describes the relocation of Vunidogoloa (called Kenanivou, "New Canaan," in its new setting) in "The Leaders of These Sinking Countries Are Fighting to Stop Climate Change. Here's What the Rest of the World Can Learn," *Time*, June 13, 2019, https://time.com/longform/sinking-islands-climate-change/.

The goal of this project was never to provide a comprehensive analysis of all sermons preached on climate change in the Fijian context. Such a survey would have needed to include sermons from a variety of Oceanic denominations and ethnicities—including a range of sermons preached by Indo-Fijians.[7] Cross-denominational comparisons are useful in the statistical analysis of broader trends and can illumine all sorts of important insights. But they can also obscure nuances in local conversations around climate change, place, and preaching that point to distinct communal and denominational framings. Christians from different communities bring different assumptions to similar texts of Scripture, even as they use similar language and preach on similar themes.[8] By investing in a smaller set of sermons preached from the pulpits of villages under threat, this project focuses attention on how displacement is impacting the preaching of *iTaukei* communities in particular.[9] It looks at the hermeneutical moves these pastors make, the questions they ask, the theological themes they resource, and the rhetorical strategies they leverage. The goal is to pay attention to how these Fijian Indigenous pastors are conceiving of and preaching about *place* from the specific places they serve.

Fijian Methodist sermons are rarely written as full manuscripts, though many Methodist pastors keep regular sermon diaries where they record the theme, Scripture, and sermon outlines they have preached. A number of

[7] One significant Indo-Fijian voice included in chapter 4 is that of Rev. James Bhagwan, current general secretary of the Pacific Council of Churches and former communications secretary for the Methodist Church in Fiji. Rev. Bhawan is an ardent regional spokesperson for ecological justice.

[8] One limitation of this study is that, in spite of specific invitations to women ministers, only one *iTaukei* woman submitted a sermon: Rev. Makelesi Watikaicolo, a chaplain posted to Nakauvadra High School. Further research on this gap is necessary. The percentage of women ministers in Fiji is small compared to their male counterparts. It may also be that there are fewer posted to rural settings vulnerable to relocation. For a project that foregrounds the biblical interpretation of *iTaukei* women in ways that connect creation's thriving with gender equity, see Jerusha Matsen Neal, "Decolonizing 'Dominion': Oceanic Methodist Women Read Genesis 1," in *Decoloniality and Methodism: A Commitment to Ecological Justice*, ed. Upolu Lumā Vaai (Oxford: Oxford Institute of Methodist Studies Publications, forthcoming 2024).

[9] Despite all the demographic shifts of recent years, Methodism continues to be a significant aspect of rural *iTaukei* life. The U.S. State Department notes influence of the Methodist denomination among the country's traditional *iTaukei* chiefs and its predominance in rural areas, where 44 percent of the population lives: Office of International Religious Freedom, "2020 Report on International Religious Freedom: Fiji," U.S. State Department, May 12, 2021, https://www.state.gov/reports/2020-report-on-international-religious-freedom/fiji/.

these outlines are detailed. The majority of the sermons we received came to us in this format: partial manuscripts or detailed outlines from sermons preached in a minister's local context. These documents were sent by post or, if Internet was available, by email. One of the requirements for sermon submission was that the sermon needed to have been preached in a particular location on a particular date (i.e., not written in the abstract or written only to be distributed and read). This criterion meant that most of the sermons we received were originally preached in *iTaukei*, and I was grateful for Bale's careful translations. It also meant that follow-up interviews were necessary to unpack how these Fijian Methodist preachers were thinking about the relationship between the biblical texts and their local contexts, how they envisioned their jobs as preachers, and how they understood their sermons' impacts in their ministry settings. Together, Bale and I chose six sermons from locations across the country to follow up with extended interviews. These recorded, semi-structured interviews took place over several months through a video-conferencing platform. Bale was present when translation assistance was necessary. In multiple cases, preachers traveled by bus or boat to reach a city large enough to have a stable Internet connection. In at least two cases, this required an overnight trip.

Because these preachers would be revealing the places and dates where their sermons were preached, anonymity would be impossible. To speak about the specifics of local villages and histories meant that the identities of these preachers would be clear to anyone familiar with knowledge of the Fijian Methodist Church, regardless of name changes. Therefore, the study was designed to provide preachers the opportunity to review the reflections attributed to them prior to publication. This in-person review process occurred after the pandemic restrictions were lifted and this chapter was written—underscoring the research's collaborative commitments.[10] The feedback from participants was, in part, an effort to clarify any mistakes or mistranslations and ensure that the broader context of their sermons had been adequately described. But it was not simply

[10] Luke Eric Lassiter's *The Chicago Guide to Collaborative Ethnography* (Chicago: University of Chicago Press, 2005) provides a basic summary of these collaborative commitments, drawing on the work and method of Glen Hinson. They include a reciprocal "reading alongside" one's research subjects and establishing aims of research that resonate with the expressed goals of the researched community. Centering the felt needs and agenda of the researched community is also a recurring emphasis of Linda Tuhiwai Smith's *Decolonizing Methodologies: Research and Indigenous Peoples*, 2nd ed. (London: Zed Books, 2012).

a matter of getting details right; it was also about discerning with these local faith leaders if the research resonated with the resistance efforts of local communities. These pastors did not cross open expanses of ocean to share their stories and sermons for the sake of disinterested observation. They offered their words as testimonies intended to serve God's restorative work in the world.

Resonant and Divergent Witnesses

These fifteen preachers do not approach the topic of climate change in lockstep. The sermons deal with scientific data, governmental responsibility, and the relationship between climate change and global industrialization differently. Some draw on scientific models and definitions, and others find such definitions suspect.[11] Some advocate for building seawalls, and others find such projects futile.[12] Climate change and ecological devastation impact different communities in Fiji in distinct ways, so it is no surprise that pastors bring diverse priorities to the work of restoration and resilience. Even within this single denomination, there is a wide range of theological perspectives. Preachers differ not only in how they define faithful response, but also in how they describe the crisis' causal root.[13]

[11] Rev. Samuela Delai, for example, draws on global-warming science taught in his daughters' secondary school to explain the importance of the ozone layer and notes how "gas and smoke from factories" is destroying it ("Spiritual Views on Climate Change: John 3:16," sermon preached for the Saqele Circuit, Fiji, in honor of Fiji Day, October 10, 2021). In contrast, Rev. Makelesi Watikaicolo describes how the expertise of scientists and politicians is increasingly "worthless" in addressing the crisis ("Climate Change: Genesis 3, 2 Chronicles 7:14, Revelation 3:20," sermon preached at Nakauvadra High School, Rakiraki, Fiji, September 2021).

[12] Rev. Iliesa Nasedra argues that "faith in God is [the village's] sea wall" (personal interview, January 18, 2022). Rev. Eparama Ligavatu, on the other hand, encourages his village to overcome their resistance to building a seawall and celebrates that after preaching on climate change in the village of Navalau, the village leaders decided to do their own fundraising for the project, rather than "waiting for the government" (personal interview, January 20, 2022).

[13] Certain preachers emphasize a link between the climate crisis and a village's "selfishness" and lack of robust spiritual practices (Inoke Tuiuruone, "Weakening Relationship Is One of the Major Causes of Climate Change: John 15:1–8," sermon given for the Oneata Circuit, Fiji, September 2021). Others see global capitalism as the culprit, placing blame at the feet of "giants who dominate us in terms of economy and technology, as the Israelites were in the midst of the Egyptians" (Netani Bau, "Respect and Sustain the Work of God's Hands: Luke 18:35–43," sermon preached in Drana village, Fiji, May 19, 2019). Still others see climate change as part of the mysterious providence of God, "renew[ing] the earth" and "inviting [humans] to return to Him"

In spite of these divergences, three assertions have striking, unanimous agreement:

First, the negative impacts of climate change and unsustainable development are undeniable. Rev. Iliesa Nasedra, in his sermon to the Malake Methodist Church, describes the rapid sea-level rise in the village over the past three years, serious enough to collapse a seawall recently built by an overseas donor.[14] Rev. Inoke Tuiuruone, from the Oneata Methodist Circuit in the remote islands of Lau, describes how the marine species in coral reefs are disappearing because of careless fishing and the warming of the ocean. As coral reefs languish and sea levels rise, he states, "sand on our beaches . . . washes away."[15] Preachers like Rev. Savenaca Kamikamica, preaching to the Vitawa Methodist congregation in Ra, describes increasing frequency and severity of cyclones and floods.[16] Rev. Samuela Delai cites data from Fiji's Minister of the Environment at the final parliamentary session of 2021 that identifies forty villages as needing relocation because of ecological crisis.[17] Delai, serving the Saqele Circuit in the Vuda Division of the Methodist Church, also notes how the building of seawalls—an expensive stopgap measure for stopping erosion—is placing the nation in greater debt. For Delai, this is particularly frustrating, as Fiji releases so few of the world's carbon emissions into the atmosphere.[18] These grassroots observations correlate with scientific studies describing "widespread shoreline erosion, groundwater salination and . . . coastal flooding" as a result of rapid sea-level rise.[19]

But rising tides and warming oceans are only part of Fiji's climate struggle. Rev. Eparama Ligavatu underscores how coastal erosion and

(Iliesa Nasedra, "Jesus Christ Is Our Hope: Mark 4:35–41," sermon preached in Malake Methodist Church, Fiji, October 2021).

14 Nasedra, "Jesus Christ Is Our Hope."

15 Tuiuruone, "Weakening Relationship."

16 Savenaca Kamikamica, "God Continues to Restore and Renew His Creations: Genesis 6; Mark 5:35–41," sermon preached in Vitawa village, Fiji, March 3, 2021.

17 Delai, "Spiritual Views."

18 Fiji's carbon emissions for 2021 account for only .004 percent of the total world carbon output. See Hannah Ritchie, Max Roser, and Pablo Rosado, "CO_2 and Greenhouse Gas Emissions," Our World in Data, accessed January 19, 2023, https://ourworldindata.org/co2-and-other-greenhouse-gas-emissions.

19 Piérick C. M. Martin et al., "Responding to Multiple Climate-Linked Stressors in a Remote Island Context: The Example of Yadua Island, Fiji," *Climate Risk Management* 21 (2018): 7–15 [8]. See also Jonathan Woodruff et al., "Coastal Flooding by Tropical Cyclones and Sea Level Rise," *Nature* 505 (2013): 44–52.

decreasing fish populations in the Navolau Circuit have exacerbated unvirtuous cycles in local communities' broader approaches to conservation. To compensate financially for the decreased fish populations, a local village began selling the sand and gravel on its beachfront to development companies, making the erosion problems worse.[20] Delai notes that a lack of oversight and accountability in industrial pine-timber harvesting in the region of Vuda has resulted in "huge landslides" and diminished "water resources on the highlands."[21] Rev. Esala Nacuva speaks of a landslide in Saumakia village caused by development companies dredging the river further downstream. For Nacuva, this raises deep theological questions: "What is of more value? . . . The life of a few people or the life of one whole village? What is of more value to the government and to the provincial administrators? Is life of more value? . . . land of more value? . . . Or is money more valuable today?"[22]

Nacuva's framing points to the second point of agreement between these preachers. For these ministers, *climate change is fundamentally a spiritual crisis*. Rev. Makelesi Watikaicolo, a chaplain at Nakauvadra High School, frames the issue in terms of a broken "cord of relationship" between human and nonhuman creation. This "breach in relationship"[23] is echoed by many preachers, drawing on understandings of *veiwekani* (relationship) that ground *iTaukei* understandings of the human. But this "weakened relationship"[24] is not just a relinquishment of cultural roles and responsibilities. It points to and amplifies a spiritual sickness. For Rev. Viliame Donu, preaching to the Nakorotubu Circuit, this lack of relationship with creation grows out of a weakened "relationship with God," who "no longer dwells in [the communities'] lives."[25] Spiritual lethargy has material consequences for the land, feeding a spirit of "greed and

20 Ligavatu, personal interview, January 20, 2022.

21 Samuela Delai, personal online interview, January 23, 2022.

22 Esala Nacuva, personal online interview, January 17, 2022. The breadth of climate stressors challenging the tropics is underscored by Sophie Chao and Dion Enari in "Decolonising Climate Change: A Call for Beyond-Human Imaginaries and Knowledge Generation," in "Tropical Imaginaries and the Climate Crisis," special issue, *eTropic* 20, no. 2 (2021): 32–54, http://dx.doi.org/10.25120/etropic.20.2.2021.3796.

23 Watikaicolo, "Climate Change."

24 Tuiuruone, "Weakening Relationship."

25 Viliame Donu, "Profound Spirituality Illuminates Our Perception of the Natural World: Job 1:6–22," sermon preached for the Nakorotubu Circuit, Fiji, February 2021.

egocentricity."[26] Donu describes an interconnected understanding of the relationship between humans, nonhuman creation, and God in which one set of relationships impacts the strength of the other connections.[27] Weakened relationships between human persons and God lead to weakened relationships between human persons and nonhuman creation. Conversely, neglect of nonhuman creation leads to (and reflects) a weak relational covenant with the divine. The material consequences of climate change may be concrete and urgent, but in the words of Rev. Nasedra, the climate crisis is first and foremost a "spiritual struggle."[28]

It is no surprise, then, that all fifteen preachers argue that *advocacy around environmental justice is central to the church's mission*. A healthy relationship with nonhuman creation is not simply portrayed as one issue among many; it is grounded in core theological claims. Delai describes the life-giving relationality of creation as a sign of the relationality of the Trinity, where it is possible to "stand together in unity" even when "various distinctive roles" are played.[29] Drawing on Genesis 1:28, Rev. Tevita Kete notes that the church's vocation as "servant of the land" has been neglected, and "we must reclaim the role assigned to us."[30] The playing of one's appropriate role in the ecclesial community and in creation is also a significant part of Ligavatu's understanding faithfulness to the divine covenant. He underscores the link between the role of *siga dina* (confirmed church member) and the role of caretaker of the land.[31]

But these preachers not only make a connection between Christian doctrine, proper ecclesial practice, and climate advocacy. They also argue that addressing the impacts of climate change has broad ethical implications for human relationships. Over a third of the sermons collected describe

26 Tevita Kete, "Your *Yalava*: Keep It or You Will Destroy It: Genesis 1:26–28; Genesis 2:15–17," sermon preached for the Korotubu Circuit, Fiji, June 7, 2015.

27 Though Donu doesn't use Christopher Wright's language of "ecological triangle," his description resonates with Wright's description of the concept in *Old Testament Ethics for the People of God* (Leicester: InterVarsity Press, 2004). Hilary Marlow notes that adaptations of Wright's reflections on the interrelation of God, humanity, and nonhuman creation continue to bring "fruitful insights" to the fields of practical theology and biblical studies: "Creation, Humanity, and Hubris in the Hebrew Bible," *Practical Theology* 15, no. 5 (2022): 491–503 [493].

28 Nasedra, "Jesus Christ Is Our Hope."

29 Delai, "Spiritual Views on Climate Change."

30 Kete, "Your *Yalava*."

31 Eparama Ligavatu, "Keeping the Navolau Circuit *Drua* in Good Shape to Withstand Rough Weather: It Depends on Our Diligence in Keeping the Covenant: Genesis 6:5–18; 9:8–11," sermon preached for the Navolau Circuit, Fiji, April 7, 2019.

how neglect of nonhuman creation leads to and amplifies the neglect of human persons—particularly women and children. For these preachers, climate change is about more than the weather. Watikaicolo connects the exploitative attitudes that have fueled the climate crisis to increases in "violent, brutal behaviors"[32] between humans. Rev. Eric Baleloa agrees, noting how a disregard for the land has accompanied "rapid increases in . . . particular crimes against women and children."[33] Rev. Jeremia Waqainabete preaches his sermon on climate justice during the church's annual "Break the Silence" commemoration, intended to decrease gender violence in the community. He uses Genesis' description of humans made in God's image (Gen 1:26–28) to call his community to faithful relationship with the land—as well as faithful relationship between the genders.[34] Climate advocacy, for these preachers, is central to the church's vocation and doctrine, its covenantal relationship with God, and its ethical responsibility for other humans. In the words of Rev. Netani Bau, climate advocacy is key to "our relationship with Jesus . . . uphold[ing] our Christian obligations, disciplines, and values."[35]

It is worth noting how this braid of Fijian Methodist affirmations cuts across and complicates Western categories of theological response to the climate crisis. These are not preachers who deny climate change's reality or its relevance for a community's spiritual calling. In reflecting on the sermons collected, Bale makes this point explicitly. Unlike preachers primarily focused on individual conversion or spiritual ecstasy, Bale notes that these Fijian Methodist sermons do not only dwell on "hell and the

32 Watikaicolo, "Climate Change."

33 Eric Baleloa, "What Must I Do to Ensure Blessing upon My *Vanua*: Luke 15:8–10," sermon preached to the Buca Circuit, Fiji, February 2, 2020. Climate science is paying increasing attention to the link between gender violence and climate change. Bharat Desai and Moumita Mandal describe the impact of climate change on gender-based violence in "Role of Climate Change in Exacerbating Sexual and Gender-Based Violence against Women: A New Challenge for International Law," *Environmental Policy and Law* 51 (2021): 137–57. Baleloa makes what is, perhaps, a more provocative suggestion: that violence against women also exacerbates climate catastrophe.

34 Jeremaia Waqainabete, "Sermon for the Commemoration of National Fisheries Day, Genesis 1:26–28," sermon preached at Epworth Hall, Suva, November 21, 2016. Many thanks to Rev. Tui Nuku Smith for his translation of this sermon. Waqainabete expands on the links between ecotheology and gender justice in his book *Christian Environmentalism: An Ecotheological Approach to Earthkeeping* (Suva: WWF Pacific, 2018), https://wwfasia.awsassets.panda.org/downloads/theological_mindset_english_2018_sep.pdf.

35 Bau, "Respect and Sustain."

afterlife." They engage "the reality of destruction taking place here and now."[36] In these ways, the Fijian Methodist sermons gathered here differentiate themselves from the priorities of newer ecclesial movements in the region (movements often associated with globalized, Pentecostal worship practices). These newer movements, which continue to grow in urban centers, have been critiqued by more traditional, regional churches for their lack of engagement around the complex socioeconomic issues facing the Pacific, accepting "the political status quo and preach[ing] an individual, but not social ethic."[37]

In contrast, protecting creation through regional ecclesial partnerships and collaborations with the scientific community is a long-standing commitment of the Fijian Methodist Church. Through ecumenical groups like the Pacific Council of Churches, Fijian Methodists have been involved in local climate advocacy efforts for decades.[38] These efforts include local preservation projects, like planting mangrove trees or preserving fish populations through seasonal *tabu* restrictions. But they also include partnerships with international and governmental relief organizations relocating climate refugees.[39] The Fijian Methodist Church has added legitimacy to new communities, like the relocated Banabans, welcoming them into

36 Marika Baleidelabaci "Bale" Tamanikaigavo, personal online interview, January 25, 2022.

37 Rusiate Tuidrakulu, "Fiji," in *Christianity in Oceania*, ed. Kenneth Ross, Katalina Tahaafe-Williams, and Todd Johnson (Edinburgh: Edinburgh University Press, 2021), 91–99 [95].

38 Kate Wheeling, "How the Methodist Church in Fiji Is Helping to Lead on Climate Change," *Pacific Standard*, November 13, 2017, https://psmag.com/environment/cop23-methodist-church-helping-lead-climate-action-fiji. More broadly, the Pacific Council of Churches has been centering this concern in regional churches for decades. Rubow and Bird note that "*The Otin Taai* Declaration (2004) makes explicit a hermeneutical connection between Biblical teaching and traditional Pacific teaching, which is then used as the basis for encouraging human beings to exercise good stewardship by living respectfully, humbly, caringly, and reciprocally on/with the Earth." They also note the Pacific Council of Churches' call to "produce, provide, and share theological resources" for local faith communities in the region (Cecilie Rubow and Cliff Bird, "Eco-theological Responses to Climate Change," *Worldviews: Global Religions, Culture, and Ecology* 20 [2016]: 150–68).

39 Tevita Nawadra Banivanua describes the church's fasting from endangered fish species like *kawakawa* and *donu*, "Mapping the Lako Yani Vou [New Exodus]," keynote address at the Divisional Superintendents and Divisional Stewards Conference, Centenary Church, Suva, March 10, 2015. Wolfgang Kempf describes the Fijian church's support for the resettled Banabans in "Climate Change, Migration, and Christianity in Oceania," in *Climate Change and Human Mobility*, ed. Kirsten Hastrup and Karen Fog Olwig (Cambridge: Cambridge University Press, 2012), 243–47.

their ecclesial structures and theological institutions.[40] It has also served as a bridge between scientific organizations and village populations. Kete names the ways that various NGOs have depended on his assistance in "spiritualizing" their local education efforts, recognizing that "their message will get across easier if I go around with them."[41]

But significantly, these Fijian Methodist sermons also differentiate themselves from the climate approaches and epistemological framings of secular Western environmental advocacy groups. These sermons do not merely attend to material realities; they insist on the spiritual significance of these material particulars.[42] More than this, they insist that spiritual practices like prayer, fasting, and church membership have material impacts on environmental health. In the polarized, politicized conversations around climate change in U.S. faith communities, the affirmations that these sermons share—that climate change is real, that it is fundamentally a spiritual crisis, and that climate advocacy is central to the church's vocation—bring discomfort to multiple ideological and theological camps.[43]

40 Kempf also notes that in various political seasons, the Methodist Church's relationship to the Banaban community has been complex. It has historically provided welcome, but it has also harbored ethnonationalist sentiments that have "caused Banabans to fear that full recognition of belonging to the present postcolonial state of Fiji might be withheld." See Elfriede Hermann and Wolfgang Kempf, "Introduction to Relations in Multicultural Fiji: The Dynamic of Articulations, Transformations, and Positionings," *Oceania* 75, no. 4 (2005): 319–20.

41 Tevita Kete, personal online interview, January 27, 2022. Waqainabete also notes the critical role that churches have played in stopping deforestation in local circuits like Seaqaqa when local populations were resistant to the Ministry of Forestry's recommendations ("Sermon for the Commemoration of National Fisheries Day").

42 Piérick C. M. Martin et al. note that the spirituality of Pacific communities "is frequently overlooked [by environmental scientists] and little attempt made to engage with it: something likely to explain why so many climate-change adaptation initiatives over the past three decades in the Pacific islands have failed" ("Responding to Multiple Climate-Linked Stressors," 13). See also Patrick D. Nunn et al., "Beyond the Core: Community Governance for Climate-Change Adaptation in Peripheral Parts of Pacific Island Countries," *Regional Environmental Change* 14, no. 1 (2013): 221–35.

43 These ideological camps, despite their differences, can evoke similar skepticisms in many Oceanic communities. Jione Havea, while emphasizing the reality of the climate crisis, notes that both Western climate activists and Western climate deniers are influenced by political and economic factors. "Who benefits from climate change and who benefits from talking about climate change?" he asks. "There are financial benefits on both fronts, with structures of corruptions in both." ("The Politics of Climate Change: A *Talanoa* from Oceania," *International Journal of Public Theology* 4 [2010]: 345–55 [352]).

These sermons refuse to bifurcate the spiritual and the material, the individual and the communal, or the secular and the religious, in ways that seem intellectually incoherent to some and ethically dangerous to others. The integration of these categories, however, is consistent with Indigenous Fijian Methodist understandings of *place*—and with the questions raised by lived experiences of displacement.[44]

The answers that the preachers of these sermons offer may not always coincide, but the questions they foreground as central to the climate catastrophe resonate with the questions of other displaced communities, including the questions of the biblical prophet Ezekiel. Similar to the biblical prophet, specters of punishment, purity, and passivity haunt these sermons' theologies of climate catastrophe, but these specters are not ends in themselves. They point to deep, nuanced questions around a community's negotiation of *sovereignty*, *belonging*, and *loss* in its theology of place. These questions illumine the significance and limits of identity, agency, and eschatology from the ground of apocalyptic disruption, and as I have argued, they voice more than a people's grief. They also challenge colonial specters of place that continue to haunt the environmental movement, the theological education classroom, and the ecclesial pulpit.

Reframing Sovereignty

Rev. Iliesa Nasedra has traveled in a small motorboat from the island of Malake to the town of Rakiraki to find a suitable Internet connection. We have agreed to meet together through a virtual conferencing platform across datelines and time zones, with Bale's translation assistance. Nasedra is the Methodist minister for the Malake Circuit in the division of Ra, so he is a busy man. But he is keen to give me his testimony: *a yearslong prayer and fasting program has halted the rise of the sea level on Malake Island*. It is a testimony that echoes the sermon of Apenisa Waqalerua, described in chapter 2, but unlike Waqalerua, Nasedra is no young seminary student. He has been a *talatala* (an ordained Methodist minister) for many years—as his white hair attests. He speaks with pastoral authority and gravitas, wearing a formal tie and button-down

[44] Chao and Enari specifically ground this integrative approach to Pacific understanding of *place* that, "as Pacific cosmologies teach us, are always at once physical, ecological, historical, cultural, and spiritual" ("Decolonising Climate Change," 41). See also T. Banivanua Mar, "Settler-Colonial Landscapes and Narratives of Possession," *Arena Journal* 37/38 (2012): 176–98. I have argued that there is a similar integrative understanding of place in the prophet Ezekiel.

shirt with his traditional *sulu*. He has disrupted his schedule and traveled across the water to discuss serious matters.

Nasedra's sermon and his subsequent testimony provide an especially fearless and explicit affirmation of the sovereignty of God in the face of rising tides. But this affirmation is not unique. Assertions that "Christ alone is our hope . . . and will renew creation,"[45] that "rooting one's faith in Christ . . . will save the land from the effects of climate change,"[46] and that "God will save us and our house" from destruction[47] echo throughout these preachers' proclamations. The theme of God's sovereignty over nature is also one of the aspects of Pacific Christianity that secular scientists find most troubling.[48] Scientists worry about communities' "lack of exposure to climate-change science . . . and their privileging of spiritual over secular beliefs, especially those relating to causation."[49] An ethnographic study of the perspectives on climate change within the Christian village of Denimanu, a village in the process of relocation, struggles with the "often aired" suggestion that "environmental changes had been 'sent by God' to punish the people for the indifferent nature of their worship."[50] To be fair, Nasedra's response to that statement would be complex as well. One of the reasons that Bale and I chose to engage Nasedra in further conversation was that his sermonic affirmation of God's sovereignty pays attention to this claim's ethical trajectory and pastoral impact. More than this, it reframes the concept of sovereignty

45 Watikaicolo, "Climate Change."

46 Esala Nacuva, "The Holy Spirit Is Our Source of Strength," sermon preached in Nadaravakawalu Circuit, Saumakia village, Fiji, June 2019.

47 Kamikamica, "God Continues to Restore."

48 Characteristically, Simon Donner notes that "in most Pacific Island nations, educators argue that the commonly held belief that the Christian God controls the weather is the greatest obstacle to educate people about climate change": "Domain of the Gods: An Editorial Essay," *Climate Change* 85 (2007): 231–33 [233]. Divine authority over the weather is not just a recurring theme in Pacific Christianity. It is also a common theme in Hebrew Scripture. James Luther Mays argues, for example, that the affirmation "the Lord reigns" is the Psalter's "theological center." This "sphere of dominion . . . includes the world and all its constituent elements" (i.e., Pss 93:3–4; 95:4–5; 97:4–5). *The Lord Reigns: A Theological Handbook to the Psalms* (Louisville: Westminster John Knox, 1994).

49 Martin et al., "Responding to Multiple Climate-Linked Stressors," 13. See also Natasha Kuruppu and Reenate Willie, "Barriers to Reducing Climate Enhanced Disaster Risks in Least Developed Country-Small Islands through Anticipatory Adaptation," *Weather and Climate Extremes* 7 (2015): 72–83.

50 Kuruppu and Willie, "Barriers," 72–83.

itself. Nasedra's assertion of God's sovereignty is engaged in complicated contextual work.

Nasedra's sermon "Jesus Christ Is Our Hope" is a reflection on Mark 4:35–41—Jesus' calming the storm while traveling with his disciples on the Sea of Galilee.[51] In the face of the disciples' fear, Jesus addresses the wind and waves directly, saying, "Quiet! Be still!" and the wind and waves obey. For Nasedra, the passage exemplifies the "authority Jesus holds over creation, including the weather or climate." Even in Malake, where "the sea level has reached our homes," Nasedra insists that Jesus can "help us overcome the crisis." Nasedra references other Scripture passages to emphasize God's sovereignty over nature: the creation of dry land through the Red Sea (Exod 14:29) and God's stopping of the sun (Josh 10:12–15). He juxtaposes faith in God with the panic and fear of the disciples (Mark 4:38), and he addresses the congregation with Jesus' words, "Why are you so afraid? Do you still have no faith?" (v. 40).

If Nasedra is attempting to quell the fear of his congregation, however, he goes about it in an odd way. His assertion of God's sovereignty over the waves is not paired with assurances that the sea will do what the community hopes. With stark matter-of-factness, he states, "The sea will continue to rise in years ahead." When he references the relocation of the Kadavu village of Narikoso in response to rising tides, he does not imply that the relocation is an act of divine punishment or that it might have been prevented. He tells his congregation, "It might also happen to us." What is strikingly absent from Nasedra's sermon are quid pro quos between faithful practice and material blessing—the transactional approach to faith often associated with "prosperity gospel" theologies.[52] Even as he asserts God's power over nature and the capacity of faith to "change what is happening around us," it does not follow that the community's faith will make the sea retreat.

This is a strange tension given the prayer and fasting program that the Malake community has sustained over years in response to the environmental crisis in their community. In his interview with Bale and myself, Rev. Nasedra explains the details of the program he instituted. *Mataqali* (extended family) groups meet for three hours of prayer and fasting each Wednesday. They pray, sing, and share from the Bible from noon until

[51] Nasedra, "Jesus Christ Is Our Hope."

[52] Kate Bowler explores the origins, unifying themes, and global reach of the American "prosperity gospel" in *Blessed: A History of the American Prosperity Gospel* (Oxford: Oxford University Press, 2013).

three o'clock—the time period of the day in which Jesus was on the cross. They began the program two years prior and plan to continue it for three more years, covering "the whole five-year span of [Nasedra's] term in Malake."[53] It is a remarkable commitment undertaken by the entire community, engaging those of different denominations and even those who do not regularly attend church services. And while Rev. Nasedra testifies to its material benefit in providing the community with a spiritual "sea wall of faith,"[54] his sermon makes no promises in this regard. His language about faith's ability to change the trajectory of climate devastation is carefully broad. "Our faith in Christ can guide us to attain new life," he says. "It can lead us to new perception and vision to sustain our land and environment." This new perception, vision, and spiritual commitment in seeking God's intervention is how "the impact of climate change on our land can be overturned," in part because the new eyes of faith reveal "our status as part of creation and not the creator."[55]

It is here that the ethical implications of Rev. Nasedra's sermonic assertions of divine sovereignty come into focus. He describes an island owned by foreign investors that is a common tourist destination in Fiji's sunny Mamanuca archipelago. "I have witnessed the development taking place on [that island]," he states. "The owner has divided the island into several smaller islands and upset the life on it." Extrapolating later, Nasedra describes how "in an attempt to beautify" God's creation, the island's owners dug out land to create a new coastline, "destroying God's perfect relation."[56] For Nasedra, this human co-optation of God's role as creator is directly related to the climate crisis. It also drives his insistence on the importance of faith in divine sovereignty. Faith keeps us from "overlook[ing] our limitations," he states, which is finally what is at the heart of the careless "overseas development pervading our land today." It does not seem accidental that both examples of "overlooking our limitations" provided by Nasedra—a crumbling seawall built by "a European lady," and foreign investors who destroy fragile island ecosystems for the benefit of tourists—are examples of those working outside the limits of their own country. Implicit in Nasedra's insistence on the limits of human agency is a critique of those who have ignored the limits of place in their

53 Nasedra, personal interview, January 18, 2022.
54 Nasedra, "Jesus Christ Is Our Hope."
55 Nasedra, "Jesus Christ Is Our Hope."
56 Nasedra, personal interview, January 18, 2022.

wielding of economic power, taking on the role of "creator" outside their own borders.[57]

Mark 4 makes a Christological assertion grounded in the assumption that creator and creation have distinct identities and levels of authority (i.e., "Who then is this, that even the wind and the sea obey him!" Mark 4:41). The disciples' terrified amazement at the end of the Gospel passage grows from their conviction that only the creator can address nonhuman creation as Jesus does and expect obedience. The passage, in other words, is about more than weather. It is about Jesus' identity discerned through the prism of his authoritative relation to the created world. Mark's claim assumes that *who one is* (and isn't) has a direct bearing on one's role, responsibilities, and capacity in relation to place. For Nasedra, this link between identity and role has ethical consequences as well. The distinction between creature and creator challenges development strategies, foreign and domestic, that overlook creaturely limitations and usurp a role that is not legitimately theirs. For all of Nasedra's emphasis on faith, faith is not a mechanism for controlling God. It prevents human creations from playing God.

At the close of his sermon, Nasedra makes an argument that seems to directly contradict his prophetic critique of exploitative foreign development. He argues that climate change is not only the result of human sin; it is also part of God's intentional work of renewal. Climate change, he argues, is part of God's plan, "invit[ing] us to return to Him and reaffirm our faith in Him." He moves from Mark 4 to Revelation 20, noting the passing away of the first heaven and the first earth. "The melting ice in the north pole and the rising sea level are all part of the passing away of the first earth," he states. "Climate change," he concludes, "is part of God's process to renew the world."

One can read this claim in a way similar to Brad Kelle's reading of God's destruction of land in Ezekiel: as the affirmation of a traumatized, displaced community that, in spite of all, God is still in control. Perhaps Nasedra wants his listeners to know that the hubris of international development does not finally control the world—even if that means accepting climate displacement as part of God's design.[58] I admit that I

57 Nasedra, "Jesus Christ Is Our Hope." J. Singh connects the concept of human "mastery" over nature to the colonial mindset, particularly when nature is understood as a passive object that only has meaning in its utility to humans: *Unthinking Mastery: Dehumanism and Decolonial Entanglements* (Durham, N.C.: Duke University Press, 2018).

58 Hilary Marlow underscores the Hebrew Bible's critique of human hubris "not only at the individual level, but . . . in national and global contexts" as well ("Creation, Humanity, and Hubris," 499).

struggle to follow this section of Nasedra's sermon. The cost of this affirmation of God's sovereignty—even for the purpose of destabilizing the sovereignties of global capitalism—seems far too high. Renewal through destruction implies the same sort of utilitarian approach to creation used by Western developers. It treats the first creation as expendable in the creation of the new.

The first time I read the conclusion to Nasedra's sermon, I scribbled my frustration in the margins of the sermon transcript. "Which is it?" I wrote. "Is climate change the result of sin or the result of God?" His assertion of divine sovereignty resulted, for me, in an unresolved theodicy puzzle. But significantly, this is not Nasedra's question. When I asked him about the cause of climate catastrophe directly in our interview, he named God's providence and human sin to both be at work, and this caused him little concern. What did cause him concern was articulating the proper relationship and role of each player in the created order: God, human creatures, and nonhuman creation. For Nasedra, God's renewal of the world was less about the exertion of absolute power and more about the leveraging of divine power to reestablish the limited creaturely sovereignties necessary for genuine relationship.[59] This is what the renewal of the earth required. Nasedra's sermon was doing more than reclaiming the sovereignty of God; it was reframing the meaning of sovereignty in the context of divine, human, and nonhuman relationship.

Nasedra's ministry illumines this reframing work in at least two ways. The first is through his village's extended, consistent attentiveness to creation in their prayer and fasting program. His insistence on divine sovereignty has not led his village to an ambivalent view of human action; it has pressed the community into a five-year discernment of their role in the creational order. When I ask Nasedra what his people are praying for during their weekly three-hour fasts, he responds with a ready answer: "We pray to know who God is. And we pray to know the purpose of our lives."[60] The community, in other words, is not praying for an easy solution that absolves them of responsibility. Trust in God's sovereignty has encouraged greater emphasis on human accountability and an articulation of human purpose in relation to climate catastrophe.

59 Faafetai Aiava notes the colonial roots of theological approaches that stress "control" over "connection," as well as the dangers of approaches that "intentionally distort connection [into] conflation": "Talking God in a Divided House: Rebuilding Spirituality in God's Pacific Household," paper presented at Iliff School of Theology, Denver, Colo., October 22, 2021.

60 Nasedra, personal interview, January 18, 2022.

The community's prayers have, in fact, led to material action. When asked if there are other initiatives that his community is undertaking to protect the environment, Nasedra produces an extensive list. They are planting mangrove trees; they have stopped disposing waste at sea. They are particularly careful with the proper disposal of batteries, recognizing that battery fluid destroys marine life. There has been a whole community campaign on the subject. Belief in God's sovereign authority over weather has not led to the community's lethargy, but rather to their active engagement in climate care.

The second way that this reframing of divine sovereignty is illumined in Nasedra's leadership is in his treatment of nonhuman creation in his sermon itself. For Nasedra, God's sovereignty does not imply place's objectification and irrelevance in God's plan. It points, instead, to God's active relationship with place, understood as having its own agency and role in God's ecology.[61] In Nasedra's sermon, place may be provisional—but it is no commodified object. It suffers, struggles, listens, and responds.

In this, Nasedra takes his cue from the biblical text. Western readers often miss the radical "subjectification"[62] of nature in Mark 4. Jesus addresses the wind and waves as if they are actors in the narrative, with the capacity to hear and obey. Nasedra echoes this approach. "All creations," he states, "have ears. They were created by [God]; they came into being through his plans and words. They will always listen to Him."[63] There is a stark difference between this biblical depiction and conceptions of place as inert capital, bought and sold at will, carved by heavy machinery for the benefit of a few. Nasedra conceives of creation—land and sea together—as having a role to play in God's covenantal story.

[61] Through their research on indigenous geographies, Soren C. Larsen and Jay T. Johnson argue for the agency of place in "The Agency of Place: Toward a More-Than-Human Geographical Self," *GeoHumanities* 2 (2016): 149–66. For Abraham Bradfield, "Places are agents that speak, create, and teach; they have autonomy that is equal to the diverse beings that live and interact with it and ultimately bring place into being" ("Decolonizing the Intercultural: A Call for Decolonizing Consciousness in Settler-Colonial Australia," *Religions* 10, no. 469 [2019]: 1–21 [9]).

[62] To borrow the term that Wojciech Pikor uses to describe Ezekiel's treatment of the land in "The Subjecthood of the Land of Israel in the Book of Ezekiel," *Biblical Annals* 3, no. 1 (2013): 29–46.

[63] Nasedra, "Jesus Christ Is Our Hope."

The importance of fulfilling one's proper role in relation to other creatures and in relation to God occurs repeatedly in the sermons gathered.[64] Performing one's proper role is, in fact, central to these preachers' definition of sovereignty. Rev. Petero Lolohea's sermon is a prime example. Preaching in Nasinu village on the island of Ovalau, Lolohea's exegesis of Psalm 104 addresses a community still struggling to rebuild after a serious cyclone.[65] He notes that the psalm speaks of rain and sun "as if they are beings who have purpose" with a specific role to play. He envisions these creaturely roles as territory, explaining that "[nonhuman creatures] are on one side and we—human beings, the last in the order of creation—are on the other." Proper relationship between creatures requires a respect for boundaries and limits, and without such respect, "our relationship is wrecked." The climate crisis, in Lolohea's view, "comes from the trespassing of boundaries" and from humans not knowing "what is enough for us."

Drawing on the psalm, Lolohea lifts up nonhuman creation and nonhuman creatures as exemplary teachers of God's relational design. In "accompanying the human," they "remain in their positions and never attempt to usurp others' status or place." Within their various "jurisdictions," they are wise, "faithful to their purpose and role." He notes that animals like fish "use [this] wisdom conscientiously." They "detect imminent dangers and upcoming weather disruptions and prepare." They understand the interconnectedness of the world and their dependence on other creatures.[66] In contrast, Lolohea states, the "global trend of becoming independent . . . is the main cause of the [climate crisis]." Creation is "speaking to us in a different language . . . and we have separated ourselves

[64] Samuela Delai notes that the traditional roles assigned to the seven clans (*mataqali*) in an *iTaukei* tribe (*yavusa*) describe the interconnected dependence of human and nonhuman creation, underscoring the significance of the performing one's "proper role" in *iTaukei* village life (personal email correspondence, June 29, 2023).

[65] Petero Lolohea, "Respect and Care for Our Environment," sermon given in Nasinu village, Ovalau, Fiji, June 4, 2000. Originally preached prior to Cyclone Winston, Lolohea has continued to build on these sermonic themes in the aftermath of increasingly frequent and violent storms (personal interview, July 7, 2023).

[66] Lolohea's argument for the epistemological wisdom of the animal kingdom sounds similar to Dame Anne Salmond's charge for climate-change science to "think like a fish" in its preparation for and adaptation to cataclysmic storms: "Think Like a Fish: Pacific Philosophies and Climate Change," in *Pacific Climate Cultures: Living Climate Change in Oceania*, ed. Peter Rudiak-Gould and Tony Crook (Warsaw: De Gruyter, 2018), 155–59.

from them as if they are not a part of us." Separation, however, is only the start of the problem. Humans are also "fond of invading the world of birds, animals, and even fish, . . . destroying their inheritance."[67]

Lolohea's sermon describes a jurisdictional sovereignty between creatures, using the metaphor of border-crossing, trespassing, and even invasion when describing human exploitation. But the purpose of these sovereign spheres is not to isolate creatures from each other. Boundaries are meant to enhance dependence and relationship; they are meant to reveal "the value, purpose and contribution" of each creature. In Lolohea's words, humans must "know their boundaries" in order to "live for God and our [nonhuman] neighbors to whom we are intimately connected."[68]

It is no great stretch to see metaphors of colonial usurpation in Lolohea's description of human exploitation of nature. Fiji knows something of invasions that result in stolen inheritances—where the value, purpose, and contribution of cultures that speak other languages are not acknowledged. This sort of colonial "sovereignty" disregards the God-given roles and wisdoms of others.[69] But Psalm 104 leads Lolohea to a different model—a relational, participatory sovereignty that allows nonhuman creation to "remind us of the greatness of God" as it "accompanies" human life.

In Fiji, the question of sovereignty is sharp. The nation is one of few places in the world where the lands of Indigenous communities are constitutionally protected. Since the late 1980s, the fear that such protections will be lost has fueled ethnonationalist rhetoric, the displacement of vulnerable ethnic minority populations, and a string of political coups.[70] In Indigenous Fijian life, identity and role relate in specific ways. There are particular relational ties between people and places, between people and animals, and between different groups of people themselves. When Kete discusses the *vanua* in the Fijian context, he continues to maintain the importance of these particularities. The *vanua*, he states, has been God-given to the *iTaukei* "for their use," and the relationship between an *iTaukei* person and their particular village is a sacred trust—even if one moves

67 Lolohea, "Respect and Care."

68 Lolohea, "Respect and Care."

69 Aboriginal scholar Irene Watson describes how colonial understandings of "sovereignty" differ from Indigenous conceptions of sovereignty, which emphasize a deeper "law" growing out of creation's interconnection: "Buried Alive," *Law and Critique* 13 (2002): 253–69.

70 Jacqueline Ryle gives a succinct history of this period in *My God, My Land: Interwoven Paths of Christianity and Tradition in Fiji* (Surrey: Ashgate, 2010), 43–54.

away.[71] Land cannot be treated as a consumable, transferable object. But, says Kete, the sovereignty of an *iTaukei* community over their land is not one of authority. It is one of "responsibility." He settles on this word in his fleshing out the nature of the relationship between human and nonhuman creation in Genesis 1:26–28. Only God has authority over creation, he argues. When humans take an attitude of authority over the land, "the *vanua* cries out."[72] In using the language of "responsibility" to describe the relationship between *iTaukei* communities and their land, Kete attempts to describe the importance of jurisdictions in decision-making, without succumbing to Western models of exclusion and exploitation. He continues to claim *iTaukei* sovereignty over their village ancestral land, even as he reframes the meaning of the word for the purpose of relationship and responsible use. Kete's sermon deals primarily with what faithful responsibility looks like in the relationship between the *iTaukei* community and their village territory. He brackets the conversation of what this relationship with land implies for *iTaukei* communities' relationships with other ethnicities, though that critical question is never far from the surface.

For a nation that continues to grapple with the impact of its colonial history and globalized economy while negotiating a multiethnic state, there is concrete significance to the ways that divine and human sovereignty are described. Indigenous leaders ministering in villages under threat leverage assertions of God's sovereignty to destabilize the power of the industrialized world. More than this, they disrupt insular, expansionist, or absolutist conceptions of sovereignty—conceptions reminiscent of colonial presumption—through an affirmation of divine relational design. For Nasedra, the sovereignty of God assumes the limited, responsive sovereignty of creation and its creatures; it assumes that there are borders to creaturely roles and responsibilities. Similar to Lolohea, he asks how human communities might learn from the wind and waves, attending more carefully to the voice of a common creator. How might humans live in a way that takes seriously the sovereignty of creation's role in God's larger purposes? And how might taking seriously the sovereignty of God and the sovereignty of place decenter anthropocentric models of sovereignty—models often associated with Western economic, military, and political force—that conveniently "overlook [their] limitations" and dismiss all claims to sovereignty but their own?

71 Kete, personal online interview, January 27, 2022.

72 Kete, "Your *Yalava*."

In a number of the sermons collected, the question of sovereignty—what it means and how a community faithfully lives out its sovereign responsibilities in relation to land and God—remains unresolved. There is a divide between the ideal of what should be and the complexity of what is. Delai preaches his climate-change sermon during a combined service of churches in his circuit on October 10, 2021.[73] It is Fiji Day, a day that celebrates the nation's political independence. It is also the first day in months that the church has been open for worship because of the COVID-19 pandemic. Despite the day's commemoration of national autonomy, Delai's mind is troubled. He fears that Fiji's sovereignty is at risk. "I was asking so many big questions that day," he reflects later in our interview. "What does it really mean to be independent? . . . When we look at the carbon emissions of First World countries—are we really free from their abusive power? We are carrying the full brunt. . . . We are paying the price. . . . Where is the freedom? What freedom?"[74] For Delai, political independence has marginal relevance in a world where the global influence of certain countries makes other countries "highly vulnerable" to climate catastrophe. True sovereignty, he argues, requires more than independence. It requires the performance of proper relationship—between humans, creation, and God.

Negotiating Belonging

For Rev. Eparama Ligavatu, a life of relationality and responsibility is given form through the language of "covenant."[75] It is a theological theme that he places at the center of his sermon on creation care. Ligavatu is preaching at the monthly combined service of his Navolau Circuit in Ra. He has recently been appointed to the area and seen firsthand the dangerous impacts of rapid coastal erosion. He has also seen the community avoiding difficult but necessary climate interventions. He had tried to address the topic at a village meeting with no result. The community worried that the planting of mangroves would muddy the beach. They resisted fundraising for the cost of a seawall, and they were resigned to selling their sand and gravel to development companies—even as the coast was being washed away. Ligavatu decided that preaching would be a significant tool in furthering conversation, and his instincts were right. Following his sermon, he noticed a number of evident changes in community response.

73 Delai, "Spiritual Views on Climate Change."

74 Delai, personal online interview, January 23, 2022.

75 Ligavatu, "Keeping the Navolau Circuit *Drua* in Good Shape."

The village moved forward in their plans to plant trees along the coast and placed a ten-year *tabu* (ban) on the selling of gravel. A local leader also informed him that the community would begin fundraising for a seawall instead of "waiting for the government."[76] Ligavatu's sermon addresses none of these issues in the particular. He does not speak of mangroves, seawalls, or gravel. Rather, taking Genesis 6:5–18 and 9:8–11 as his texts, he uses the story of Noah to speak to the significance of "diligence in keeping covenant."

Pacific interpretations of the Noah narrative have been widely discussed by ethnographers who study Oceanic theologies of climate crisis. Wolfgang Kempf describes how popular church songs based on the Genesis narrative make contravening theological claims in relation to the contemporary crisis of rising sea levels.[77] Some look to God's promise that a flood will never again destroy the earth as a proof of protection, regardless of the predictions of climate scientists.[78] Others point to Noah's diligence in preparing for the flood, arguing that rising tides are God's way of telling the community to get ready for a crisis that is on the way.[79] But Ligavatu follows neither of these approaches. He focuses instead on the consequences to the land if God's covenant is not followed.

Defining the Hebrew word for covenant (*berit*) as "a step taken to strengthen a relationship," Ligavatu spends a significant portion of the sermon exegeting sin as a lack of diligence in covenantal commitment. He describes sin as "missing one's target," "failing to meet divine requirements," a "crossing agreed-upon boundaries," and "deliberate acts of rebellion." There are consequences to such sin, he argues, not the least of which

[76] Ligavatu, personal interview, January 20, 2022.

[77] Wolfgang Kempf, "Climate Change, Christian Religion and Songs: Revisiting the Noah Story in the Central Pacific," in Dürr and Pascht, *Environmental Transformations and Cultural Responses*, 19–48.

[78] One set of lyrics that exemplifies this approach: "It was God alone who created the earth, as well as all that is on it and He alone rules everything. No one else can do so in this world. The scientific theory which came to the public's notice unsettled us deeply about what will happen to the people living here . . . If it is God's will that the flood will cease, who other than He who knows all could predict these things?" (Kempf, "Climate Change, Christian Religion and Songs," 37).

[79] A counter-narrative to the above lyric, also expressed in song: "How amazing events are, day by day, the global changes, the rising of the sea level, is affecting Kiribati, If it is really going to happen why don't we get prepared? The flood in Noah's time acted as a warning . . . What are our responsibilities? We have to be prepared" (Kempf, "Climate Change, Christian Religion and Songs," 41).

is that the land itself will need "cleansing" (*vakasavasavataki* in *iTaukei*). The idea that the land must be cleansed from human sin, or even from ancient associations with pre-Christian spirits, is a controversial issue in the Fijian Methodist Church.[80] But it is particularly arresting in connection with the Noah narrative and the rising tides of climate catastrophe. Climate change, for Ligavatu, is a mechanism through which God cleanses creation from the pollution of sin—not unlike the Genesis account (Gen 6:5–7). "We did not obey or honor the covenant that God gave us to be in his likeness," Ligavatu explains. "We took it for granted."[81]

This connection between covenantal faithfulness and the flourishing of nonhuman creation is a deeply rooted biblical idea. Richard Middleton notes the clear link between the "moral and cosmic orders" as early as "Genesis 3, where human sin results in a curse on the ground, explained as a resistance of the ground to human efforts to eke a living from it (v. 17)."[82] Ligavatu traces the idea that humans impact creation for good or ill to humanity's creation in God's image (Gen 1:26) and to the explicit theme of "covenant" in the context of his sermon text. God's covenant in Genesis 9 is extended to all creation, giving human fidelity ramifications for the whole created community.[83]

When the stakes for faithful covenantal practice are this high, sin is a serious matter. Alongside "serious crimes like rape, murder, and violence,"

[80] Jacquelyn Ryle's discussion of "spiritual warfare" in her chapter on "Healing the Land" teases out the complexity. At stake is the relationship between pre-Christian Indigenous practices, ancestral spirits, and Christian practices. On the one hand, rituals like *ibulubulu* rites, "chain prayers," or exorcisms can bring reconciliation between land and people and are part of a long Indigenous tradition that recognizes the communal nature of sin and its impact on the land. On the other hand, such rituals of cleansing can challenge "traditional values and mainstream [Fijian] Christian beliefs and practice." The Methodist Church in Fiji head office has "distanced itself" from such cleansings, even as many church members participate in them: *My God, My Land*, 65–82 [78–80]. See also Matt Tomlinson's discussion in "Sacred Soil in Kadavu, Fiji," *Oceania* 72, no. 4 (2002): 237–57, and "Ritual Risk and Danger: Chain Prayers in Fiji," *American Anthropologist* 106, no. 1 (2004): 6–16.

[81] Matt Tomlinson describes the significant role that Gen 1:27 and the concept of the *imago dei* plays in Fijian preaching in *In God's Image: The Metaculture of Fijian Christianity* (Berkeley: University of California Press, 2009).

[82] J. Richard Middleton, "The Image of God in Ecological Perspective," in *The Oxford Handbook of the Bible and Ecology*, ed. Hilary Marlow and Mark Harris (Oxford: Oxford University Press, 2022), 284–98 [293].

[83] See Randy Woodley's treatment of creation as "community" in *Shalom and the Community of Creation: An Indigenous Vision* (Grand Rapids: Eerdmans, 2012).

Ligavatu describes the dangers of "syncretism" as a danger to the purity of covenantal life—"blending together" the worship of different gods at the same time.[84] This "blending together" does not only endanger the purity of doctrine and ecclesial practices. It also endangers the relational ties of the community and its connection to the land. Covenantal loyalty, for Ligavatu, consists in "maintaining the things that God has given us—like culture, like our beach, like our land, like our customs. This is covenant, and if we neglect these things, climate change will come and take them away."[85] The ease with which Ligavatu moves between valences of culture and faith reveals the complex relationship in *iTaukei* Fijian communities between the *vanua* (the land and the Indigenous culture it sustains) and the *lotu* (the church). These spheres of influence are distinct in the understanding of many Indigenous Fijian Christians, but they are also inextricably related.[86] Tuwere describes an "inseparable union" between three strands in the "Fijian collective consciousness": the *vanua*, *lotu*, and *matanitu* (the government). "Their union is so complete that if one is affected, the whole is affected."[87] Ligavatu's sermon exemplifies the overlapping purviews of the *lotu* and *vanua* in many of the climate-change sermons preached in the Indigenous Fijian context.

Some Fijian preachers locate the danger of syncretism in an inappropriate blending of pre-Christian *iTaukei* practices with those of the Christian *iTaukei* community. Eric Baleloa, for example, connects increases in gender-based violence in his circuit with a lingering disregard for women rooted in certain pre-Christian *iTaukei* attitudes.[88] He calls for a "complete cleansing" of the community—both for the sake of women and for the

84 In his sermon, Ligavatu uses the struggle between Israel's YHWH and the Canaanite fertility god, Baal, to make his point.

85 Ligavatu, personal interview, January 20, 2022.

86 Tomlinson describes the "tensions or frictions between *lotu* and *vanua*," as well as the "productive power they generate interactively," particularly around questions of "the sanctification of territories, the rhetorics and rituals of nation making, and the politics of tradition": "Sacred Soil," 237, 252–53.

87 Quoted in Ryle, *My God, My Land*, 55.

88 Matt Tomlinson brings complexity to this characterization, noting the ways that Fijian theologians are attempting to reconcile contextual theologies of *vanua* and Indigenous culture with "quasi-feminist" theological understandings. His article also notes the lived experiences of discrimination encountered by female Fijian clergy as a result of *vanua* protocols: "Gender in a Land-Based Theology," *Oceania* 85, no. 1 (2015): 79–91. Sociologists like Akanisi Sobusobu Tarabe would complicate Eric's characterization as well, emphasizing the ways that missionary Christianity weakened the role and authority of women in Indigenous contexts: "Re-assessing Gender Roles:

sake of the larger cosmic order.[89] When I ask Ligavatu for specific examples of "syncretism" in his context, he does briefly mention several ways that *iTaukei* culture might conflict with Christian practice. For example, he mentions the dangers of drinking too much *yaqona*—a mild narcotic and significant symbol in *iTaukei* ceremonies and celebrations. But *yaqona* is not his real worry.

Ligavatu is more concerned about the impact of religious and cultural others on communities of *iTaukei* Christians. The question of how those of other faiths relate to faithful Christian covenantal practice is a fraught topic in Fijian history,[90] and it is also a fraught topic in Ligavatu's village. Naimuimada does not only contain Fijian Methodists. He tells me that "there are many Hindus and Muslims living here." Given the connection that Ligavatu's sermon makes between a people's covenantal fidelity to YHWH and the thriving of place, he struggles to imagine what purity of practice looks like in this pluralistic context.

Crucially, he does not suggest that people of other faiths be removed from the community.[91] But he does emphasize the importance of the *iTaukei* community retaining legal responsibility over their land. If the strength of one's relationship with one's land indicates the strength of one's relationship with God, the most concrete piece of advice he has for his Indigenous Methodist congregation is stop leasing their land to others—be they outside the faith community or outside developers. "The land is covenant," he tells them. "It is for your children that will come after you, and it will strengthen your relationship with God."[92] For Ligavatu, there is "a difference between giving the land [to ethnic/religious outsid-

---

A Story of Indigenous Fijian Women's Role in Funerals" (master's thesis, School of Social Science, University of the South Pacific, Suva, September 2015).

[89] Baleloa, "What Must I Do."

[90] Lynda Newland describes this conversation through the lens of Fiji's adoption of a secular constitution in 2013, in "Imagining Nationhood: Narratives of Belonging and the Question of a Christian State in Fiji," *Global Change, Peace & Security* 25 (2013): 227–42.

[91] This is particularly significant given the "anti-Indian rhetoric that promoted the images of Indo-Fijians as *vulagi* (i.e., foreigners)" during the 2000 coup—the nation's most violent. "Hand-in-hand with *taukei* assertions of 'Indigenous rights' were condemnations of non-Christians and calls for Fiji to restore its status as a Christian state" (Susanna Trnka, *State of Suffering: Political Violence and Community Survival in Fiji* [Ithaca, N.Y.: Cornell University Press, 2008]). Ligavatu, for all of his concern about "blending," works to distance himself from such condemnations in our conversation.

[92] Ligavatu, personal interview, January 20, 2022.

ers] through lease—and giving it for them to stay." Land leases in Fiji are traditionally ninety-nine years long. "When you lease land for ninety-nine years, you don't know if that will come back or if it has left you," Ligavatu explains. The community also relinquishes responsibility for the land's care over that time period. "But if you give the land for them to stay . . . it is like love (i.e., *loloma*)."

Ligavatu recognizes that something is not right in the relationship between the *iTaukei* village and its non-*iTaukei* neighbors. It does not sit well with him that "our relationship with [religious others] is just like a smiling of faces—but in our hearts there is . . . [*he grimaces*]." What Ligavatu leaves unspoken is Fiji's struggle with ethnonationalist ideologies of belonging, exacerbated by long-standing policies of colonial segregation.[93] Theologies of place that weave together identity and land through covenantal performances of faith often struggle with this challenge—especially when fears of displacement are at the door.

In Ligavatu's view, relational tensions between those of different faiths and ethnicities continue to negatively impact the area in which he serves. First and foremost, they affect the climate of trust in the community. He worries, for example, when he hears village people say things like, "Why are these people living here? They should leave this place."[94] The sentiment is lacking in Christian love, as well as the prized *iTaukei* virtue of hospitality. More practically, it hinders the ability of the community to collaborate in its decision-making. It doesn't matter if the Indigenous community makes decisions to protect the land if outsiders in the community increase their rates of development. Place connects the community, regardless of their religious differences.

93 Marika Bale's history of the Methodist Church in Fiji's relationship to racial identity celebrates the vision of unity and interethnic cooperation championed by Rev. Setareki Tuilovoni, the first MCIF president following the church's independence in 1964. He contrasts that vision with that of the Taukei Movement—an ethnonationalist political movement that emerged shortly after the defeat of the Alliance party in 1987. He notes the opposition of President Josateki Koroi and other key MCIF leaders to this entrance of "*taukeism* philosophy into the church": "Racial Integration in Transition," 44–45. Lynda Newland notes the ways that the "colonial project" also exacerbated a "rigid and more politicized . . . basis of Fijian identity": "Religion and Politics: The Christian Churches and the 2006 Coup in Fiji," in *The 2006 Military Takeover in Fiji: A Coup to End All Coups?* ed. Jon Fraenkel, Stewart Firth, and Brij V. Lal (Canberra: ANU Publications, 2009), 189. Chapter 4 further engages this history through the lens of the MCIF's "New Exodus."

94 Ligavatu, personal interview, January 20, 2022.

In articulating his perspective, Ligavatu returns to the logic of covenant. Ligavatu argues that those of other faiths worship other gods, so the strictures of Christian covenant need not apply to them. "We serve a universal God," he reasons, ". . . who made different kinds of humans with different kinds of faiths." But for Ligavatu, covenant creates a common responsibility. "To keep the covenant means we should go in with a very wide and open mind to understand that the world God gave us is not only for us, for *iTaukei*. It is for all of us. We need to understand them, live with them, talk to them. Pushing them aside or leaving them, this is not covenant."

Mark Brett notes that in Ezekiel's text and the Holiness School associated with his tradition, there is an imagining of "ways in which permeable social boundaries need not undermine the core identity and vocation of Israel."[95] He notes that "strangers" (*gerim*) continue to have responsibility in maintaining the land's holiness (i.e., Lev 20:2–5), but "the full range of obligations that bind Israelites are not . . . imposed on the population in the same way."[96] More significantly, he notes the assumption that the *gerim* will be living alongside the covenantal community—and that the primary concern of the text is on appropriate behaviors of relationship with land and neighbor, rather than bloodlines. In Ligavatu's tensive description of the differences between his *iTaukei* Christian village and its Hindu and Muslim neighbors, as well as the universality of God's covenantal call to relationship, one hears a similar sort of negotiation. Ligavatu leaves space for difference in definitions of faithful covenantal practice; significantly, his understanding of a "universal" God does not assume that conversion precedes relationship. And yet his understanding of God's universality does call the Christian community into authentic relationships with those who are outside its covenantal practices.

Ligavatu is trying to imagine a new way—or more accurately, he is articulating an older, Indigenous way. It is significant that Ligavatu's concern about belonging is not simply *who* belongs to place, but *how* that belonging is grounded and lived out in time. He is interested in establishing relationship "in the manner of the land" (*i tovo vakavanua*), rather than "doing things 'in the manner of money' (*vakailavo*)."[97] Sociologists

95 Mark Brett, *Political Trauma and Healing: Biblical Ethics for a Postcolonial World* (Grand Rapids: Eerdmans, 2016), 109.

96 Brett, *Political Trauma and Healing*, 109.

97 Jara Hulkenberg describes an Indigenous Fijian opposition between these understandings in how kinship ties are established: "Fijian Kinship: Exchange and

like Unaisi Nabobo-Baba have described the concrete ways that "belongingness," "relatedness," "boundaries," and "cooperation" are given form in *iTaukei* villages.[98] Belonging is practiced, not simply conferred. There is a performative nature of Fijian identity and its participatory relation to the *vanua*. It is not a category that is stable and fixed.

This provisionality means that what Ligavatu is describing would very likely be unsatisfying to his Hindu and Muslim neighbors. To Western ears, it sounds naïve.[99] He knows this. When I tell him that what he describes is "well said—but hard to do," he nods in silence. But it is worth noting that this vision of hospitality is different than an ethnonationalist's vision, where one's belonging to land is authenticated by one's bloodline. It is also different from models of belonging based on economic transaction and secular contracts. The syncretism that Ligavatu finds dangerous is not simply the "blending" of Christian practices with those other religious traditions. There is another form of syncretism that he is pressing against: the "blending" of covenantal commitments to place with noncovenantal models of belonging, be they rooted in blood or money. He stretches to articulate a belonging negotiated through practices of shared responsibility and limits, and he is not discouraged by the challenge. This kind of belonging, growing out of communally recognized covenantal relation, is not unprecedented in Fiji.

In *Vanua: A Fijian Theology of Place*, Tuwere describes occasions where land is gifted in order to mark or strengthen relationships of kinship: *veiwekani*.[100] Expanding on Tuwere's work, Nabobo-Baba notes that gifts of land have also been given to "create a new settlement for a displaced population" and "as a gift to the state to benefit the wider community, province,

---

Migration," in *Living Kinship in the Pacific*, ed. Christina Toren and Simonne Pauwels (New York: Berghahn Books, 2015), 60–85 [60].

98 Unaisi Nabobo-Baba, *Knowing and Learning: An Indigenous Fijian Approach* (Suva: University of the South Pacific Press, 2006).

99 The issue is not only unconvincing idealism. Martha Kaplan notes the problematic ways that *vuravura vakaloloma* (i.e., world in the way of love) and *vuravura vakailavo* (i.e., world in the way of money) have "roots in [first British governor of Fiji, Sir Arthur] Gordon's separation of Fijians from the market with objectifications of nineteenth-century exchange practices": "Promised Lands: From Colonial Lawgiving to Postcolonial Takeovers in Fiji," in *Law and Empire in the Pacific: Fiji and Hawai'i*, ed. Sally Engle Merry and Donald Brenneis (Santa Fe, N.Mex.: School of American Research Press, 2003), 153–86 [166].

100 Ilaitia S. Tuwere, *Vanua: A Fijian Theology of Place* (Suva: Institute of Pacific Studies Press, 2001), 39–41.

or nation."[101] Such gifts reframe the definition of "kinship" through the gift of land itself. Nabobo-Baba describes such gifts of land (*i solisoli*) "as instances of *veiwekani vakaturaga*" (chiefly behavior) that make evident values of sharing, reciprocity, and "distributive justice."[102] Behavior that is *vakaturaga* is behavior that is "so steeped in respect for others that . . . [it] acts as if others were of more importance than themselves."[103] Nabobo-Baba emphasizes that "when land is gifted (*i solisoli*), the original owners . . . [continue] to be recognized and blessed in all things that relate to the land that was given." They do not relinquish responsibility for the land and "are mentioned and recognized even if they are not present to remind all, especially those who do not know, who the landowners are."[104] None of the examples Nabobo-Baba provides are gifts of land to Indo-Fijians. But as I listen to Ligavatu, I wonder if he is imaging just such a reframing of relationship. In the context of climate crisis, new models of reciprocity and cooperation will be increasingly necessary with those outside the Christian Indigenous community. Ligavatu's description of a covenantal belonging, for all of its fear of "syncretism," recognizes this. He attempts to posit a belonging across difference that requires more of his community than a "smiling face" and a lease. He envisions a belonging that will strengthen his community's faith commitment and his community's Indigenous commitment to place rather than dilute it. He also knows it will not be possible without *loloma* (love) and a new orientation of the heart.[105]

Covenantal negotiations of belonging are critical in the climate conversations of Fijian preachers. If faithful covenantal relation is God's desire between human and nonhuman relation, then the question of who is responsible for a given place is key—as is the question of what those obligations look like in lived practices of commitment. The character and scope of belonging becomes more complicated to negotiate given colonial legacies that have displaced those with very different understandings of

[101] Nabobo-Baba names such a land gift "a i solisoli ni koro vou" (the gift of a new village) and "a i solisoli nib ula torocake raraba" (a gift of life for the good of all): "The Mutual Implication of Kinship and Chiefship in Fiji," in Toren and Pauwels, *Living Kinship*, 33.

[102] Nabobo-Baba, "Mutual Implication," 17.

[103] Nabobo-Baba, "Mutual Implication," 17.

[104] Nabobo-Baba, "Mutual Implication," 19.

[105] The necessity of "a new heart and a new spirit" in order to keep the covenant properly is also a focus of Ezekiel (36:26).

identity and piety, but in the wake of climate catastrophe, such negotiations are increasingly urgent. This is not only because diverse communities are needing to work together; it is also because Indigenous communities themselves are losing their land and needing to establish new belonging in new places.[106] In a recent article outlining the challenges faced by villages needing to relocate because of climate stressors, lack of land and tensions with donors who had gifted land were both listed as thorny challenges.[107] The examples provided show these tensions within the *iTaukei* community itself. Given these challenges, the stakes of Ligavatu's insistence that belonging be rooted in faithful covenantal performances are clear. To frame belonging through covenantal connections with land leaves open both the possibility of genuinely hospitable relationships with strangers and new connections of belonging formed in new places. Theologies of covenant allow Indigenous Fijian Christians to do more than plait their faith together with cultural practices performed *i tovo vakavanua* (in the manner of the land). They also create the potential for holding land with an open palm, received and offered as gift, even as they assert the crucial spiritual ties between people and place.

Of course, in the biblical text and contemporary life, potential is not always realized. As with the question of sovereignty, the question of belonging continues to be actively negotiated in Fiji—particularly in the political sphere. The constitution may apply the name "Fijian" to all its citizens, but there are still political parties which debate that designation for non-*iTaukei* citizens.[108] Love is notoriously difficult to legislate. It is no surprise that many Indo-Fijians prefer the protections

[106] Wolfgang Kempf and Elfriede Hermann describe the "repositionings and multiple belongings" that are part of how Indigenous communities are "coping with the consequences of climate change and sea level rise." These repositionings are, in the authors' words, "permeated with ambivalence": "Uncertain Futures of Belonging: Consequences of Climate Change and Sea-Level Rise in Oceania," in *Belonging in Oceania: Movement, Place-Making and Multiple Identifications*, ed. Elfriede Hermann, Wolfgang Kempf, and Toon van Meijl (New York: Berghahn Books, 2014), 189–213 [194].

[107] Kate Lyons, "How to Move a Country: Fiji's Radical Plan to Escape Rising Sea Levels," *Guardian*, November 8, 2022, https://www.theguardian.com/environment/2022/nov/08/how-to-move-a-country-fiji-radical-plan-escape-rising-seas-climate-crisis.

[108] For history of the debate around the designation of "Fijian," see Newland, "Imagining Nationhood." For an extension of this debate into the 2014 elections, see Lynda Newland, "From the Land to the Sea: Christianity, Community and State in Fiji—and the 2014 Elections," in *The People Have Spoken: The 2014 Elections in Fiji*, ed. Steven Ratuva and Stephanie Lawson (Canberra: ANU Press, 2016), 109–34.

of law (extended under the purview of the government, i.e., *matanitu*) to the benevolences of the *vanua*. Jacqueline Ryle describes the relief that the constitutional designation of "Fijian" gives a priest from the Arja Samaj Hindu organization: "There's assurance that should anyone disturb you in your prayers, you . . . have recourse to the law . . . that the law should protect you . . . [should there be] any . . . misbehavior against any religion."[109]

What does Christian covenantal commitment require of the *lotu* in such a circumstance? Many *iTaukei* preachers are still discerning. But within that necessary labor, a subversive challenge to neocolonial conceptions of place finds voice. Rather than conferring belonging through financial transaction or ethnicity, Indigenous Fijian preachers describe a belonging grounded in fidelity to emplaced relationships. Like Ligavatu's understanding of the *imago dei*, such living relationships with place cannot be taken for granted. They result in a belonging that is precious, costly, and elusive.

Facing Loss

I sit with Taniela Balenaikorodawa on the floor of his Nausori home reflecting on a sermon he preached two years before. Pandemic protocols have lifted, and I am grateful to be with my friend in person as he shares the story of Cogea village. "The morning after Cyclone Yasa," he recalls, "the village members returned to their *vanua*, . . . and the wailing began." His eyes get teary, and his voice becomes soft. "Eighteen homes were washed away."[110] Cogea is the *vanua* of Balenaikorodawa's mother, and his memories of the place go deep. When he was a young boy in the early 1970s, he walked five miles each week with his grandmother's brother, a landowning unit chief named Simione Vunibaka, looking for mangrove seedlings to replant along the coast.[111] Those decades-old trees were torn from their roots when Cyclone Yasa, a Category 5 storm, hit the Vanua Levu shoreline. But the community's griefs were not over. Two weeks later, Category 3 Cyclone Ana made landfall, confronting the already battered

109 Jacqueline Ryle, "Religious Freedom," in Ross, Tahaafe-Williams, and Johnson, *Christianity in Oceania*, 309–20 [315].

110 Taniela Balenaikorodawa, personal online interview, January 3, 2023.

111 Taniela Balenaikorodawa and Jerusha Matsen Neal, "Anti-colonial Homiletic Strategies for Combating Climate Change in the Fijian Methodist Church," paper presented at the World Christianity Conference, Princeton Theological Seminary, March 6, 2021.

village with fierce wind, drenching rain, and severe flooding. When Ana subsided three days later, the village was uninhabitable.

Balenaikorodawa had come to Cogea before the storm for the funeral of his cousin, Tamai Bobby.[112] Cyclone Ana had torn through the village over the course of the funeral, stranding over twenty members of his extended family (not including children). It was the night after the funeral, and the family had asked Balenaikorodawa to bring a sermon to their family worship service. His text was Genesis 1:1–5.

Balenaikorodawa begins his sermon by noting the preciousness of the land on which they sit, a land where "our ancestors lay buried with the roots of the *baka* tree."[113] The burial imagery is particularly poignant given the funeral of the day prior. He reiterates the living connection between the community and their land: "The Taukei land, rivers, sea, reefs, creeks, animals, fish, flowers, waterfalls, and natural surroundings are closely connected to our people's identity. [They] reveal who we are." The "topography and geography and relationships (*veiwekani*)" of this place are "woven into the lives of our community." In this opening paragraph, Balenaikorodawa doesn't explicitly name the unvoiced question that hangs in the room, but it is clearly implied. *What if this land—so central to the village's identity—is lost?*

Ellen Davis describes the biblical exile as taking place "at the end of known history."[114] In this sense, the exile takes on the character of an apocalypse—an end that one cannot see beyond, when the natural relations of the world are undone, and where the ground of one's identity slips through one's fingers like sand.[115] Balenaikorodawa's sermon posits

112 Balenaikorodawa uses a respectful form of address, calling his cousin by the name of his eldest son. He explains that this address, "Tamai Bobby" (*father of Bobby*), honors the relationships established in marriage, not only of the husband and wife, but also of their respective *vanua*.

113 Taniela Balenaikorodawa, "God the Creator, Recreating," sermon preached in Cogea village, January 29, 2021.

114 Ellen Davis, "Seeing God in Exile," public lecture at New York Avenue Presbyterian Church, Washington, D.C., October 21, 2018, https://www.youtube.com/watch?v=n6V_BJWLjSU.

115 Elizabeth Phillips distinguishes between "apocalyptic*ism*," which she describes as a worldview which "sees little value within history" and "apocalyptic . . . visions of the future which function to jolt us out of our status-quo perceptions and see current social reality for what it truly is in relation to God." She goes on to "consider more closely the darkness of the apocalyptic" and its connection in communal narrations of catastrophic events: "Narrating Catastrophe, Cultivating Hope: Apocalyptic Practices and Theological Virtue," *Studies in Christian Ethics* 31, no. 1 (2018): 17–33 [21–22].

just such a crisis, and no simple affirmations of community resilience or pragmatic troubleshooting feel sufficient to the challenge. Instead, he names the community's grief and suffering through a linguistic analysis of the *iTaukei* word for weather: *draki. Dra* is the word for blood, he explains, and *ki* is a preposition that indicates direction and place. Weather, for Balenaikorodawa, "reveals itself [as] the blood or sacrifice that the land and people encounter on their journey through . . . place." He uses this observation to make a direct connection between the suffering of creation after a cyclone, the suffering of his mother's village, and the suffering of Christ on the cross. "Creation sacrifices itself in times of cyclones such as this, [just as] Christ sacrifices himself for the salvation of humanity."

Understanding the death and resurrection of Christ as an apocalyptic interruption of history is familiar interpretive strategy for New Testament scholars. Paul describes his own transformation and vocational calling in relation to God's revelation (*apokalypsis*) of Christ (Gal 1:16).[116] In this case, the language of "apocalypse" does more than describe the end of known history. It also implies an "unveiling" of truth that has consequence in the present—an epistemological shift that allows one to see the world laid bare.[117] The "apocalypse of God"[118] revealed through the cross exposes the lies and false promises of the powers and principalities (Col 2:15), just as it reveals God's intimate solidarity with a suffering world. Balenaikorodawa's sermon makes a similar shift. Rather than simply naming *draki* as catastrophe, he wonders what "role *draki* plays in our journey." He looks into the devastation of the storm and attempts to name a deeper providential reality. Scripturally, he also makes a shift, moving from the cruciform images of the New Testament to the imagery of Genesis. Sitting in a village cut off from electricity and outside communication, grieving the death of a beloved family member, and contemplating the loss of ancestral land, Balenaikorodawa's sermon works to unveil the "darkness" that covers "the face of the deep" (Gen 1:2).

[116] Martinus C. De Boer, "Paul, Theologian of God's Apocalypse," *Interpretation* 56, no. 1 (2002): 21–33 [29ff.].

[117] J. Louis Martyn's classic interpretation of 2 Cor 5:16–17, for example, speaks of the apocalyptic struggle of the "turn of the ages" as an "epistemological" reframing of the world "by the cross" (*kata stauron*): "Epistemology at the Turn of the Ages," in his *Theological Issues in the Letters of Paul* (Nashville: Abingdon, 1997), 89–110 [108].

[118] De Boer, "Paul," 24. De Boer's use of this phrase draws on J. Louis Martyn's discussion in *Galatians* (New York: Doubleday, 1997), 97–105.

Following Cyclone Winston in 2016, a social media hashtag swept the country: #StrongerThanWinston. Originating in urban centers, the tag was meant to communicate community resolve in overcoming the storm's devastation. The sentiment resonates with Western ears. But it also created anxiety among many traditional *iTaukei* Christians, who experienced the tag as exhibiting a dangerous pride and oppositional approach to creation itself.[119] It is striking how differently Balenaikorodawa conceives of the storm that has destroyed his mother's village. He notes that "darkness" is not an anomaly in created order, that "darkness has always been part of creation [from] the beginning." More than this, he underscores that even when God created light, God did not "abolish darkness. . . . [It] is made to have a function in creation." Describing the linguistic connections between the *iTaukei* word for darkness (*butobuto*) and the word for umbilical cord (*buto*),[120] Balenaikorodawa wonders if the storm might strengthen "cords of connection that will seed new life." At stake for Balenaikorodawa is a conception of God's Spirit that "cannot be boxed by our human minds, which may not fully comprehend the wonders of creation, their relatedness and interconnectivity."

Balenaikorodawa does not keep silent about the losses the community has suffered, including the "millions of dollars" that the lost houses and crops will cost. He notes that the community has been "broken" by an event "beyond our control." It is a night, he says, to "lament our loss." And yet in the same paragraph, he notes that in the creation story, God does not "delete darkness," but uses it for "renewal" and a "rekindling of relationship." Insisting on God's providence when faced with the *butobuto* (darkness) of climate disaster might be perceived as dangerous fatalism, but Balenaikorodawa uses the theological affirmation to a different end. He describes a hope that is not "stronger than" creation—but rooted in the community's ties to place and their trust of place. Creation suffers with them, as Christ suffers with them. And creation will renew itself through God's creative, connecting power.

There is a significant pastoral consequence of this approach. Balenaikorodawa does not blame his community for the storm. Even as he stresses their relationship to the land and to each other, he reiterates that

119 John Cox et al., "Disaster, Divine Judgement, and Original Sin: Christian Interpretations of Tropical Cyclone Winston and Climate Change in Fiji," *Contemporary Pacific* 30, no. 2 (2018): 380–410 [389].

120 Balenaikorodawa notes that this linguistic link is particularly clear in *iTaukei* dialects from the island of Kadavu.

they are not central in this story. They did not cause the storm, nor can they turn back the world to the way it was before the cyclone made landfall. "Always," he protests, "the blame is placed on the people" when natural disasters strike—but what if *draki*, in all of its fury, is part of creation's process of recovery and renewal? What if "*butobuto* (darkness)" is unveiled as "the *buto* (creative connection)" that creates relationship and life?

This is a place of notable difference between Balenaikorodawa's sermon and Ezekiel's approach to suffering and punishment. Ezekiel counters Exodus' warning that the "iniquity of the parents" will be "visited on the children" (Exod 34:6–7) with a new word: "Only the person who sins . . . shall die" (Ezek 18:4). Such framing emphasizes the link between sin and consequence, and Ezekiel uses the affirmation to assure his people that their righteousness matters to God; their fate is not determined by the decisions of generations prior. Similar to a number of Fijian Methodist sermons described in this chapter,[121] Ezekiel argues that present righteousness has material impact (18:5–9), as does unrighteousness (18:10–13). Suffering and sin are linked. But Balenaikorodawa refuses to read his community's loss in this way, and he points to Christ's crucifixion and Genesis' relational depiction of darkness and light to make his point.

It is a mysterious and courageous word that has a lasting effect on his community. In January 2023, Vani Catanasiga, the executive director of the Fiji Council of Social Services, visited the village in preparation for the community's decision-making process around relocation. The village had been told on multiple occasions that the government would break ground on a relocation site, but there had been little follow-through. Many were still living in tents. "They have had so many false promises," Catanasiga notes.[122] The community approached her to see if she might help them organize a relocation themselves, and she had come to talk logistics. Surprisingly, Catanasiga—an employee of a secular social service agency—found herself talking about sermons instead. The villagers told her of sermons from outside church leaders that had blamed the com-

[121] Many of the sermons collected arch toward this view, some focusing on a neglect of spiritual practices and others on the dangers of consumerism and pride. Savenaca Kamikamica, for example, warns that "technology and advanced scientific discoveries seem to drift [our] attention far away from God . . . and so God calls [us] back through climate change" ("God Continues to Restore"). Cliff Bird, a Solomon Island theologian, bemoans and complicates the framing of natural disasters as a divine "wake-up call" in "God and Natural Disasters," Facebook, Methodist Church in Fiji and Rotuma, February 29, 2016, https://www.facebook.com/notes/685264578781425/.

[122] Vani Catanasiga, personal interview, Suva, January 11, 2023.

munity for their losses. Then, in contrast, they told her about a sermon "from a Rev. Balenaikorodawa."[123] To the director's surprise, the community remembered the sermon as a source of strength and comfort, though it had been preached two years prior. Balenaikorodawa's proclamation, which emphasized the community's lack of control over the weather and lamented the difficult road ahead, had increased the community's reserves of agency and hope.

Emmanuel Katongole has argued that lament is about more than existential release; it "deepens and intensifies engagement with the world of suffering, . . . [inviting] us into deeper political engagement."[124] More than this, communities who engage in lament—naming inscrutable, inequitable loss—perceive that more than "legal and administrative adjustments within the [current] framework" are necessary for human flourishing. Instead, they "invoke a totally new vision of society."[125] Rather than asserting a simplistic list of #StrongerThanWinston solutions to an intractable problem, Balenaikorodawa's sermon invites the community into deeper reflection and engagement with the land. Such engagement takes time, but that is not an unfamiliar concept in *iTaukei* village life. Balenaikorodawa's approach resonates with Tuwere's description of the *vanua*'s rhythms. "Time," Tuwere argues, "is not so much something to be used as it is *a place* in which one waits—for the birthing of the fish, for the flowering of the *drala* tree, or the appearing of the *balolo* [an edible sea worm] on the reefs."[126] Even as these rhythms are disrupted by climate catastrophe, Balenaikorodawa refuses to sever the relational dependence that grounds his people's conception of place. Creation, he insists, is not the enemy. Rotuman playwright Vilsoni Hereniko describes how the material losses of climate change can unravel Indigenous intimacies with places themselves. When "'the deep, beautiful, but dangerous sea' is no longer a [source] of regional strength and unity, but a threat that will engulf,"[127] the trust that

123 Vani Catanasiga, personal correspondence, January 16, 2022.

124 Emmanuel Katongole, *Born from Lament: The Theology and Politics of Hope in Africa* (Grand Rapids: Eerdmans, 2017), 261.

125 Katongole, *Born from Lament*, 262–63.

126 Tuwere, *Vanua*, 39, quoted in Rubow and Bird, "Eco-theological Responses," 154–55 (emphasis original).

127 Vilsoni Hereniko, "The Deep, Beautiful but Dangerous Sea," program for the film premiere of *Moana: The Rising of the Sea*, University of the South Pacific, Suva, 2014. More information on the film can be found at: https://catalog.library.vanderbilt.edu/discovery/fulldisplay/alma991043676880703276/01VAN_INST:vanui. Quoted in Maebh Long, "Vanua in the Anthropocene: Relationality and Sea Level Rise in Fiji," *symploke* 26, nos. 1–2 (2018): 51–70 [60].

sustains communities in times of crisis is threatened. Balenaikorodawa's sermon laments loss even as it rejects fear, and in this way, it is more than an act of pastoral care. It is an act of resistance.

Rev. Esala Nacuva had been serving the village of Saumakia for two years when a massive landslide threatened the village in 2018. The government issued a directive for relocation, and the community was stunned with grief. The instability of their land had never been an issue in communal memory, but severe storms and the dredging of the river by development companies downstream had changed the force of their river's current. A water source that had provided life, transportation, and food was becoming increasingly dangerous. Nacuva recounts the fear in the village on the day when he preached his sermon "The Holy Spirit Is Our Source of Strength."[128] It was the week of Pentecost, and so it is no surprise that, like Balenaikorodawa, Nacuva focuses on the creative capacity of the Spirit and the life that "sprouts" in the Spirit's presence. Birth imagery also makes an appearance in his sermon. The Feast of the Visitation, when the church remembers Mary's visit to Elizabeth early in her pregnancy (Luke 1:39–45), brought to Nacuva's mind the promise of the Annunciation: "Nothing is impossible with God" (Luke 1:37). Remembering the sermon, Nacuva notes the "low spirit" of the community on that day—"no hope. The government says . . . you won't be staying here, and you have to move. As a Fijian, with your connection to the land, with the memories of your ancestors, with your connection to them in that particular place—[*a pause*] . . . at that very moment, I feel what they feel."[129]

The sermon that he preaches is not an affirmation of strength; it is a call to humility. He asks his community to approach the church's altar "with humbleness," trusting that God can accomplish what the government cannot. To his surprise, the chief who sits in a special place of honor and "normally doesn't stand . . . comes down. . . . He moves from his place . . . to be at the same level as the villagers. . . . No rank. Just a normal human being in need of God's power." Nacuva becomes quiet as he remembers the sermon's impact. "There were tears flowing," he tells me.

In "The Role of *Vanua* in Climate-Related Voluntary Immobility in Fiji," the authors note that "decolonizing climate change from a Pacific perspective" will have to take into account "local voluntary immobility."[130]

[128] Nacuva, "Holy Spirit Is Our Source of Strength."

[129] Esala Nacuva, personal online interview, January 17, 2022.

[130] Merewalesi Yee et al. "The Role of *Vanua* in Climate-Related Voluntary Immobility in Fiji," *Frontiers in Climate* 4 (December 2022): 1–19.

In other words, it will have to take seriously the crisis of loss that relocation creates for culture, livelihoods, and social networks. Communities that "resist mobility" often understand something about place that governmental agencies tasked with relocation miss. In the case of Saumakia, the village's "humility" in relation to God and in relation to each other led them to reject the governmental relocation directive and instead look for ways to press the provincial administrator for greater attention to the impacts of regional development. In Nacuva's words, we need "them to witness [the problem], so they can feel what we feel."[131]

In scientific descriptions of religious approaches to climate change in the Pacific, a binary is regularly described between a narrative of "religiously informed . . . adaptation-oriented environmental stewardship, . . . stressing human responsibility in the face of climate change" and a "counter-narrative" that considers "climate change as God's business."[132] Such a binary does not fully account for the scope and urgency of the questions local communities are asking—particularly around questions of *loss*. When the apocalyptic realities of climate catastrophe are taken seriously, the line between spiritual action and material action blurs. Reflecting on what made it possible for the village of Cogea to move toward a just, community-led model of relocation, Catanasiga notes the significance of "prayer and fasting."[133] Spiritual practices increase community cohesion, and they inoculate communities from the panacea of empty promises by human institutions who have much less to lose than local communities themselves. They also unveil the limits of governments, institutions, and technologies to bring restoration. Given the extent of what is broken and all that is at stake, a greater Source of salvation will be necessary.

Like the questions of sovereignty and belonging, the question of how best to respond to the loss of place in climate catastrophes is unresolved in Fijian Methodist preaching. The question intersects an evolving, tensive theology of hope. Indigenous Fijian conceptions of place resist models of hope grounded in subduing, controlling, or abandoning creation. They press toward a hope that is communal, humble, and relational. Such relationality is a source of strength, even as it leaves the community vulnerable to tragedy. Delai tells the story of a family whose home collapsed on top

131 Nacuva, personal online interview, January 17, 2022.

132 Elodie Fache and Hannah Fair, "Turning Away from Wicked Ways: Christian Climate Change Politics in the Pacific Island Region," *Anthropological Forum* 30, no. 3 (2020): 233–53 [233].

133 Catanasiga, personal interview, January 11, 2023.

of them in the middle of Cyclone Yasa. The father was able to shift enough debris for his children to scramble free, but he and his wife were trapped under a fallen wall. He pleaded with his children to leave them and find shelter elsewhere. "No, Dad," his thirteen-year-old daughter replied. "We will not go; we will die together."[134] The father would later recount "a surge of pure energy coursing through his veins as he pushed the wall and freed himself and his wife."[135] But not all stories have such endings. The family would later discover that, unknown to them at the time, a different relative had died under the very same wall.

Development continues on the river that cuts into the hillside of Saumakia. The villagers of Cogea struggle with trauma, despite their resilience. Catanasiga tells me she will bring a counselor to the village when she returns, as the pain runs deep.[136] The question of how to sustain hope in the midst of a crisis one cannot control is real—as is the question of how to sustain connection to land and community in ways that allow for adaptation and relocation. How is hope of new life seen, protected, and nurtured in the "darkness" of increasingly fierce storms and rising tides? Once again, the witness of Indigenous Fijian pastors echoes Ezekiel when asked, "Can these bones live?"

"O Lord God, you know" (Ezek 37:3).

Places of Reckoning and Resistance

Catanasiga has spent her career at Fiji Council of Social Services thinking about how to activate the agency of rural Fijian villages and increase their access to trustworthy information when facing rapid climate changes. She is aware of how even the work of well-meaning NGOs can exacerbate narratives of victimization, disempowering Indigenous communities. "We give a handout—which is good for NGOs and raises money for NGOs—and we seem to be the hero in the situation."[137] She continues, "I think the messaging should change. God has created us strong . . . and we can fail to acknowledge what communities bring on board." In her relocation efforts, she insists on centering the decision-making processes and resources of the local community, with outside agencies providing assistance upon request from the margins. In this way her work resonates

[134] Delai, "Spiritual Views on Climate Change."

[135] Luke Rawalai, "TC Yasa: Father's Will to Survive," *Fiji Times*, December 31, 2020, https://www.fijitimes.com/tc-yasa-fathers-will-to-survive/.

[136] Vani Catanasiga, personal correspondence, January 16, 2022.

[137] Catanasiga, personal interview, January 11, 2023.

with environmental activists who maintain the "epistemological privilege of subaltern communities."[138]

Theological education has its own versions of playing the "hero." Particularly in the West, practical theologians can become especially invested in assessing, fixing, or diagnosing their interlocutors from the global south.[139] I have suggested that these prescriptive, objectifying postures are related to conceptions of place that overemphasize its universality, insularity, or constructed character at the cost of interdependence. Such postures can be observed in biblical hermeneutical projects that marginalize conceptions of place that decenter the human subject. When this decentering of the human occurs in scriptures or communities marked by traumatic displacement or colonization, it can be construed as pathology.[140] I have discussed certain ecological critiques of Ezekiel in this vein. Certainly, the traumatic effect of displacement is real and may lead to ethically and pastorally dangerous discourses: tragedies can be interpreted as punishment; communities might exhibit exclusionary preoccupations with purity or mask passive resignation with language of

[138] George Zachariah, "Colonization of the Watersheds and the Green Politics of Hagar," in *People and Land: Decolonizing Theologies*, ed. Jione Havea (Minneapolis: Fortress, 2019), 141–57 [144].

[139] Even the language of "decolonization" can be co-opted by the Western theological academy in this way. Robert Heaney warns against a "cheap postcolonialism" that adopts postcolonial language without commitments to postcolonial practice: "Prospects and Problems for Evangelical Postcolonialisms," in *Evangelical Postcolonial Conversations: Global Awakenings in Theology and Practice*, ed. Kay Smith, Jayachitra Lalitha, and L. Daniel Hawk (Downers Grove, Ill: IVP Academic, 2014), 29–44 [38]. Eve Tuck and K. Wayne Yang deliberately extend Aimé Césaire's critique of colonialism to describe what "decolonization" is not: "neither evangelism, nor a philanthropic enterprise, nor a desire to push back the frontiers of ignorance, disease and tyranny," in "Decolonization Is Not a Metaphor," *Decolonization: Indigeneity, Education and Society* 1, no. 1 (2012): 1–40 [21]. For the original quote, see Césaire, *Discourse on Colonialism*, trans. Joan Pinkham (New York: Monthly Review Press, 2000), 32.

[140] Hannah Fair, "Three Stories of Noah: Navigating Religious Climate Change Narratives in the Pacific Island Region," *Geo: Geography and Environment* 5, no. 2 (2018): 1–15, expands on Kempf's research, describing and problematizing numerous scientific engagements with Oceanic faith communities that call for "an increase in scientific information and a rejection of religious knowledge." Fair connects this posture with Bruno Latour's broader discussion of "purification of scientific and religious knowledges, and epistemological hierarchization" in *We Have Never Been Modern* (Boston: Harvard University Press, 1993), 6. Kempf notes that religious knowledge is regularly associated with "deviation, ignorance, passivity, and maladaptivity" ("Climate Change, Christian Religion, and Songs," 30).

faith. These specters do appear in Fijian Methodist sermons on climate catastrophe. The sermons collected in this chapter, however, are marked by more than trauma. They are also marked by anti-colonial resistance. They raise different questions about *place* from displaced ground, rejecting the underlying assumptions of a "colonial gaze" that "*thingified* and commodified"[141] place, denying the land's relational claims and sacral mediation. I have characterized these questions as negotiations of sovereignty, belonging, and loss.

The Fijian Methodist preachers described here are not invested in transcending the particularity of their places. They are working, instead, to protect that particularity, even as they insist on place's interrelation with a complex, often dangerous world. To image *sovereignty* in this relational way names place as both bordered and connected, resisting Western assimilations and exclusions.[142] It also raises the question of a community's role in relation to God and its responsibility to the place it lives. In contrast to Western models of sovereignty that stress an insular, self-determinative agency, Fijian Methodist preachers reframe self-determination in the context of divine vocation and relationship.

Because of this, these Fijian Methodist preachers do not dwell on how the self relates to place. Instead, they grapple with the question of how emplaced *belonging* is shared and if that belonging can be extended to others. The question is not only *who* belongs—but also *how* belonging is maintained and gifted. Drawing on covenantal language that emphasizes spiritual fidelity and commitment, the community images place as more than a commodity or an inheritance. Fijian preachers who emphasize covenantal practices as the basis of belonging embrace a performative understanding of belonging, prioritizing a community's relationship with place, while leaving open the possibility of the community's embrace of the stranger.

Finally, these Fijian Methodist preachers are ambivalent about the human ability to restore places once they have been lost. Human technologies that attempt to control, subdue, or insulate communities from

[141] Zachariah, "Colonization of the Watersheds," 142 (emphasis original).

[142] Lisa Thompson gives helpful, accessible examples of these assimilations and exclusions from the U.S. context, concluding, "We have not lived well and responsibly in this interconnected space": "Struggles for Belonging: Environment, Land, and Borders," in *Preaching the Headlines: Possibilities and Pitfalls* (Minneapolis: Fortress, 2021), 65–84 [72].

creation are seen as part of the problem, rather than the solution.[143] The human capacity to "make" places elsewhere is also suspect. Relocation often exacerbates social vulnerabilities and can ask communities with the least culpability for the crisis to pay the highest price. The urgency and scope of the questions raised by place's *loss* amplify hope's fragility. When one is standing at the edge of known history, questions of *who is responsible* and *who belongs* are critical. But neither are as fraught as the question, *who saves?*

Fijian preachers are not creating systematic theologies of place in their sermons; they are responding to particular texts of Scripture in particular moments of community crisis. And yet careful attention to their priorities reveals a startling challenge to the philosophical assumptions of an unsustainable global economy.[144] These sermons resist Western objectifications of place, even as they reveal relational tensions in the ways place is understood in Fijian Indigenous communities. Place is particular and permeable, changing and continuous, shaped by memory and hope alike. It mediates divine presence, even as it is encountered and transformed by that presence. It is holy ground.

143 Upolu Lumā Vaai agrees, arguing that a broken relationship with creation has led to "unlimited production without ethical limits," ". . . reinforced by a human-centric neo-liberal capitalist system": "'Others Are Enjoying Life from Our Death': Eco-relational Theology and a Methodist Ecological Revolution in Oceania," in *Methodist Revolutions: Evangelical Engagements of Church and World*, ed. Joerg Rieger and Upolu Lumā Vaai (Nashville: Wesley's Foundery Books, 2021), 173–96 [175].

144 These priorities resonate with Upolu Lumā Vaai's work on "Dirt Theology"—a theology foregrounding God's solidarity with dirt. Vaai describes the radical implications of this theology for economic development in "A Dirtified God: A Dirt Theology from the Pacific Dirt Communities," in *Theologies from the Pacific*, ed. Jione Havea (Cham, Switzerland: Palgrave Macmillan, 2021), 15–30.

4
Reframing Exodus
Decolonizing Place in the Fijian Methodist Church

Ezekiel's opening vision identifies *place*—its character, its loss, and its relation to God—as a central concern in the book's theological storm. In words that pile up like cumulus clouds, the prophet stands in carefully described dislocation among the exiles residing near the Chebar Canal (v. 1) and claims a difficult promise: "the appearance of the likeness of the glory of the Lord" moves (v. 28). It is a hard word, but not because it is hard to believe. As Moshe Greenberg notes, assertions of God's freedom are not "revolutionary"[1] in Israelite thought. "YHWH is nowhere in Scripture anything less than a god of universal dominion."[2] It is a difficult promise because God's presence in a foreign land is a promise that cuts multiple ways. It testifies to more than God's agency and authority; it testifies to place's fragility and loss.

There's a sound to testimony that cuts multiple ways. Richard Lischer describes it as a resonant chord of theological "tones": judgment, grace, and a call to action. Charles L. Campbell adds a fourth tone to Lischer's list—a world-upturning dissonance of expectation.[3] Ezekiel's vision

[1] Greenberg cites D. M. Stalker's *Ezekiel* (London: SCM Press, 1968) as representative of this view, alongside G. Fohrer's *Ezechiel* (Tübingen: Mohr, 1955) and Walther Zimmerli's *Ezekiel*, trans. Ronald E. Clements, 2 vols. (Philadelphia: Fortress, 1979). See Moshe Greenberg, *Ezekiel 1–20* (New York: Doubleday, 1983), 58–59.

[2] Greenberg, here, is not denying the broader scriptural conversation concerning the relation between God's revelation and the land of Israel. He notes that Ezekiel does push back against a "Jerusalemite claim that the exiles are removed from God" because of their displacement (*Ezekiel 1–20*, 59). He questions depictions of an oppositional binary between God's particular attention to people and place and God's universal dominion, as well as interpretations of biblical history that assume a linear progression from the former to the latter.

[3] Charles L. Campbell, "Preaching Gospel: Four Theological Tones," in *Preaching Gospel: Essays in Honor of Richard Lischer*, ed. Charles L. Campbell et al. (Eugene, Ore.: Cascade, 2016), 1–15. Campbell describes these four tones as judgment, grace,

of God's glory, carried by four creatures moving atop "a wheel within a wheel" (v. 16), vibrates with this multivalence. The vision is weighted with awe, leaving Ezekiel groping for adequate words. Greenberg stresses the vision's individual significance for a prophet called to a painful vocation.[4] It is, in one sense, Ezekiel's burning-bush affirmation that God will accompany him in the difficult work of preaching judgment (Exod 3:12). But the vision is no simple word of comfort. God's glory has followed God's displaced people into exile, echoing their experience of deportation—and raising a wind-whipped question: *What does divine mobility mean for the land Ezekiel loves and for the place of worship that shapes his religious imaginary?* Ezekiel's vision in chapter 1 does not necessitate the catastrophic exit of God's glory from the Jerusalem temple in chapter 8. But it does make that exit imaginable. Whatever YHWH's "universal dominion" means for God's people, it does not imply a taken-for-granted guarantee of presence. It also raises the gutting prospect of absence. The glory of the Lord moves. It is no wonder that Ezekiel describes the splendor of this mobile God as "the bow in a cloud on a rainy day" (v. 28), simultaneously referencing the assurances of the Noahic covenant (Gen 9:13) and the grief-soaked ambivalence of that narrative. Ezekiel has been through floods of violence, and the place that matters most to his identity is hidden under those waves. Visions of glory notwithstanding, Ezekiel will not stand on that ground again.

There is one interpretation of God's movement foreclosed by Ezekiel's broader witness. Some scholars sever the movement of God's glory in Ezekiel 1 from YHWH's covenantal connection to place. God's mobility, in this view, suggests an untethered freedom, no longer "bound to the holy land."[5] But in Ezekiel, YHWH's freedom aches with ties that bind. The land's restoration is a central concern for the prophet (Ezek 34:26–29; 36:8–15), as is the land's purification (39:11–16). The book's final chapters attend to emplaced practices of faith in minute detail (Ezek 40–48), and the book ends with careful descriptions of the land's reallocation (Ezek 47:21–48:29). Wojciech Pikor argues that moving manifestations of God's glory (*kavod*) in Ezekiel point to "God's immanence in the world

---

a new imperative, and "ha!" For Lischer's discussion of the gospel's multivalent "tones," see *A Theology of Preaching: The Dynamics of the Gospel* (Nashville: Abingdon, 1981), 46–65.

4 Greenberg, *Ezekiel 1–20*, 80–81.

5 Stalker, *Ezekiel*, 49.

without . . . violating God's transcendence,"[6] a generative tension familiar in Scripture and certainly as old as the pillars of fire and cloud in the Exodus wilderness narrative (Exod 13:17–14:29). For Pikor, the Exodus account is Ezekiel's most significant conversation partner, shaping the book's structure around a "new exodus" of reformation, with Judah's land playing a central narrative role.[7] Greenberg also emphasizes Ezekiel's reframing of the exodus story. Much has been made of the Babylonian and Assyrian iconography in Ezekiel's vision of God's glory,[8] but Greenberg emphasizes a different antecedent. He notes that the "appearance of the majesty (*kavod*) of YHWH in cloud and fire but without visible shape is a feature of the wilderness narratives of the Pentateuch" and connects Ezekiel's winged creatures to the "small cherubs on the desert ark," between whose "outspread wings . . . God 'met' Moses to speak to him" (Exod 25:22).[9] When place has been lost, the way forward often requires a reclaiming of foundational knowledge and, as these interpreters suggest, an ancient story.

When Tevita Nawadra Banivanua was elected president of the Methodist Church in Fiji (MCIF) at the 2014 Annual Conference, his church's political, cultural, and theological skies were also roiling with storm. It was the church's Jubilee Year, its fiftieth year of autonomy after claiming ecclesial independence from the Australian Methodist Church in 1964. Such a significant anniversary invited assessments of the church's complex past and its next fifty years, conversations that were not easy. The former president, Tuikilakila Waqairatu, had been dead for only six months, and Banivanua was tasked with casting the vision for his predecessor's program of ecclesial reform: the New Exodus (*Na Lako Yani Vou*). The prior eight years had been years of painful upheaval and dislocation for the Methodist Church. Once power brokers in a nation where a vast number of the Indigenous population identified as Methodist, the church was coming to terms with a starkly diminished political role. Voreqe Bainimarama,

6 Wojciech Pikor, *The Land of Israel in the Book of Ezekiel* (New York: T&T Clark, 2018), 175. While Pikor distinguishes Ezekiel's theology of place from a strict Zion theology that confines YHWH to land and temple, he underscores the significant role that land and place have in the book's theological arc (177ff.).

7 Pikor, *Land of Israel*, 178.

8 Margaret Odell, for example, notes similarities between Ezekiel's description and the representational art in Assyrian palaces: *Ezekiel* (Macon, Ga.: Smyth and Helwys, 2005), 22–32.

9 Greenberg, *Ezekiel 1–20*, 54.

the military dictator who seized power in 2006, had done all he could to thwart the church's influence, dismantling ecclesial funding structures and canceling denominational meetings necessary for church governance. The Annual Meeting itself had not been allowed to convene since 2008.[10] The first democratic election since the coup was just months away, but distrust of the government continued to be high in Methodist circles. This was a government that had jailed church leaders and restricted ecclesial criticism of its policies.[11] Early in 2015, Banivanua would address the church leadership engaging this difficult history while reporters for the government-friendly *Fiji Sun* took notes on Banivanua's address from the back of the room. In the days to come, they would publish a distorted, partisan account of Banivanua's exhortation, claiming that this new church leader had named Bainimarama's rise to power as "God's will."[12]

In reality, Banivanua was doing something far more complex. The church had interior work to do if it was going to survive the rapidly changing world in which it found itself, and it couldn't do that work if it expended all its energy battling the government. Banivanua was not only interested in asserting the church's authority in the general culture; he was interested in constructing a theological alternative to the church's colonial inheritance. He was interested in calling the Fijian church to repentance. Banivanua's call unfolded in a way that was less than self-evident to my Western Christian understanding, but his approach would have made sense to Ezekiel. Over the course of his next three years in office, through image, illustration, and Scripture, Banivanua would center the question of *place's* relation to the promise and purpose of God. Drawing on the biblical text and Fijian *iTaukei* tradition, the MCIF leader and his team would craft language for an Indigenous, exilic theology of place shaped by the wilderness narratives of the Pentateuch. To speak of place (*vanua*) in the Fijian context is to wrestle with nothing less than the identity of the Fijian Methodist Church. It is a conversation with critical political, cultural, and ecclesial implications. Banivanua—and the *Lako Yani Vou*—refused to let

10 David Crampton, "Fiji's Government Again Cancels Methodist Conference," *Christian Century*, August 25, 2011, https://www.christiancentury.org/article/2011-08/fiji-government-again-cancels-methodist-conference.

11 Sam Sekek, "Police Again Detain Fiji Church Leaders," ABC News, July 22, 2009, http://www.abc.net.au/news/2009-07-22/police-again-detain-fiji-church-leaders/1363198.

12 Ana Sovaraki, "Act of God," *Fiji Sun*, March 11, 2015, https://fijisun.com.fj/2015/03/11/act-of-god/.

Western assumptions or the intimidations of a secular state frame that conversation's terms.

I had just arrived in Fiji in 2014 and knew far too little about the fears and frustrations of the church that had welcomed me, but there were tonal resonances to President Banivanua's calls for ecclesial reform that my preacher's ear recognized. I knew the sound of a preacher being brave. Judgment, grace, calls to action, and overturned expectations vibrated in Banivanua's scriptural choices and his framing of the future. His proclamation resisted easy binaries and digestible sound bites, relying instead on careful engagements with narrative and image. One of those images was the church's new ecclesial logo, designed to articulate its New Exodus vision: an ocean-faring vessel blown by the wind of the Spirit. Like Ezekiel's image of God's glory, it was an image that cut in multiple directions. For Banivanua, the image was more than a symbol of a moving church; it was an icon of a moving God. From displaced ground, the image reclaimed foundational knowledge and, alongside Banivanua's call to a New Exodus, reframed an ancient story.

The first three chapters of this book have argued that the witnesses of displaced, Indigenous communities facing climate catastrophe illumine shortcomings in Western philosophies of place marked by colonial history. These communities ask questions that many climate advocates in populations more insulated from the crisis skirt: *who is responsible*, *who belongs*, and *who saves?* These questions disrupt climate solutions that assume universal relevance, reinscribe insularity, or prioritize technologies of control.[13] Instead, they highlight the deep challenges of relationally engaging with place when one's place is under threat. Much of my argument to this point has been deconstructive in character, framing the approaches to place taken by Ezekiel and Indigenous Fijian Methodist ministers as resistance strategies against imperial usurpations. These resistance strategies are significant interventions in their own rights. Despite Fiji's political independence in 1970, vestiges of colonial power continue to find expression in globalizing economic and political encroachments, threatening the work of climate justice. Resisting such present-day objectifications of place is crucial prophetic work. *What would it mean, however, to listen to the witness of displaced communities constructively?* What would it mean to hear a displaced community's

[13] As a practical theologian, I note that these questions disrupt preaching practices that do the same.

theology of place as more than a rejection of colonial hubris, and instead attend to it as a contextually specific attempt to center relationality as a defining characteristic of emplaced life?

In the years between 2014 and 2017, the leadership of the Methodist Church in Fiji prioritized this very theological project. This Fijian theology of place took a different form than that found in Tuwere's classic 2001 monograph.[14] It was embodied through a review of the ecclesial constitution, engagements with secularism, adaptations to theological education, and conference decisions about church finances. It also rearticulated the church's commitment to the work of climate justice. Significantly, the New Exodus—as it was called—found expression in sermons. During my three years at Davuilevu, easily half the sermons preached ended with a closing paragraph about the sermon's implications for the church's New Exodus. The metaphor became a regular feature of Fijian sermons during Banivanua's tenure. Indeed, one cannot adequately interpret the sermons described in the last chapter without this broader context. In a climate of apocalyptic loss, the reframing of exodus gave the church an image and a story to navigate its need for continuity alongside its call for change. It also named the faith community's performed relationship with place as a site of divine mediation. The church's New Exodus leaders were deconstructing colonial approaches to place that had been internalized by the Fijian Methodist Church, reorienting its collusions with Western essentialisms and exclusions.[15] But they were doing more than that. They were also envisioning a decolonized, ecclesial future, grounded in *iTaukei* interpretations of the biblical text and a covenantal commitment to creation. I will argue that this vision charts a trajectory for preachers around the world engaging the work of climate justice.

[14] Ilaitia S. Tuwere, *Vanua: A Fijian Theology of Place* (Suva: Institute of Pacific Studies Press, 2001).

[15] Drawing on Carolyn J. Sharp's description of colonized theologies that "collude with a colonizer's exercise of power" in "Sites of Conflict: Textual Engagements of Dislocation and Diaspora in the Hebrew Bible," in *Interpreting Exile: Displacement and Deportation in Biblical and Modern Contexts*, ed. Brad E. Kelle, Frank Ames, and Jacob Wright (Atlanta: SBL, 2011), 365–76 [365]. Namsoon Kang also acknowledges the "colonialism-in-here" that can exist in previously colonized contexts: "Envisioning Postcolonial Theological Education: Dilemmas and Possibilities," in *Handbook of Theological Education in World Christianity: Theological Perspectives, Ecumenical Trends, and Regional Surveys*, ed. Dietrich Werner et al. (Eugene, Ore.: Wipf and Stock, 2010), 30–41 [33].

Reframing an Ethnonational Exodus?

Mark Brett's *Political Trauma and Healing: Biblical Ethics for a Postcolonial World* notes the complicated role that the biblical exodus has played across colonial history, fueling both the colonial imagination and anti-colonial liberation movements in turn. "The exodus yields different meanings," he explains, "depending on the social location of the reading, and part of the task of the reconciliation within the church is to understand why this is so."[16] In the language of this book's introduction, Brett argues that knowing where one stands in the world is critical for interpreting where one stands in the biblical text. Drawing on examples from the Australian context and the Aboriginal church, Brett notes how "in the hands of powerful churches and nations, the exodus material is potentially damaging, whereas in the hands of weak groups, it can inspire well-founded resistance."[17] Brett makes his point convincingly from Australian ground, but the place of Fiji is more complex. The lines between powerful and weak are less clear in a nation where the political agendas and constitutional rights of Indigenous Christians often clash with those of religious and ethnic minorities—particularly those of Indo-Fijian Hindus and Muslims. It is the very reason why the New Exodus offers such a compelling example of a theological and ecclesial engagement with *place*. In interpreting the church's responsibility and vocation in relation to the losses of its colonial history, military dictatorship, and ecological devastation, the MCIF leaders asserted the importance of place for the Indigenous church, even as they called that church to repentance for displacing and excluding others. It is a tension that the following excursus on Fiji's recent political history brings into relief, emphasized by the church's strategic, costly use of the word "new" in its New Exodus vision.

In the Methodist Church in Fiji, it is no simple thing to call something "new." Given Fiji's long memory of colonial usurpation and the contemporary speed of globalizing change, novelty is often suspect. *iTaukei* culture's appreciation of continuity grows, in part, from a widespread lament that *mana* (spiritual power) from the past has been lost.[18] But more than this,

16 Mark Brett, *Political Trauma and Healing: Biblical Ethics for a Postcolonial World* (Grand Rapids: Eerdmans, 2016), 160.

17 Brett, *Political Trauma*, 160.

18 Matt Tomlinson, *In God's Image: The Metaculture of Fijian Christianity* (Berkeley: University of California Press, 2009). Tomlinson argues that a sense of loss undergirds much of Fijian Christianity, described as a lost spiritual potency (*mana*).

there is the real fear that the multicultural hybridity of Fiji's urban centers is hastening the extinction of *iTaukei* culture in the present.[19] This fear is given concrete form through the physical displacements described in the last chapter. Culture is not the only thing disappearing; land itself is washing away. Given this unstable ground, a proposal for MCIF reform regularly gains legitimacy through its rootedness in the past, not its "newness."[20] And yet, for all of the ways that Banivanua and Waqairatu drew on an ancient narrative in charting the church's path forward, they insisted that this exodus was "new"—an insistence born from the fraught arc of the nation's struggle with ethnonationalism and a series of coups that have disrupted its politics.

Much Western academic conversation about Fiji over the past twenty years has circled around this history. If a Western academician knows something of the Fijian context—and particularly something of the Methodist Church in Fiji—this is probably the history she knows. The church's complicity in these coups, with the marked exception of Bainimarama's military overthrow of the government in 2006, has been well documented.[21]

Following the 1987 election of the multiethnic Labour Party, Methodist lay preacher and military general Sitiveni Rabuka became the mouthpiece for *iTaukei* ethnonationalists, stoking fears that the Fijian constitution's protection of tribal lands would be rescinded. Much of Rabuka's rhetoric was incendiary and unsubstantiated, finding voice in sermons steeped in biblical references to exodus and exile. "Our property is in the hands of strangers," he preached, weaving the contemporary context into his reading of Lamentations 5. "Foreigners are living in our homes."[22] Significantly,

19 Manfred Ernst describes the impact of a globalized culture on local ecclesial practice in *Globalization and the Reshaping of Christianity in the Pacific Islands* (Suva: Pacific Theological College, 2006).

20 Matt Tomlinson discusses the ambivalence that Fijian Christianity has for the "new" in "The Generation of the Now: Denominational Politics in Fijian Christianity," in *Christian Politics in Oceania*, ed. Matt Tomlinson and Debra McDougall (New York: Berghahn Books, 2013), 78–102 [79].

21 Lynda Newland gives a succinct account of this history in "Religion and Politics: The Christian Churches and the 2006 Coup in Fiji," in *The 2006 Military Takeover in Fiji: A Coup to End All Coups?* ed. Jon Fraenkel, Stewart Firth, and Brij V. Lal (Canberra: ANU Press, 2007), 187–207. See also Jacqueline Ryle, *My God, My Land: Interwoven Paths of Christianity and Tradition in Fiji* (Surrey: Ashgate, 2010).

22 Lynda Newland, "From the Land to the Sea: Christianity, Community and State in Fiji—and the 2014 Elections," in *The People Have Spoken: The 2014 Elections in Fiji*, ed. Steven Ratuva and Stephanie Lawson (Canberra: ANU Press, 2016), 109–34.

Rabuka was not referencing British colonial history in this sermon, but rather the nation's growing Hindu and Muslim citizenry. This theological framing heightened the ethnic tensions already present in the country, seeding two political coups in quick succession. Rabuka would be democratically elected as the nation's prime minister in 1992. The Methodist president at the time, Josateki Koroi, strongly critiqued Rabuka's ethnonational version of *iTaukei* Christianity and was himself replaced through an ecclesial "coup."[23] Koroi's ousting from church leadership led to a sea change in the Fijian Methodist approach to racial unity in the country. In decades prior, the Methodist Church in Fiji had followed the leadership and vision of its first president, Setareki Tuilovoni—a stalwart supporter of racial cooperation. The MCIF had even elected an Indo-Fijian Methodist, Daniel Motilal Mastapha, as its president in 1976.[24] In contrast, the years following Rabuka's coup (1987–2000) saw a number of ethnonationalist conservatives directing the church's head office.[25] A third coup with ethnonational overtones would follow the election of Indo-Fijian Mahendra Chaudhry as prime minister in 1999. While the Methodist Church was not officially involved in this action, the muted response of the head office was significant and troubling. The suffering and fear visited upon

---

Excerpts from Rabuka's sermon can be found in Claire Walmsley's documentary "Paradise in Peril," *Everyman*, BBC TV, December 20, 1987, https://www.youtube.com/watch?v=23A4S1ZBmgM.

[23] In 1992, Koroi would give a keynote address to the Fiji Teacher's Union in which he critiqued the new church leaders for bringing a "*taukeism* philosophy into the church" (quoted in Manfred Ernst, *Winds of Change: Rapidly Growing Religious Groups in the Pacific Islands* [Suva: Pacific Council of Churches, 1994], 208). Marika Baleidelabaci (Bale) Tamanikaigavo provides one of few available historical accounts of the takeover of the MCIF presidency written by an *iTaukei* Methodist: "Racial Integration in Transition: A *Magimagi* Historical Assessment of the Methodist Church in Fiji's Response to Identity and Race Issues in Fiji" (unpublished master of theology thesis, Pacific Theological College, September 2021).

[24] Rev. Dr. Jolame Lasawa, church historian and current principal of Davuilevu Theological College, stresses the church's bold promotion of racial unity prior to the 1987 coup (personal interview, Nausori, Fiji, January 5, 2023).

[25] A significant exception is the presidency of Dr. Ilaitia S. Tuwere (1996–98). Tuwere's paper "The Church-State Relation in Fiji—a Reflection" was presented at the 1997 MCIF Annual Conference (Centenary Church, Suva) and argued that a Christian state goes against principles of religious freedom. The history of this time period is still difficult for many *iTaukei* Methodists to discuss. Jolame Lasawa's doctoral dissertation outlining this history is currently quarantined at the University of the South Pacific because of fears that the information is too divisive to be made public.

the nation's Indo-Fijian population during this third coup, regardless of religion, was widespread. Indo-Fijian families that had farmed land for generations were kicked off their property. Acts of physical and sexual violence were reported, as was the widespread looting of Indo-Fijian businesses in urban areas.[26]

During these difficult years, progressive voices in the Methodist Church spoke out against violence and ethnonationalism, particularly through the academic platforms provided by Pacific Theological College and ecumenical institutions like the Fiji Council of Churches.[27] But Matt Tomlinson notes a different kind of grassroots protest among the laity. Indigenous membership in the Methodist Church in Fiji fell significantly during these years, many joining Pentecostal congregations characterized by globalized worship practices and an otherworldly spiritualism.[28] The growth of the Pentecostal church in Fiji is the result of many complex factors,[29] but in Tomlinson's view, one of those factors was distaste among the laity for the church's interventions in the nation's political life. When Bainimarama seized power in 2006, he leveraged this growing quietism for his own political purposes, co-opting the language of secularism to silence critique and consolidate power.[30] It was the fourth coup the country had seen in twenty years.

[26] Ethnographic evidence of these attacks is preserved in Susanna Trnka, *State of Suffering: Political Violence and Community Survival in Fiji* (Ithica, N.Y.: Cornell University Press, 2008).

[27] Ryle's final chapter of *My God, My Land* catalogs the progressive streams in the MCIF leadership, ecumenical organizations, and institutions of theological education during this period (199–231).

[28] Tomlinson, "Generation of the Now," 79.

[29] Ernst, for example, correlates Pentecostalism's rapid rise with the nation's increasing cultural and economic globalization: *Globalization and the Reshaping of Christianity in the Pacific Islands.*

[30] Despite Bainimarama's self-proclaimed support of secularism, Wadan Narsey argues that Bainimarama's reasons for rejecting the draft constitution prepared by Yash Gai in the lead-up to the 2014 election was not because it violated secular, multiracial ideals. Rather, the draft was scrapped because it required the Bainimarama government to step down six months prior to elections and did not provide members full legal immunity from human-rights abuses. Bainimarama, in other words, championed secular tolerance when politically expedient. See "Why the Fiji Regime Rejected the Draft Constitution," in The Interpreter (blog), Lowy Institute for National Policy, Sydney, January 15, 2013, https://www.lowyinstitute.org/the-interpreter/why-fiji-regime-rejected-draft-constitution. Bainimarama's exile of renowned academic and long-standing advocate of democracy in Fiji Dr. Brij Lal, following Lal's criticism of the Bainimarama government in 2009, reinforces this portrait. See Christine Rovoi,

The story of the church's response to Bainimarama's military takeover and his marginalization of the church's influence has been described by multiple Western academicians, but rarely by Fijian Methodists themselves.[31] I trace this history with caution, as outside voices can be quick to evaluate and advise, assuming correspondence between diverse sociocultural contexts and ignoring the West's own role in the crisis.[32] I am also cautious, however, because Fiji's negotiation between race, land, faith, and government continues to unfold. The Bainimarama government was officially elected as the country's majority party in 2014 and again in 2018 following eight years of military rule (2006–2014)—bookending the Methodist Church's "New Exodus" reform movement and Banivanua's tenure as MCIF president. But in 2022, as this chapter was being researched, the Bainimarama government lost narrowly to the People's Coalition, a party created through a multiracial political collaboration among those determined to remove him from power. This ethnically and religiously diverse collective was headed up by none other than Sitiveni Rabuka, the Methodist military general who spearheaded the nation's first coup in 1987.[33] Having apologized for his prior actions, Rabuka now committed himself to cooperation across diverse communities. Whether this commitment stemmed from an awakened appreciation of religious tolerance or the fact that Rabuka needed Indo-Fijian votes to retain political control is an open question. When I visited the country following the 2022 election, the Methodists I spoke to expressed widespread relief that the Bainimarama

"Fiji Should Be Ashamed, Says Ali," RNZ News, December 30, 2021, https://www.rnz.co.nz/international/pacific-news/458794/fiji-should-be-ashamed-says-ali.

31 Bale's thesis, "Racial Integration in Transition," is an exception, as is Laisenia Qarase's self-published autobiography, *Prisoner 302* (self-pub., Amazon Digital Services, 2022), Kindle.

32 The West's role in the contemporary crisis has been emphasized by scholars like John D. Kelly, who describe how colonial legal arrangements undergirding Fijian land distribution contributed to the coups and instability of the postcolonial period; see his "Gordon Was No Amateur: Imperial Legal Strategies in the Colonization of Fiji," in *Law and Empire in the Pacific: Fiji and Hawai'i*, ed. Sally Engle Merry and Donald Brenneis (Santa Fe: University of New Mexico Press, 2003), 61–100. But Western influence is also a contemporary phenomenon. Martha Kaplan notes that the catalyst for George Speight's 2000 coup was a government-thwarted scheme to sell a forest of mahogany to a U.S. buyer: "Promised Lands: From Colonial Lawgiving to Postcolonial Takeovers in Fiji," in Merry and Brenneis, *Law and Empire in the Pacific*, 153–86 [154].

33 Natasha Frost, "In Fiji's Election, One Former Coup Leader Replaces Another," *New York Times*, December 20, 2022, https://www.nytimes.com/2022/12/20/world/asia/fiji-election.html.

government had been removed from power. Whether conservative or progressive, *iTaukei* or Indo-Fijian, many expressed an ardent hope that free speech, financial transparency, and an accountability to local communities would improve under the newly elected leadership. When asked about Rabuka's apparent change of heart, however, there was shared circumspection. Some alluded to God's power in calling the former coup leader to repentance, but more commonly, there was a pragmatic gratitude that Rabuka had won the election by so few votes. The fragility of his majority would require cooperation across religious and ethnic lines.

The church is actively negotiating what righteousness requires in such a season, which makes the question of the church's relationship to place as relevant today as it was in the tumultuous years after Bainimarama seized power. It also gives new urgency to the conversations that framed the New Exodus vision. Between the years of 2006 and 2014, the Fijian Methodist leadership was faced with a similarly complex theoethical challenge. They believed in the necessity of the church's witness in the public sphere. Whether the issue was poverty, government corruption, or unsustainable development practices, the church's witness was vital. To embrace a faith that impacted only a person's individual spiritual reality was to submit to a colonial reduction of the gospel. And yet asserting the church's public witness required that it face its ugly history of ethnonational co-optation. It required, in fact, that the church see this ethnonational movement not as a safeguard of Indigenous land and community, but as another colonial distortion of *place*.

Following the 2022 election, Banivanua would reflect on the church's crafting of the *Lako Yani Vou* (the New Exodus) through this broad historical lens.[34] More than a repudiation of the Bainimarama government's intimidation or a translation of progressive, democratic norms into a biblical vernacular, the MCIF's New Exodus was a deep interrogation of colonialism's hold on the church's self-understanding. Describing the generations of church leadership that he had seen over the course of his lifetime, Banivanua noted that, despite the extensive changes that had occurred in the Fijian church, the malformations of its colonial history had never been fully addressed. The "generation of colonial time" gave way to "a generation of the independence time," he explained, but the colonial mindset had never completely gone away. "The generation of the coup" might have claimed the preservation of *iTaukei* communities as its goal, but its

34 Tevita Nawadra Banivanua, personal interview, January 2, 2023.

mentality did not align with Indigenous values. Its goal was to "take whatever it wants."[35] In Banivanua's view, righteousness required the Methodist Church in Fiji to a deeper work around the nature and call of a decolonized church.

For Banivanua and his predecessor Waqairatu, the *Lako Yani Vou* was "new" in at least two respects. It was new in its repudiation of earlier Methodist leaders who had supported the coups of the past. "We were . . . a new group who believed that we should work . . . under the law."[36] This was not, in other words, Rabuka's 1987 "exodus." As if to underscore the point, Waqairatu and Banivanua would entrust an Indo-Fijian Methodist minister with articulating their vision, appointing the Rev. James Bhagwan as MCIF secretary for communications and overseas mission between the years of 2013 and 2017.[37]

More radically, the New Exodus differentiated itself from the culminating conquest of the *biblical* exodus, emphasizing instead the wilderness formation—and reformation—of God's people that make up the core of the Pentateuch narratives. Like other Indigenous biblical interpreters, these MCIF leaders were troubled by the account of Canaanite displacements and genocides in the book of Joshua.[38] "The Exodus story ends with a conquest," Banivanua reminded me in 2022, his voice serious and searching. "People pushed out and killed. . . . We wanted to avoid violence at all costs." Can an exodus movement have a "a nonviolent destination?"[39] he wondered. For Banivanua, this is a biblical question. But deeper still, it is a question of theological and anthropological ontology. It is a question about the nature of land, place, and human creation in the wake of colonial and environmental apocalypse.

The ambition of this ecclesial project should be underscored, if not abundantly clear. The New Exodus attempted to reframe, or perhaps reclaim, the meaning of *vanua* in a decolonized church—no slight or simple labor. Such "newness" would require a deep remembering. In the

35 Banivanua, personal interview, January 2, 2023.

36 Banivanua, personal interview, January 2, 2023.

37 Livai Tuisaravere, an instrumental lay leader in the *Lako Yani Vou*, was also a key voice of support in Bhagwan's appointment (James Bhagwan, personal email correspondence, June 20, 2023).

38 For example, Anna Kasafi Perkins, "Resisting Definitive Interpretation: Seeing the Story of the Exodus through Caribbean(ite) Eyes," *Caribbean Quarterly* 51, no. 2 (2005): 53–66.

39 Banivanua, personal interview, January 2, 2023.

iTaukei language, to refer to the past (*gauna mai liu*) is to refer literally to "the time in front," and to refer to the future (*gauna mai muri*) is to refer literally to "the time behind."[40] Banivanua and Waqairatu understood that an Indigenous theology of place would require just such a dialectic. Indeed, Fiji's present political shifts require the same. A "new" way forward for this multiracial, multireligious nation—beyond Western models and military coups—depends on it.

Navigating *Vanua*: The Church as *Drua*

This overview of the political context of the MCIF's New Exodus illumines the costly ethical inventory often required when *place* becomes a category of theological reflection—even for Indigenous communities themselves. The wounds of empire go deep, in the histories of colonized peoples and in the biblical text. If one wants to lance and sterilize those wounds for the sake of a sustainable, just world, preachers must be willing to face failings in themselves and their communities, unburied bones of generational grief, and the revelation of an uncontainable God. They must be willing to stand in *places* that shift underfoot and recommit to that risky, God-given relation. For the prophet Ezekiel, the risks of relating to God and place are articulated through an inaugural vision that recollects God's ancient presence in the Pentateuch wilderness narratives and calls the displaced priest to a new vocation. The image of a moving God and a moving church would also be critical to the New Exodus of the Methodist Church in Fiji, its Indigenous theology of place, and its decolonized ecclesiology.

Nearly a year before Banivanua's election, James Bhagwan—MCIF secretary of communication at the time—described a new church logo to the *Fiji Sun*.[41] The image was more than updated branding. It was a form of crisis theology, responding to a season of profound change and uncertainty in the MCIF's life. It also reaffirmed the resilience, creativity, and ecclesial significance of Indigenous approaches to place. Carefully designed through the collaborative efforts of church leaders that were headed up by Waqairatu prior to his passing,[42] the logo had been

40 My thanks to Fijian sociologist Akanisi Sobusobu Tarabe for this linguistic insight (personal interview, Suva, December 29, 2022).

41 Losalini Rasoqosogo, "Renewed Symbols for Church," *Fiji Sun*, August 22, 2013, https://fijisun.com.fj/2013/08/22/renewed-symbols-for-church/.

42 Bale describes, for example, the 2012 Secretariat Retreat that took place at Nataleira Eco Lodge in Tailevu North following the government's truncation of the church's Annual Conference. The retreat was a key moment in the development of the New

approved by the church's Standing Committee—the executive council responsible for directing the church between conferences. The image, in other words, had considerable institutional support.

The MCIF's original logo was a fishing boat, a *takia*, built to sail in protected lagoons and the relatively calm waters between islands. The new logo was a *drua*, an ancient two-hulled ocean-faring vessel—designed to carry entire communities across the sea. In Bhagwan's description, the *drua*'s size spoke to the church's desire to be an "inclusive church for all" and its commitment to "go outside its comfort zone and connect with others outside its usual sphere."[43] The complexity of sailing such a vessel required the coordination of a team. Bhagwan noted that the church would need lay and clergy leadership working together to navigate "the winds of change" impacting the twenty-first-century church.

The shift in imagery was striking enough that sociologists like Lynda Newland who study the impact of religious rhetoric on Fijian politics took note, describing the logo as a significant shift in ecclesial self-understanding and emphasis: "from the land to the sea."[44] But, reflecting on the power of the logo some years later, Bhagwan suggests something more crucial.[45] The logo was a re-remembering of pre-colonial understandings of *vanua* and emplaced life. It did not simply pair *vanua* (land) and *moana* (sea) as complementary contrasts. It reframed the *vanua* itself as informed by and informing *moana*, recognizing the ways that place itself shifted through space, time, and performances of relation. Maebh Long's "Vanua in the Anthropocene" notes that Indigenous understandings of the term *vanua* are multivalent. The term distinguishes land from sea, even as it describes the relation between both elements. Linguistically, the word *vanua* can "operate simultaneously on three . . . conceptional levels," Long notes, as "section [distinct from ocean], totality [inclusive of ocean], and as structure of relation [between the two]."[46] Indeed, this epistemological fluidity is part of what makes it possible for Indigenous conceptions of place to disrupt and resist colonial reductions. Place is not

---

Exodus vision and the working document of the MCIF's Connexional [*sic*] Plan, "The Way Forward" (*Racial Integration in Transition*, 82–85).

43 Rasoqosogo, "Renewed Symbols for Church."

44 Lynda Newland, "From the Land to the Sea: Christianity, Community and State in Fiji—and the 2014 Elections," in Ratuva and Lawson, *People Have Spoken*, 109–34.

45 James Bhagwan, personal interview, Suva, January 7, 2023.

46 Maebh Long, "Vanua in the Anthropocene: Relationality and Sea Level Rise in Fiji," *symploke* 26, nos. 1–2 (2018): 51–70 [64].

an empty space to be filled, an object to be possessed, or a commodity to be manufactured. For those familiar with Indigenous traditions of sea travel, the *drua* connotes this fluidity, reframing place's relation to movement, its relation to time, and—for the Fijian theologians of the New Exodus—its relation to divine presence.

Place and Movement: A Relational Journey

Reflecting on the importance of the *drua* as an ecclesial symbol, Bhagwan notes that "before the time of Christ, we were a people of the ocean. . . . At a time when the biblical view of the ocean was one of fear, the unknown, the depths that held the Leviathan—it was our highway."[47] Contrary to stereotypes of Indigenous cultures that foreground conservative stasis, in Indigenous history, place traveled. But this was not a culturally unmoored wayfinding. In the midst of migratory movement, there was continuity and stability in communal identity. "For voyaging communities," Bhagwan explains, "the *drua* represents the *vanua*."[48] The two hulls of the boat echo the layout of a traditional Fijian *bure*, or home. More than this, "in northern Pacific cultures of wayfinding, it was believed that the vessel stands still, and the ocean brings the islands to *it*." The *drua*, then, speaks to the "fixed and fluid" nature of Indigenous understandings of place, muddying simple binaries between *vanua* and *moana*—land and ocean.[49] Similar to Long's description of the *vanua*, the *drua* provides a way for a community to speak of "grounding" through "aquatic mutability."[50]

Reflecting on Anishinaabe intellectual traditions in North America, Kyle Whyte describes the importance of adaptive capacity in the continuance of communal Indigenous traditions.[51] Such adaptation is not change for its own sake. Change makes possible a community's lived relationship to place, engaging communal wisdom to meet present realities rather

47 Bhagwan, personal interview, January 7, 2023.

48 Bhagwan, personal interview, January 7, 2023.

49 Long, "Vanua in the Anthropocene," 51. Long's language here draws on a poem by Upolu Lumā Vaai, "We Are Because We Don't Have," in *The Relational Self: Decolonizing Personhood in the Pacific*, ed. Upolu Lumā Vaai and Unaisi Nabobo-Baba (Suva: University of the South Pacific Press, 2017), 283–84 [283]. For one of the most influential theological reflections on the significance of *moana* for Oceanic theology, see Winston Halapua, *Waves of God's Embrace: Sacred Perspectives on the Ocean* (Norwich: Canterbury Press, 2008).

50 Long, "Vanua in the Anthropocene," 67.

51 Kyle Whyte, "Settler Colonialism, Ecology, and Environmental Injustice," *Environment and Society: Advances in Research* 9, no. 1 (2018): 125–44 [140].

than preserving wisdom in museum-like stasis. Historically, for example, continuity in a community's relationship with natural ecosystems often required the migration of Indigenous people.[52] Whyte argues that colonial "sedimentations" of place and identity undermine these adaptive strategies.[53] The logic of colonial acquisition requires a rigid mapping of "immovable territorial boundaries,"[54] reinforced by what Willie James Jennings calls the "terrible choice" of the colonial moment: "either you own it—or I will own it."[55]

Relational notions of place did not create a pre-colonial utopia. Fijian chiefs used violent means to vie for power and territory, as any number of political and military leaders have done over the course of human history.[56] But there was relative stability in the waxing and waning centers of chiefly authority, rooted in long-standing relationships between communities themselves and between communities and land.[57] These relationships frayed—and at times, unraveled—when confronted with European demands for a government that could guarantee the security of European property and financial investment. Jane Collier notes that the "first laws western colonizers pressed chiefs to pass and enforce were laws designed to facilitate the expansion of liberal capitalism by creating responsible individuals and private properties."[58] Rather than emphasize the self's

52 Jione Havea, for example, notes that migration has always been a necessary Indigenous Pacific response to changing weather conditions in "The Politics of Climate Change: A *Talanoa* from Oceania," *International Journal of Public Theology* 4 (2010): 345–55. He extends his analysis on the relationship between sea and land and the intersecting, shifting borders of Oceanic identity in "Drifting Homes," in *The Future of a Biblical Past: Envisioning Biblical Studies in a Global Key*, ed. Fernando Segovia and Roland Boer (Atlanta: SBL, 2012), 195–206 [202].

53 Whyte, "Settler Colonialism," 140.

54 Whyte, "Settler Colonialism," 139.

55 Kenan Institute for Ethics, "Facing the Anthropocene: A Conversation with Willie Jennings," March 11, 2021, https://www.youtube.com/watch?v=SGMXeE6p-tQ.

56 Sanjay Ramesh describes the regular warfare of pre-colonial Fijian history: "State Hegemony and Ethnicity: Fiji's Problematic Colonial Past," in *The Palgrave Handbook of Ethnicity*, ed. Steven Ratuva (Singapore: Palgrave Macmillan, 2019), 247–63. Jane Collier notes that Indigenous notions of place do not necessarily correspond with reductive tropes of "helpful, generous and cooperative [natives] in contrast to selfish, greedy, competitive Westerners": "A Chief Does Not Rule Land: He Rules People (Luganda Proverb)," in Merry and Brenneis, *Law and Empire in the Pacific*, 35–60 [56].

57 Aubrey Park, *Degei's Descendants: Spirits, Place and People in Pre-Cessation Fiji*, ed. Matthew Spriggs and Deryck Scarr (Canberra: ANU Press, 2014), 88.

58 Collier, "Chief Does Not Rule Land," 52.

permeability to place, these laws emphasized place's differentiation from self and community—a recurring characteristic in various streams of Western political philosophy.[59] Collier notes the profound effect that such laws had on chiefly authority. Prior to the imposition of colonial law, chiefs ruled people; after cessation, chiefs ruled land.[60] The impact on the environment was extreme. Within ten years of European contact, "all sandalwood stands were logged from Fiji."[61] This new approach to territory served European economic interests, but it also served to reinforce an increasingly rigid, hierarchical understanding of race and ethnicity. In John D. Kelly and Martha Kaplan's chilling summary, "the constituting moments in colonial history . . . were designed, not just to keep races in their places, but in fact, to make places for races," regardless of the impacted communities' consent.[62]

This colonial shift toward bounded identities and control of territory continues to manifest itself in contemporary Oceanic contexts. Upolu Lumā Vaai describes an "imperial theology" that marks much Pacific Christianity, sedimenting a single way of engaging God and reading Scripture.[63] Newland notes the fervor in many nineteenth-century Western missionary writings for mapping and stabilizing Fijian genealogical origins, a fervor echoed in contemporary Fijian ethno-religious nationalistic movements.[64] Martha Kaplan describes how early decisions around land ownership by the British colonial government "created colonial order meant to deny competition, . . . to allow local elites to maintain their rank, and above all, to create separated racial spheres . . . especially 'Natives' separated from others."[65] Bhagwan worries that these rigid approaches to place have eroded "our sense of mobility" and replaced it with a colonial desire

[59] For all of their differences, Collier notes the self-sufficient anthropology undergirding the political philosophy of Thomas Hobbes, John Locke, and Jean-Jacques Rousseau ("Chief Does Not Rule Land," 42).

[60] Collier, "Chief Does Not Rule Land," 57.

[61] Ramesh, "State Hegemony and Ethnicity," 251.

[62] John D. Kelly and Martha Kaplan, *Represented Communities: Fiji and World Decolonization* (Chicago: University of Chicago Press, 2001), 196–97.

[63] Upolu Lumā Vaai, "Va'atapalagi: De-heavening Trinitarian Theology in the Islands," in *Colonial Contexts and Postcolonial Theologies: Storyweaving in the Asia-Pacific*, ed. Jione Havea and Mark Brett (Melbourne: Palgrave, 2014), 41–53 [42–43].

[64] Lynda Newland, "The Lost Tribes of Israel—and the Genesis of Christianity in Fiji: Missionary Notions of Fijian Origin from 1835 to Cession and Beyond," *Oceania* 85, no. 3 (2015): 256–70 [263].

[65] Kaplan, "Promised Lands," 182.

for permanence. Echoing Whyte's description of North American Indigenous adaptive strategies, Bhagwan understands the Pacific's "diasporic and migratory" character to be part of how long-standing relationships with land are maintained, alongside inclusive understandings of the "extended, huge family" of Pacific relationships. "We have a hangover from the colonial experience [that describes] extinct tribes or empty land—land where nobody lives," without recognizing the long histories that connect places and people groups.[66] As an example, Bhagwan describes how communities that now occupy far-flung eastern Fijian islands are connected to land in the Viti Levu interior, hundreds of miles away. Indigenous approaches to place honor the oral histories that preserve these ties. "This over there is your people's land," returning islanders are told by present-day elders from the interior—not because the connections between land and people do not matter, but precisely because they do.[67]

It is not just the ties with the land that are at stake; it is also the ties with the sea. The trope of "empty land," a common historical justification for colonial acquisitions and exploitations,[68] has sinister, contemporary relevance in Pacific conversations around deep-sea mining. "Most of the focus of the environment is on the land," Bhagwan states, ". . . but we can't see the tears of the ocean."[69] Banivanua agrees. As international development companies vie to make the ocean floor financially productive, local faith communities are increasingly wary of the unknown costs and losses of speculation.[70] "Development means you dig up things," Banivanua tells me, "and that's not what we need. There are many ways to get jobs, but we cannot create fish."[71] In this case, the relationship between land and sea imaged in the *drua*'s movement connotes more

66 Bhagwan, personal online interview, September 28, 2021.

67 Bhagwan speaks specifically of the migratory histories connecting the people currently residing on the islands of Lau with those living in the Tailevu Province.

68 Santiago Slabodsky describes the "trope of empty land" as a colonial narrative that justified the seizing of territory that was not being "productively" managed by Indigenous populations: "Empty Land: Righteous Theology, Sneaky Coloniality," in *People and Land: Decolonizing Theologies*, ed. Jione Havea (Minneapolis: Fortress Academic, 2019), 51–64 [53].

69 Bhagwan, personal interview, January 7, 2023.

70 The focus of the 2022 G20 Interfaith Forum was on "Ecological Racism and Deep Sea Mining in the Pacific," February 11–12, 2022, https://www.g20interfaith.org/2022-webinar-deep-sea-mining-and-ecological-racism/. Thanks to G20 Interfaith communications consultant JoAnne Wadsworth for making this webinar available.

71 Banivanua, personal interview, January 2, 2023.

than the relationally fluid boundaries of landed places. It also calls the church to reinforce its relational ties to aquatic places, protesting the ways that colonial constructs of "emptiness" continue to justify the destruction of ecosystems for economic gain. In Bhagwan's words, this is an "exodus that includes the ocean."[72]

It is no surprise that the image of an ocean-faring *drua* evokes a renewed ecological awareness. Fijian Methodist historian Jolame Lasawa notes that Waqairatu's theology of New Exodus was always tied to his theology of a "green church."[73] But Lasawa notes that Waqairatu's definition of a green church wove together concepts of nature and culture, connecting the health of environmental and political ecosystems. Given Fiji's colonial history of foreign powers creating "places for races," the church's relational ties to a moving *vanua* cannot be bracketed off from its practices of ethnic inclusion. There are political implications to embracing the movement of place, with corresponding implications for how the Bible is read and interpreted in preaching. For Bhagwan, the migratory character of Oceanic life is a significant scriptural hermeneutical lens. Alongside Exodus stories of liberation are "the stories from the Torah . . . that share how to live with people who are alien in your land. [There are laws about] gleaning fields . . . and radical hospitality. Being an inclusive society."[74] The differences between this scriptural understanding and most Western accounts of how the Bible has been leveraged to promote exclusion and stasis in Fijian Christianity are stark.[75] But the theologians of the New Exodus insist that the Bible can be a resource in helping Fijian Christians negotiate the cultural, theological, and political movement that emplaced relationality requires. Perhaps this emphasis seems a natural fit for Bhagwan, an Indo-Fijian Christian who has been negotiating categories of

[72] Bhagwan, personal interview, January 7, 2023.

[73] Lasawa, personal interview, January 5, 2023.

[74] Bhagwan, personal online interview, September 28, 2021. Bhagwan refers here to passages like Lev 23:22, "When you reap the harvest of your land, you shall not reap to the very edges of your field, or gather the gleanings of your harvest; you shall leave them for the poor and for the alien: I am the LORD your God," and Exod 22:21, "You shall not wrong or oppress a resident alien, for you were aliens in . . . Egypt."

[75] Lynda Newland, for example, acknowledges the Bible's use in movements of colonial resistance, such as Navosavakadua's rebellion in 1877—but much of her discussion centers on how Scripture was leveraged by missionaries and Fijian *iTaukei* communities to connect "authentic" Fijian clans to Israel, legitimating "power regimes and alliances and . . . justify[ing] boundaries around Fijian land": "Lost Tribes of Israel," 267.

insider and outsider in relation to the Fijian Methodist Church over the course of his lifetime. But he is not alone in bringing this mobile understanding of place to the reading of Scripture. Place's movement—and its implications for lived relationship—is a homiletic and biblical touchstone to which *iTaukei* MCIF president Tevita Banivanua repeatedly returns over the course of his 2014–2017 tenure.

In his first address as president to church leadership in 2015, Banivanua foregrounds themes of journey, movement, and preparation for the "process of birthing" that God is doing in the MCIF. His sermon "Mapping the *Lako Yani Vou* [New Exodus]" makes no secret of the "demoralizing and humiliating" treatment of the church by the military government of the time. From 2009–2013, "all the MCIF systems of governance were strangled and shut down."[76] But rather than focus on the government's failings, Banivanua wonders how this period of hardship is calling the church "to start anew and think anew," drawing on the portions of the exodus story that emphasize the wilderness reordering of God's people. In a "time of disorder," Banivanua notes, God's people are given "the law that governed societal living for the people of God."[77] This biblical narrative of dislocation and re-creation is then brought into dialectical relation with the image of the *drua*. Banivanua's homiletic interplay between text and image clarifies that the migrating movement of the church is not meant as a promise of escape from the challenges of the present, nor does it lionize a leaving behind of communal identity. Instead, at the center of the New Exodus journey Banivanua places the work of "restored relationship" (*veiwekani*)—a core *iTaukei* value. He discusses the need for restored relation with the government, with other denominations, and with the church's Hindu and Muslim neighbors, framing each in light of the larger effort required to restore relationship with the *vanua* itself.

It is a world-upturning juxtaposition—a simultaneous return to traditional Indigenous values and a call to risky transformation. Banivanua reframes each in reference to the other, defining ecclesial movement as a commitment to deeper relation. It is a paradox embedded in the image of the *drua* itself. Movement on a *drua* requires the careful negotiation of roles and responsibilities. It requires a carefully woven sail and a strong, two-hulled structure. It requires the embrace of communal wisdom—not

[76] Tevita Nawadra Banivanua, "Mapping the *Lako Yani Vou* [New Exodus]" keynote address at the opening of the 2015 Divisional Superintendents and Divisional Stewards Conference, Centenary Church, Suva, March 10, 2015.

[77] Banivanua, "Mapping the *Lako Yani Vou*."

its negation. It is no accident that the first tasks to which Banivanua calls the New Exodus church are a constitutional review and the creation of a new code of pastoral conduct. This is no celebration of free-floating autonomy.[78] Instead, Banivanua follows the trajectory of the Exodus text: framing liberation as a journey toward law and framing law as a gift of covenantal relation.[79]

Banivanua's attention to place's movement is costly, and his sermon takes seriously the implications of that movement. There is a difficult journey ahead. "As Christian Fijians," Banivanua explains, "we need to be inclusive in our understanding of the *vanua* to include all those other ethnic groups that live in the *vanua*."[80] It is a sentence that depends on the fluidity inherent in *iTaukei* understandings of place, framing *vanua* both as a bounded descriptor of cultural particularity (i.e., "our understanding of the *vanua*") and as a space for pluralistic difference (i.e., "those other ethnic groups that live in the *vanua*"). Newland notes that this call to inclusivity is coherent to Banivanua's listeners not through Western models of individual rights and religious privatism, but because it is rooted "firmly in Taukei conceptions of the world."[81] I would add that it is a call firmly rooted in a biblical story interpreted through that *iTaukei* lens, rather than the sedimentations of empire. In one of Banivanua's final sermons as president, he reiterates the importance of "the stories of the Christian tradition" to the "stories of our [*iTaukei*] culture," though he notes that the "struggle over the interpretation" of those stories is "crucial to political and social life."[82] For Banivanua, the church's wilderness journey—envisioned

78 Indeed, the "Connexional [*sic*] Plan" that would serve as the church's New Exodus navigational map had no less than twelve "pillars." This was not a journey that feared structure and planning. See MCIF's "Methodist Church in Fiji Connexional Plan," accessed June 10, 2023, https://www.academia.edu/6620913/Methodist_Church_Connexional_Plan_First_Draft_. The Connexional Plan's twelve-pillared structure alludes to Exod 24:4, where Moses sets up "twelve pillars" at an altar at the base of Mount Sinai prior to receiving the law.

79 Banivanua frames the constitutional review, code of conduct, and "Connexional Plan" as "attempt by the church to bring covenantal relationship much closer to the people": Tevita Nawadra Banivanua, "The Status of the Methodist Church in Fiji," keynote address, 2015 MCIF Annual Conference, Centenary Church, Suva, August 14, 2015.

80 Banivanua, "Mapping the *Lako Yani Vou*."

81 Newland, "From the Land to the Sea," 127.

82 Tevita Nawadra Banivanua, "Staying the Course in the Winds of Change," keynote address to Davuilevu Theological College graduates, Baker Memorial Church, Nausori, Fiji, November 24, 2017.

as a migratory *drua*—calls God's people to a "prophetic tradition" that does not separate the "salvation of society from the shalom of society."[83] A "just peace" born of right relation marks the place of God's people, even when they are being buffeted by storms of change.

In a world where increasing numbers of Fijian villages are facing relocation, Netani Rika, communications director for the Pacific Council of Churches, names the practical significance of theologies of place that embrace place's migratory character. Communities "need the priest or church minister or pastor to say: look, whether you are here or up on that mountain . . . God will accompany you on this journey."[84] But there is more at stake in remembering place's fluidity than pastoral consolation. Recognizing place's movement equips communities to reject the false choice of cultural essentialism or erasure, and instead to embrace place's relational claim. In choppy seas, relationships require change even as they require continuity. And they regularly require repentance.

On May 14, 2023, the then-president of the Methodist Church in Fiji, Rev. Ili Vunisuwai, offered a public apology to the Indo-Fijian community for the wrongs committed against them by the *iTaukei* members of the MCIF during the years 1987–2006.[85] The service was held in the Suva's central arena during the nation's first celebration of the descendants of the Indian indentured laborers brought to Fiji by the British colonial government. Rev. Vunisuwai noted the "immense suffering" that the Indo-Fijian community had endured: "We firmly believe that the Lord our Creator has seen and heard your cries and sorrows." "This is a God who loves justice and hates sin," he continued, "and so the only way forward for us and for . . . Fiji is to confess our wrongdoings, to turn from committing sin, and live together . . ."[86] Apologies are notoriously insufficient in bringing about just peace. But the very fact that this formal apology took so long in coming reveals something of the depth of spiritual transformation it

83 Banivanua, "Staying the Course."

84 Kate Lyons, "How to Move a Country: Fiji's Radical Plan to Escape Rising Sea Levels," *Guardian*, November 8, 2022, https://www.theguardian.com/environment/2022/nov/08/how-to-move-a-country-fiji-radical-plan-escape-rising-seas-climate-crisis.

85 Naveel Krishant, "Methodist Church of Fiji Apologises to the Descendants of the Girmityas for the Hurt of the Past," Fijivillage, May 14, 2023, https://www.fijivillage.com/news/Methodist-Church-of-Fiji-apologises-to-the-descendants-of-the-girmityas-for-the-hurt-of-the-past-5rf4x8/.

86 Krishant, "Methodist Church of Fiji Apologises."

required.[87] Reframing *vanua* as a place of restored relation is risky, decolonial work. And it is a journey that continues.

Place and Time: A Formational Journey

The New Exodus *drua* is an intentional weaving together of remembered history and expectant hope. In this way, it is not unlike the *Uto ni Yalo*, a two-hulled sailing vessel named *Heart of the Spirit* that currently transverses the Fijian archipelago.[88] The *Uto ni Yalo* is a contemporary wind-powered canoe that calls to mind the ancient navigational histories of Pacific peoples, even as it enacts that history. Pacific climate activists use the canoe as a place for convening urgent conversations around the stewardship of the ocean.[89] Past and future intersect in such voyaging, embodying a nonlinear conception of Indigenous time. The New Exodus *drua* does something similar, harkening to a past *vanua* that points forward and a future *vanua* that looks back. The *drua*, in other words, not only reimagines place's relationship to movement. It reimagines place's relationship to time itself. Time, of course, is not unrelated to movement. Time makes movement visible. But time also introduces the question of destination, purpose, and telos. How does the place of the present intersect the place of the past and the place of the future? Do the hopes of a displaced, colonized community lie in the reclamation of pre-colonial history? Or is hope found on the shores of an unknown hereafter, far from tradition and home? As with the flawed dichotomy between "fixed" and "fluid" understandings of place, the sermons of the New Exodus suggest this dichotomy between past and future is a false choice.

And yet it is a choice codified in biblical studies' polarized theological categories for much of the twentieth century.[90] So-called "creation

[87] In a 2015 interview, Bhagwan notes that the church actively initiated grassroot conversations with local congregations on the long process of reconciliation in 2012, stressing the "spiritual journey" it would require: *Speak Your Mind*, FBCTV, episode 163, April 8, 2015, https://www.youtube.com/watch?v=hrRMrC4TNvQ.

[88] A brief history of the *Uto ni Yalo* and its mission is outlined by Andrea Egan at "Heart of the Spirit: GEF-SGP UNDP 25th Anniversary Stories," *Uto ni Yalo* Trust, accessed June 10, 2023, https://undp.shorthandstories.com/gef-sgp-heart-of-the-spirit/.

[89] Epeli Lalagavesi describes the significance of the *Uto ni Yalo* as bringing new, younger voices into participatory climate advocacy: "COP26: Calling for Greater Ambition and Climate Justice," Peace Boat, November 16, 2021, https://peaceboat.org/english/news/cop26-greater-ambition-and-climate-justice.

[90] The supposed dichotomy between Exodus salvation and Genesis creation narratives was popularized through influential twentieth-century works like Gerhard

theologies" in the scriptural text were regularly characterized as "mimetic" and conservative in character, reinforcing an idealized past.[91] Redemption theologies, on the other hand, emphasized interruption, future transformation, and liberation from the status quo. Both reductions held ethical dangers. Scholars like Walter Brueggemann worried that creation theologies were susceptible tools for imperial propaganda.[92] But redemption theologies could also be leveraged by powers that be, turning biblical faith into an "otherworldly piety, authoritarian ethical legalism, [or] a dualism regarding sacred and secular."[93] Indeed, in the U.S. ecclesial landscape, redemptive theologies that emphasize the promise of a heavenly home have fueled a dangerous quietism in relation to the environmental crisis.[94]

Terence Fretheim's influential commentary on Exodus signaled a sea change on the way. Not only did he describe the echoes of Genesis' creation account in the Exodus text, he argued that the trajectories of creation and redemption in the biblical witness were intimately connected. For Fretheim, the thriving of creation was God's redemptive goal. In the

---

von Rad's *Old Testament Theology*, trans. D. Stalker (Edinburgh: Oliver and Boyd, 1962). The bifurcation fueled a corresponding dichotomy in understandings of biblical time. In contrast, scholars like Terence Fretheim assert the "decisive" impact of creation theology on the book of Exodus; see his *Exodus* (Louisville: Westminster John Knox, 1991), 13. See, for example, Fretheim's fleshing out of the anti-creation consequences of Pharaoh's disobedience in Exodus' plague cycle: "The Plagues as Ecological Signs of Historical Disaster," *Journal of Biblical Literature* 110 (1991): 385–96.

91 J. Richard Middleton analyzes the influence of these characterizations in biblical studies in "Is Creation Theology Inherently Conservative? A Dialogue with Walter Brueggemann," *Harvard Theological Review* 87, no. 3 (1994): 257–77 [269].

92 Walter Brueggemann, "A Shape for Old Testament Theology, I: Structure Legitimation," *Catholic Biblical Quarterly* 47 (1985): 28–46 [42]. Even given the large shifts in Brueggemann's own approach to creation theology and his acknowledgment of its significance for the contemporary ecological crisis, he continues to be wary of creation theology's "powerful propensity for the maintenance of the present system." See "The Loss and Recovery of Creation in Old Testament Theology," *Theology Today* 53, no. 2 (1996): 177–90 [189].

93 Middleton, "Is Creation Theology Inherently Conservative?" 272. Indigenous theologian George "Tink" Tinker (Osage Nation) underscores that liberation theologies can be dangerous in their negation of cultural discreteness, "promising Indigenous people nothing better than continued cultural genocide": "The Full Circle of Liberation: An American Indian Theology of Place," *Sojourners* 21, no. 8 (1992): 12–17 [13].

94 Barbara Rossing describes and counters this kind of eschatological escapism in "Reimagining Eschatology: Toward Healing and Hope for a World at the *Eschatos*," in *Planetary Solidarity: Global Women's Voices on Christian Doctrine and Climate Justice*, ed. Grace Ji-Sun Kim and Hilda Koster (Minneapolis: Fortress, 2017), 325–47.

book of Exodus, the "objective of redemption is to free people to be what they were created to be. . . . Redemption is in the service of a creational end."[95] It is an argument that hinges on particular reframing of biblical time. In his reading, creation is not meant to exist in stasis,[96] repeating life's cycles as ends in themselves. But it is also more than a linear arrow, moving through history as if one's present materiality is tangential to one's destination. Biblical time, in Fretheim's description, is a "spiral"[97]—moving forward in ways that honor the relationality of community, land, and created world. Both aspects of this movement are necessary: time's hopeful arc alongside its remembered repetitions and responsibilities.[98] The book of Exodus brings this tensive understanding of time into focus. In Fretheim's words, "Passover and sea crossing keep redemption and creation firmly bound together, without collapsing them."[99]

Given the significance of the Exodus account in Banivanua's theology of place, it is no wonder that time becomes a recurring theme in his sermons. Fretheim is not the only one to notice how the wilderness of Exodus arcs toward promise and echoes God's work in creation. In Banivanua's proclamation, there is no redemption that does not honor place's created sacredness, just as there is no attention to place that does not make room for the redeeming, sanctifying work of God. This honoring of what is and has been, alongside what will be, creates an ambiguity in the New Exodus journey. Traveling connotes linearity. It moves from somewhere, to somewhere else. But Banivanua's New Exodus sermons are unspecific about where the church's movement will lead.[100] The promised land that lies on the other side of dislocation and migration remains stubbornly undefined. I have noted elsewhere that this lack of detail has left open various interpretations of the New Exodus vision—some focused on modernization, some focused on spiritual purification, some focused on reclamation of

[95] Terence Fretheim, *God and World in the Old Testament: A Relational Theology of Creation* (Nashville: Abingdon, 2005), 125.

[96] Note, for example, God's command to "multiply and fill" the earth (Gen 1:28), which implies transformation and change.

[97] Fretheim, *God and World*, 124.

[98] Tinker differentiates, for example, between "value-neutral creation theology" and theologies of community that "generate a consistent interest in justice and peace" ("Full Circle of Liberation," 17).

[99] Fretheim, *God and World*, 123.

[100] For example, Tevita Nawadra Banivanua's "Kairos, God's Plan and Wilderness Formation in the New Exodus," sermon preached at Davuilevu Theological College, Nausori, Fiji, November 2016, describes the destination only as "the kingdom of God."

heritage.[101] The promise shifts as different preachers bring different interpretive lenses to bear. The phenomenon frustrates some. One Fijian layperson expresses impatience when we speak in 2022. "I'm tired of this New Exodus," she shares, with unusual candor. "New Exodus—all the time, New Exodus. What is the destination? That's what I want to know. Where are we headed?"

I ask Banivanua this question in our interview, and it is precisely here that he voices his worry over the biblical conquest narratives. It is not only the scriptural depictions of violence that concern him; he understands the dangerous allure of the language of "promised land" in his own context.[102] The idea that the *vanua* is promised by God to *iTaukei* Christians has not always led to an increase in relationality with land and community. It has regularly fueled resentment, entitlement, and division.[103] Given this history, it is the *process* of the Exodus journey that Banivanua emphasizes in his preaching, even if "we have no idea where it will lead us."[104] In this refusal to stabilize and claim a promised destination, Banivanua refocuses attention on the present action of God in the MCIF—what he calls the work of "formation." The goal of wilderness wandering is not just to move from one place to another. It is to be formed into a new sort of community, a community with a "decolonialized mind."[105] This requires an honoring of ancestors and communal ties, even as it discerns a future on the other side of colonial loss. It embraces *iTaukei* framings of the future as "time following" and the past as "time in front,"[106] even as it envisions the intersection of past and future in a present, expectant reality. The desert wilderness of Exodus, for Banivanua, is a "dusty womb."[107]

A womb may seem a very different image than a *drua*, but both speak to life sustained in the watery resonances between memory and hope. Life is held and preserved by place, even as life is on the way—provisional and becoming. One might protest that there is an endlessly delayed

101 Jerusha Matsen Neal, "Exodus of Exile: Hermeneutic Shifts in a Shifting Fijian Methodist Church," *International Journal of Homiletics* 1, no. 2 (2019): 16–25 [20–21].

102 Banivanua, personal interview, January 2, 2023.

103 For example, Martha Kaplan describes several examples of "promised land" rhetoric undergirding Christian nationalist sentiment, land disputes, and Fiji's recent coups in "Promised Lands."

104 Banivanua, personal interview, January 2, 2023.

105 Banivanua, personal interview, January 2, 2023.

106 Referencing Akanisi Tarabe's literal translation of the *iTaukei* words for "future" and "past."

107 Banivanua, "Kairos."

eschatology to Banivanua's imagery, a promise that never reaches fulfillment. But this is to misunderstand his point. The material implications of this journey are of critical importance to him. The losses of colonialism are real, and in the wake of those losses, there are better and worse futures possible for the Fijian people.[108] What worries Banivanua more, however, is an ecclesiology that does not take seriously the ongoing quality of its formational journey. When "the church decides it's complete," Banivanua states, it becomes "dangerous."[109] It's an insight echoed by Bhagwan, who also prioritizes the church's formational process over its destination. The church is always in the process of "reformation," Bhagwan explains. If not, we are "no longer a voyaging people . . . [and] we become instead an 'Are we there yet?' people."[110]

One of the implications of taking seriously the impact of time on communal identity and place is a careful weaving together of contextual theology and liberation theology in Banivanua's sermons. Even as he centers Indigenous knowledge through his descriptions of "traditional navigators" who "knew the stars" and "listened to the ocean" to discern "not only the way to go but when to go,"[111] he also disrupts assessments of Indigenous culture as finished and complete. One example of this is Banivanua's prioritization of equity for women deaconesses and ministers in the MCIF governmental structure. In his first year as president, Banivanua worked with church leaders to revise the MCIF constitution, granting deaconesses the status of "ordained" conference members. At the 2016 Annual Conference, for the first time "sisters from the Deaconess Order were present in a Ministerial Session in recognition of the equal status of the ordinations of our male and female *Talatala* as Ministers of the Word, Sacrament and Pastoral Care and our *Vada ni Turaga* as the Order of Deaconesses."[112] Given the fact that there are still relatively few ordained women ministers in the MCIF,[113] representing the voices of deaconess alongside the voices

[108] Tomlinson documents the recurring trope of lost *mana* in Fijian sermons in *In God's Image*. In the face of such losses, Banivanua references the work of Michael Walzer, arguing that there is a "march toward a goal . . . a transformation" in wilderness wandering: Walzer, *Exodus and Revolution* (New York: Basic Books, 1986), 12.

[109] Banivanua, personal interview, January 2, 2023.

[110] Bhagwan, personal interview, January 7, 2023.

[111] Tevita Nawadra Banivanua, "Signposts of the New Exodus," keynote address, 2016 MCIF Annual Conference, Centenary Church, Suva, August 15, 2016.

[112] Banivanua, "Signposts of the New Exodus."

[113] In my final year teaching at Davuilevu Theological College, there were four women studying to become ordained clergy out of nearly seventy students.

of ministerial clergy is a move toward gender equity in the leadership decisions of the church. But it is also a move that broadens the meaning of ordination itself, recognizing that the service that deaconesses provide in schools, in hospitals, and in social service agencies has sacred, ecclesially significant weight. Banivanua celebrates the 2016 change, but he does not stop there. He presses the conference further, noting that if deaconesses share the same ordination status as ministerial clergy, they should also receive the same financial compensation.[114] The work, in other words, is not over. Banivanua's theology of ecclesial formation reinforces the importance of communal identity, even as that identity stretches and shifts.

Perhaps even more significantly, Banivanua's formational ecclesiology serves a pastoral function, moving the MCIF away from false hopes of quick fixes and toward a hope that sustains. The legacies of colonialism and the present impacts of globalization can feel insurmountable. In his sermon "Kairos, God's Plan, and Wilderness Formation in the New Exodus," Banivanua describes the work of confronting those legacies in visceral terms, "crawl[ing] forward on our hands and knees, pleading with God for His grace to help us push this heavy canoe into the water, as others watch, waiting to climb aboard only if we can prove that the *drua* will float."[115] Banivanua gives this sermon at the November graduation of Davuilevu Theological College in the year that Cyclone Winston destroyed large sections of the Viti Levu interior. Myriad church plans for investments in education, holistic mission, and grassroots outreach have been disrupted by the storm. Churches and congregations have had little to no financial resources, and for much of 2016, pastoral stipends have been in jeopardy. Banivanua is speaking to a room of future Fijian Methodist ministers who may very well be placed in communities unable to provide for their welfare. Climate catastrophe is no abstraction in this room. The increasingly strong, increasingly frequent storms that result from a warming ocean have impacted these ministers' livelihoods and the security of their families. They have also sobered early visions of New Exodus transformation. In Banivanua's sermon, there is no romanticization of the "wilderness" voyage ahead. "Excitement gives way to hard work," he tells his listeners;

114 This proposal was formally discussed at the 2017 MCIF Annual Conference, but was unsuccessful. See Iva Danford, "Discussion to Be Held on Deaconesses Being Given the Same Pay as Ordained Ministers," Fijivillage, August 21, 2017, https://fijivillage.com/news/Discussion-to-be-held-on-deaconesses-being-given-the-same-pay-as-ordained-ministers--sr2k95/.

115 Banivanua, "Kairos."

there are "storms and wind . . . but also days with no wind." Such days require something more difficult than grit. They require "waiting."[116]

Banivanua's sermon juxtaposes the Greek concepts of *kairos* and *chronos* time to underscore the difference between "God's time and human limited time." But he doesn't illustrate those concepts through philosophy or through reference to the New Testament. Instead, he draws on Exodus 13:17–18 and the "roundabout way" that God leads God's people in their wilderness journey, rather than the more direct route. God's time, in this case, was "longer than expected" and experienced as a time of delay. It is a sharply relevant word for his congregation, but his personal tone suggests that it is also a relevant word for himself in a difficult year of leadership. The "easiest route is not always the best route," he acknowledges, and it is worth discerning "the reasons for detours." There is "no wasted time in the process of formation," he continues, calling his listeners to a difficult dependence on each other and God. To "rely on strange food called manna" (Deut 8:2–3) is to rely again on God's providence and to remember what it means to live as community. This is the work of the New Exodus, Banivanua tells his listeners: "not a five-year plan, but an epic half-century voyage of reshaping a community of faith."[117]

When I speak to clergy in Western settings about climate-crisis preaching, the question of hope is raw—particularly around questions of time. The paralysis of fatalism and despair are close for these preachers, especially if they are young. They recognize that hope is necessary to fuel the faithful action required of the present challenge. But hope itself seems dangerous. Hope can become an eschatological opiate, dulling urgency with spiritual platitudes. It can also cast technological solutions as idols, providing easy answers that never address the spiritual, societal root of the crisis. Fred Iutzi describes the problem well. "The gravity of climate change tells us we don't have time to work slowly, but the gravity of the climate change situation also tells us we don't have time for 'quick fixes.'"[118] It requires action that is "both fast and slow," a hope rooted in mem-

[116] Banivanua, "Kairos." Cecilie Rubow and Cliff Bird discuss the importance of "waiting" in Indigenous understandings of the relationship between time and place, and how the value of "waiting" is being deconstructed by "commodified, urban life" in Pacific Island communities: "Eco-theological Responses to Climate Change in Oceania," *Worldviews: Global Religions, Culture, and Ecology* 20 (2016): 150–68 [164–65].

[117] Rubow and Bird, "Eco-theological Responses," 164–65.

[118] Fred Iutzi, "We Were Born for This," The Land Institute, *Land Report* 125 (Fall 2019): 4–11 [7].

ory, and a memory pressed forward by hope that sustains. In this light, Banivanua's preaching does give specificity to the church's formational journey—though his description might seem as ordinary and elusive as the "strange food called manna" in the wilderness narratives. "What are we being formed into [through] the journey of the New Exodus?" Banivanua asks. We are being formed into "a people able to respond to the prevenient, justifying and sanctifying grace of God . . . [a people able] to love God and love the stranger, friend, and enemy as a neighbor." This is the "larger, wilderness journey" of the church—and its "reformation."[119]

One of the few comprehensive sermons on the church's exodus journey given by Tuikilakila Waqairatu before his death ends at the communion table—another place that sustains life at the intersection of memory and hope.[120] The 2012 sermon is preached in the year that the New Exodus vision is being crafted, and like that vision, it is ambitious. It follows the trajectory of the Exodus narrative through attention to four themes: liberation, covenant, wilderness journey, and destiny. Waqairatu's assertive, holistic theology is on clear display in his discussion of liberation and covenant. As a preacher, he is less cautious than Banivanua in reclaiming an increased political role for the church, and he is more concrete in his calls for spiritual cleansing. But in his final point discussing the destination of the Exodus journey (i.e., its "destiny"), he destabilizes and deepens the meaning of "promised land" in ways that foreshadow Banivanua's leadership. Waqairatu reframes place through a reframing of time and through lived practices of Christian formation. For Waqairatu, the promised destination of the Exodus journey is tasted in the Lord's Supper, where God's people participate in a shared proclamation of Christ's death and a new covenant in Christ's blood.[121] This "promised land" is not a new *vanua* of the future or a reclaimed *vanua* of the past. It is a *vanua* reborn through the repetitions and responsibilities of covenantal commitment. Waqairatu does not mention the *drua* in this sermon. In 2012, conversations about the church's new ecclesial logo are still in process. And yet, the *drua* is

119 Banivanua, "Kairos." For Bhagwan, one of the distinguishing marks of New Exodus is the time that it requires: an "intergenerational voyage of many legs" (personal email correspondence, June 20, 2023).

120 Tuikilakila Waqairatu, "Na Lako Yani [The Exodus]," unpublished sermon notes, Centenary Church, Suva, November 4, 2012. My deep thanks to Radini Qase Levu Vakacegu Joana Waqairatu for access to Qase Levu Vakacegu Waqairatu's sermon notes.

121 Waqairatu, "Na Lako Yani."

visible on his theological horizon. Attending to wilderness journey and eucharistic table, Waqairatu's reframing of the *vanua*—his reframing of place—does the very work that Fretheim describes in the imagery of Exodus itself. Redemption and creation are bound together through echoes of "Passover and sea crossing," without collapse.[122]

Place and Divine Presence: A Sacramental Journey

Waqairatu's sermon makes explicit that Fijian Methodist understandings of *vanua* are fundamentally theological. Given these stakes, the New Exodus goal of decolonizing place requires more than a reorientation of place's relation to movement and time; it also requires a reorientation of place's relation to God. Similar to Ezekiel's inaugural vision of a God on the move, the New Exodus *drua* describes a God that is simultaneously free and committed to covenantal relation. In that tension, there is a critique of two colonial conceptions of divine presence and place. The first collapses divinity into ecclesial institutions—making the church the earthly representative of God. The second severs divinity from creation, desacralizing the world. But the New Exodus' vision provides more than critique. It also casts a positive vision of place's revelatory character. Place becomes the site of the church's longing for, dependence on, and encounter with divine presence.

When the New Exodus' ecclesial logo was being redesigned, one significant change was in the placement of the cross. The MCIF's prior symbol, a small fishing boat (*takia*), had included the cross on the boat's sail. For Bhagwan, this imagery suggested a colonial missionary model, where the church "championed the message of Christ," carrying his cruciform "symbol on its sail" like a banner.[123] There was, in such a model, a kind of triumphalist representation—a collapsing of divinity into the structure of the church itself. But there was also a settling for the semiotics of faith, rather than its substance. To carry a symbol across the sea is one thing. It is something different, in the words of Winston Halapua, for "Christ to be mast and Master of our canoe."[124] The New Exodus *drua* images the cross as resting on the boat, not simply represented by the boat. It is a way to envision the church's accountability to a God who is with the community, and indeed, carries the community—but cannot be replaced by the

[122] Fretheim, *God and World*, 123.

[123] Bhagwan, personal interview, January 7, 2023.

[124] Bhagwan describes the input of former archbishop of Polynesia Winston Halapua into the MCIF logo (personal interview, January 7, 2023).

community. "This is Christ being present in the canoe and master of the vessel itself," Bhagwan explains.

More obviously, the cross' presence on the *drua* is a sign that the church does not travel alone. The migrations and displacements of the community are shared by a Migrating God.[125] At the 2016 Annual Conference, the question of discerning God in the midst of climate catastrophe was a central concern of Banivanua's sermon "Signposts of the New Exodus." "Where is God?" he asks, addressing the trauma present in the room following Cyclone Winston. "God is not in the destructive winds and rain and tidal waves and storm surges," Banivanua tells the conference. Instead, "God was with our people as they sought refuge in the church, in the home of a neighbor, in a bathroom, even in a cave. God is always right where the suffering is—with us."[126] Banivanua's words push back against spiritualities that collapse God and creation.[127] Just as there are theological and ethical dangers in conflating the place of the church with the presence of God, conflations of Creator and creation are treacherous in times of climate catastrophe. Such understandings can foster a fatalistic acceptance of suffering—or even justifications of the same. In this case, God's otherness makes possible a more radical solidarity. Banivanua's claim that God is "where the suffering is" draws deeply on the promise of God's presence with God's people. Banivanua follows this affirmation with a Methodist trope based on John Wesley's final words. "Best of all," he reminds the congregation, "God is with us."

But this affirmation of God's transcendence and immanence in emplaced life is only the beginning of Banivanua's theological exploration. His deeper concern is in helping the community embrace a faithful, ecclesial response to this Migrating God. To "take seriously [Wesley's] saying in its totality," Banivanua argues, is to foreground the question of discernment. If God is other than the *vanua* and yet with the *vanua*, how does

125 Seforosa Carroll describes the significance of theologies of migration for Pacific communities in "Displacing Liturgy: A Pacific Exploration," *Liturgy* 36, no. 2 (2021): 26–35. For a discussion of "God as the primordial migrant," see Peter Phan, "God, the Beginning and End of Migration: A Theology of God from the Perspective of Migrants," in *Christian Theology in an Age of Migration: Implications for World Christianity*, ed. Peter Phan (Lanham, Md.: Lexington Books, 2020), 101–28 [120].

126 Banivanua, "Signposts of the New Exodus."

127 Banivanua's proclamation is more in line with relational understandings of Creator and creation, characterized by Indigenous theologian Randy Woodley through the biblical concept of "shalom": *Shalom and the Community of Creation: An Indigenous Vision* (Grand Rapids: Eerdmans, 2012).

one discern God's movement and what it would mean to faithfully follow? Drawing on the cloud of God's presence in Numbers 9:15–23, Banivanua places the question in the context of the wilderness tabernacle. "The strength of the tabernacle was not the tabernacle itself," he notes, ". . . but the cloud." And the presence of the cloud was not guaranteed; it had to be watched for and responded to. "The movement of the cloud determined the movement of the community. When it lifted, the people arose and followed it. . . . When it descended, the community stopped and waited."[128]

The cloud of God's presence, then, is a cloud that cuts multiple ways. It affirms God's deep commitment to the community; but it is not a cloud that can be controlled or owned. It is a cloud that calls for response, obedience, and trust. When Bhagwan describes what the New Exodus journey has been like for the Fijian Methodist Church, he describes this release of control as part of what the *drua* represents. "An Exodus that includes the ocean challenges us to go beyond the horizon of what is known to us. Voyagers would set a destination beyond what could be seen, trusting that it would be there," Bhagwan describes. When the New Exodus shifted the church's ecclesiology from images of earth to images of sea, it shifted from a "narrative of 'owning land' to a narrative of being 'of the land.'" The *vanua* becomes "something we cannot own."[129] For Bhagwan, this is a spiritual insight about what it means to be a church following a Migrating God into myriad other situations the church cannot control. "Everything [the Fijian church has faced] so far, we have tried to control it—politically, economically, socially. . . . Rather than respond to those issues, we have tried to bend everything toward us." We say, "Come to the church and the church will guide you," Bhagwan states, rather than recognizing that the church itself needs to be led "out of its depth" to a new horizon.[130]

Place, in this understanding, becomes a site of longing and dependence—a reminder of the church's need rather than an assertion of its territory. Bhagwan, who often serves as a chaplain on the *Uto ni Yalo*'s sailing voyages, describes a moment when the vessel was carrying the MCIF president and his wife from Matuku to Suva in 2017. "You have to do so much work," he explains, "to keep wind in the sail as you steer." Tracking the wind's direction and tacking to keep the boat properly positioned is difficult labor. But even with all this effort, "there are days when

128 Banivanua, "Signposts of the New Exodus."
129 Bhagwan, personal interview, January 7, 2023.
130 Bhagwan, personal interview, January 7, 2023.

the wind doesn't blow. And you must sit and wait for the wind." Such a stillness fell during the president's voyage, and his wife, Radini Qase Bale Banivanua, did what she had been trained to do by her people. Reminiscent of Ezekiel prophesying to the wind of God's Spirit (Ezek 37:9), Radini Banivanua "sang and called for the wind." There would be no movement without it.[131]

The cross is not the only reference to divine presence in the MCIF logo. The dove of the Spirit actively flies overhead, creating the momentum that propels the *drua* forward. Similar to the work required to keep a sail filled with wind, or the attention given to God's tabernacling cloud in Numbers, Banivanua notes the work necessary to discern and respond to the Spirit's leading. He argues that this discernment is central to ecclesial identity. "We, as the church, must make listening to our spiritual instinct, our following the Cloud and waiting with the Cloud, part and parcel of the journey process," he states.[132] "This means that each one of us, from our children up to the *Qase Levu* [the MCIF president], must learn to listen to the voice of God."

This listening is not simply a mystical experience; it is embedded in daily life. It occurs, in other words, in and through *place*. Like other Western polarities and false choices described in this chapter—essentialism or erasure, circular time or linear time—the polarity between natural and revealed theologies blurs in place's matrix. Banivanua has already named the suffering of the *vanua* as one of the most significant places of where the church discerns the presence of God. But there are other places where God's people "peek around the corner at eternity." "God flows into our existence, when we least expect it," Banivanua preaches, ". . . [making] inexplicable intersections with our daily walks."[133] To encounter this divine presence, discernment is required. One must "know the difference between the voice of God, the voice of our own thoughts, the noise of the world and the voice of temptation." But one must also be open to discerning God's mediation in the places we find ourselves—particularly through the natural world. Banivanua uses the migratory wisdom of marine turtles, whales, and Pacific golden plovers to remind the ecclesial community of its deep spiritual instinct—"the divinely connected instinct which is in tune with the voice of God."[134] For Jolame Lasawa,

131 Bhagwan, personal interview, January 7, 2023.
132 Banivanua, "Signposts of the New Exodus."
133 Banivanua, "Signposts of the New Exodus."
134 Banivanua, "Signposts of the New Exodus."

God's revelation in the natural world opens up possibilities for cooperative engagements between the church and secular "institutions like government." "The vision of the New Exodus is very holistic," he notes. A "broader coalition" would help the church discern God's revelatory work in unexpected places and communities.[135]

Banivanua acknowledges that there will be "days when we will doubt the Spirit of God which hovers over the water with us."[136] The risk of a God that moves is a God that can be absent. Such is the challenge of a divine presence that cannot be placed in one's pocket. But for all of Banivanua's honoring of God's transcendence, he is a preacher who understands place as a site of divine encounter. "God tabernacles with us in the wilderness," Banivanua reminds his listeners, giving the wilderness journey of God's people revelatory significance. Place has a sacred—even sacramental—function, in the mediatory definition of that term. It reveals something of the character, purpose, and direction of God. And it can also become a place of communion.

The president's *Uto ni Yalo* voyage to Matuku was the first trip that a *qase levu* had made to the remote island in over fifty years. Banivanua was coming to this far-flung area of Fiji to perform the induction service of Rev. Jope Navuki, a divisional superintendent (*talatala qase*), but he was also coming to raise awareness about climate change.[137] The island had suffered from serious drought over the past several years. Rising sea levels were also eroding its land. "Our coconut trees are in the ocean," Rev. Navuki explained.[138] The *Uto ni Yalo* had agreed to sail the president and his wife to Matuku in order to amplify conversations around sustainable sea travel and the large carbon footprint of fiberglass motorboats.[139] The event felt like a precious reclaiming of Indigenous history

135 Lasawa, personal interview, January 5, 2023.

136 Banivanua, "Kairos."

137 Methodist Church in Fiji, "Methodist President Banivanua Sails on the *Uto ni Yalo*," *Fiji Sun*, September 13, 2017, https://fijisun.com.fj/2017/09/13/methodist-president-banivanua-sails-on-uto-ni-yalo/.

138 Methodist Church in Fiji and Rotuma, "Facing Climate Change in Matuku," interview with Talatala Qase Jope Navuki, divisional superintendent of Matuku, September 25, 2017, https://www.facebook.com/watch/?v=1538505886239344.

139 James Bhagwan, "Methodist Church in Fiji President Sails on Traditional Voyaging Canoe—Reflects on Climate Change," *First Friday Newsletter: World Methodist Council*, September 28, 2017, https://firstfridayletter.worldmethodistcouncil.org/2017/09/methodist-church-in-fiji-president-sails-on-traditional-voyaging-canoe-reflects-on-climate-change/.

and wisdom. A *qase levu* of the Methodist Church had not sailed across the ocean on a traditional canoe since 1964.

Six weeks before the president's arrival, four whales entered the village harbor and rested in a particular section of the bay for four days. "Four, of course, is a biblical number that speaks to us of creation," Rev. Navuki would recount later. "For the people it was . . . a blessing of good tidings for the land." When the *Uto ni Yalo* entered the bay over a month later, it put down anchor in the exact spot where the whales had rested. "God is telling us that he is concerned about what he has created," Navuki explained to Bhagwan and Banivanua that weekend.[140] Reflecting on the story four years later, Bhagwan's eyes fill with tears. "This is how God coordinates the universe. This is how the Spirit moves, but . . . the depth is missed by so many people. How do we get the whole church to recognize it? How do we translate and contextualize?"[141]

Multiple ministers recounted this story to me over the course of my interviews, and I stretched to understand its significance. It was not until I had heard the story several times that I began to make out familiar textures of detail. Here was another vision of God's commitment to God's people in a time of displacement. Another icon of a moving God, able to move in every direction, resting atop four living creatures. A place experiencing an apocalypse of land loss had become a site of encounter between something like "the appearance of the likeness of the glory of the Lord" (Ezek 1:28) and God's people. More than this, place had become a sign, mediating God's own apocalypse—God's own revelation of divine presence and call.

Covenantal Wayfaring in a Climate Storm

One of the last sermons that Tevita Banivanua delivered in his role as president of the MCIF is a sermon entitled "Staying the Course in the Winds of Change," preached to the graduating class of Davuilevu Theological College in 2017. Climate change is a central theme in his address. To minister in the face of "sea-level rise, saltwater inundation, . . . extreme weather patterns, . . . the pollution, acidification and heating of our ocean, and [the] forced migration of our people" is to live "in solidarity with those

140 Methodist Church in Fiji and Rotuma, "Newly Inducted Talatala Qase of the Matuku Division in the Lau Group . . . Reflects on the Significance of His Induction," interview with Talatala Qase Jope Navuki, divisional superintendent of Matuku, September 19, 2017, https://www.facebook.com/watch/?v=1533311446758788.

141 Bhagwan, personal online interview, September 28, 2021.

who are suffering and . . . work for a just transition to a more sustainable lifestyle by all," he states.[142] At its core, the MCIF's reframing of place is a theological, philosophical proposal that describes what such "solidarity" with the suffering requires, particularly in an era of climatic crisis.

As in the proclamation of many displaced communities, tensions are visible in the New Exodus vision. The recurring questions around *sovereignty*, *belonging*, and *loss* outlined in the early chapters of this book are embedded in this description of church as *drua* and place as exodus journey. In spite of their lack of resolution, the questions themselves disrupt the emplaced assumptions of empire. Continuity is negotiated alongside change, communal cohesion creates an openness to difference, and agency is held together with dependence. But the Fijian Methodist Church's New Exodus does more than critique colonial reductions of place as commodities, extensions of nationalistic or individual will, or constructions of human ingenuity. It reclaims another, more ancient, way. In a season of ecclesial soul-searching, the MCIF's New Exodus engaged biblical texts and images of Indigenous ocean migration to stress place's relationality, its formational arc, and its sacred mediation of the holy. There is a biblical word for such a posture toward place. In the words of Banivanua at the church's 2015 Annual Conference, it is the "covenantal nature of the *Lako Yani Vou* church."[143]

Covenantal understandings of place have circled through the sermons of the Indigenous Fijian preachers collected in this text, just as covenantal themes recur in Ezekiel's own "new exodus" prophecies (i.e., "covenant of peace" [Ezek 34:25] and "everlasting covenant" [37:24–26]).[144] It is a concept that can seem rather ordinary—or at least familiar. Those trained in Western settings are relatively certain it's a word we understand. Given this, those who feel the urgency of climate catastrophe might wonder: "Have we traveled all this way for a word so simple, and perhaps simplistic?" But the MCIF's New Exodus suggests that there is a thinness in many Western academic interpretations of the term, in both breadth and depth. Covenantal relationship can be understood primarily in terms of con-

[142] Banivanua, "Staying the Course."

[143] Banivanua, "Status of the Methodist Church."

[144] While the imagery of a "new exodus" is more commonly associated with Isaiah, Pikor sees it as fundamental to Ezekiel's structure and covenant theology. Ezekiel's "new exodus is contingent upon the new covenant, signaled in Ezek 20:37 as 'lead[ing] you into the bond of covenant' in the 'wilderness of the peoples'" and embodied through the restoration of the land (*Land of Israel*, 113–14).

tractual obligation, with place serving as proof of contractual fulfillment. Land can be understood as little more than a transactional commodity given to God's people in exchange for faithful behavior.[145] For the New Exodus leadership, however, place (*vanua*) is something more. It is an active participant in covenantal relation to God and God's people, meaning that place's continuity and particularity matters, even as place changes over time. Place has a role to play in the formation of a community, and its primary promise lies in that formational work, not in its reification of some idealized final destination. More significant still, for the leaders of the New Exodus, place has theological significance, serving as a site of divine mediation for a living, transcendent God.

For all of the New Exodus' insistence on "newness," this embrace of emplaced covenant has deep historical and biblical roots. In Wojciech Pikor's view, covenantal conceptions of place as relational, formational, and sacramental resonate with Ezekiel's description of Israel's land. More than "a guarantee of covenant," the land in Ezekiel "constitutes the aim of the covenant as the space in which YHWH's communion with Israel is to be realized." It is "the place of God's theophany."[146] In this view, place provides more than proof of a community's purity—or even proof of divine power. It is the ground of relationship and revelation. My next chapter will use the New Exodus' framing of place as a hermeneutical lens in interpreting Ezekiel 40–48, arguing that theologies of place developed by displaced communities are critical conversation partners in interpreting the biblical text for climate-crisis preaching. The goal, of course, is not to transpose one place with another. Western pulpits are distinct from Fijian pulpits (which are themselves diverse), and both are different than the particular

[145] Norman Habel helpfully distinguishes no fewer than six biblical land ideologies, bringing complexity to this transactional understanding of land ownership as a "conditional grant," in *The Land Is Mine: Six Biblical Land Ideologies* (Minneapolis: Fortress, 1995). Ellen Davis notes that while land ownership as an issue of political sovereignty is certainly addressed in the biblical text, a "neglected aspect" of biblical scholarship is "possession 'from below' based on care" and relationship, rather than possession "'from above' based on conquest" or treaty: *Scripture, Culture, and Agriculture: An Agrarian Reading of the Bible* (Cambridge: Cambridge University Press, 2009), 107. H. Daniel Zacharias notes how the "objectification" of land in Western settings creates "non-relational readings" of biblical, covenantal texts: "The Land Takes Care of Us: Recovering Creator's Relational Design," in *Theologies of Land: Contested Land, Spatial Justice, and Identity*, ed. K. K. Yeo and Gene L. Green (Eugene, Ore.: Cascade, 2021), 69–97 [85–89].

[146] Pikor, *Land of Israel*, xix.

place that shapes Ezekiel's witness. Covenantal theologies of place, however, speak to our relationship with creation, with God, and with each other. As such, they help preachers to better own "where we stand" when we address climate catastrophe in our particular communities and better articulate the responsibilities that connect us across space and time.

There are many resonances between the New Exodus *drua* and Ingold's description of place as "wayfaring,"[147] but this attention to limits, particularities, and communal accountabilities differentiates the two. The New Exodus presses conversations about "wayfaring" past phenomenological acknowledgments of place's eventfulness and toward place's political, economic, and theological function—its covenantal role in God's larger work in the world. This means that covenantal wayfaring takes seriously the question of whose voices are centered in conversations about place and whose needs require priority. None of these decisions are easy, and all of them are provisional. But they are necessary discernments if the language of covenant is to be constituted by more than abstraction, self-legitimizing circularity, or romanticism. Covenantal wayfaring is lived out through practical obligations of reciprocity and collaborative yielding—which, sadly, are the very qualities that Bhagwan finds lacking in many of his contemporary engagements with climate-justice advocacy groups around the world.

Elected as general secretary of the Pacific Council of Churches in 2018, Bhagwan candidly describes numerous opportunities to speak at international conferences on the subject of climate justice that rarely "follow a process of partnership." "There is no consultation about a strategy or development of a campaign," he explains. "The Pacific continues to be used by some international organizations as a token of representation—and we are so desperate to get our message across, that we say yes."[148] This desire for genuine processes of partnership is why Bhagwan did not attend COP27, though he had attended COP meetings in the past and his involvement would have been paid for by international partners. He was troubled by the assumed priority of the trip over existing regional projects and funding needs. And in his assessment of his own leadership, he was aware of an

[147] I offer a respectful critique of Tim Ingold's understanding of place as "wayfaring" in chapter 1 (see Ingold, *Being Alive: Essays on Movement, Knowledge, and Description* [New York: Routledge, 2011], 148–50). I use "covenantal wayfaring" to nuance and expand his concept.

[148] Bhagwan, personal interview, January 7, 2023.

unresolved commitment to the Fijian Methodist district superintendent of Vatulele that he had made a year before.

When the island of Vatulele was hit by Cyclone Yasa at the end of 2020, the *Uto ni Yalo* brought roofing supplies for rebuilding. But these supplies only scratched the surface of the community's need. It was a message that the district superintendent reiterated with "persistence and hope" whenever he saw Bhagwan over the year to follow. "You said you'd try to do something," he would say, "so let's try to do something."[149] The community was dealing with loss and damage; but they were also focused on mitigating the next disaster. Solar-powered floodlights would be necessary for lighting the village's emergency evacuation center, and solar-powered refrigeration would help sustain the fishing economy in a crisis. The village's outboard motor had been damaged by Yasa and was not able to be repaired. The community needed a climate-induced risk assessment, including a measure of the health risks impacting the local community. The village had one of Fiji's highest rates of noncommunicable diseases (NCDs) and amputees, with numbers particularly high among women.[150]

Being "constant and consistent" is key in faithful relationship, Bhagwan tells me. It is also a challenge when needs are great and communication with remote areas is difficult. A year would pass before Bhagwan would be able to respond, but the "burden kept pricking." The Pacific Council of Churches doesn't have a "climate change" program, Bhagwan reminds me. "It has a climate *justice* program—and I had not been just in this matter. The PCC had not been just."[151] And so, Bhagwan imagined how local, ecumenical cooperation might fund a second *Uto ni Yalo* sail, delivering solar panels and a new boat motor, bringing pastors equipped to minister to communities facing climate disaster trauma, and bringing reproductive health supplies for women through the Feminist Coalition for Climate Justice. The trip was not inexpensive, but it was no more costly than a flight to a COP27 meeting.[152]

149 Bhagwan, personal interview, January 7, 2023.

150 Bhagwan connects these health problems to the long hours that women spend sitting and pounding *masi*, work that is currently necessary to sustain the local economy.

151 Bhagwan, personal interview, January 7, 2023.

152 World Council of Churches, "Pacific Conference of Churches Calls for 'Less Talk and More Action' as COP27 Begins," November 7, 2022, https://www.oikoumene.org/news/pacific-conference-of-churches-calls-for-less-talk-more-action-as-cop27-begins. The Methodist Church in Fiji, the Anglican Diocese of Polynesia, the

For Willie James Jennings, the colonial conception of place is fundamentally marked by a desire for control. In Western contexts, nationalism and individualism have been that desire's primary modes of expression. Jennings paraphrases the mindset this way: "I [as either a self or a nation-state] can only *become*, if I own, control, and extract."[153] The belief has had ruinous impacts on creation. In the undertow of this history, Jennings insists that covenantal understandings of place are far from ordinary. They offer potent, subversive alternatives grounded in communal commitments: "I choose you, and you choose me." Covenant describes more than a "freedom from." It describes a "freedom for," reorienting theology toward relationships of "belonging."[154] The vision is beautiful. But its implications should not be sentimentalized, for such choices have costs. One cost, in particular, cuts to the heart of the colonial project. To embrace relationship and responsibility is to leave control behind.

This letting go of control is seen, in part, through a decentering of climate-advocacy models that impede local relationships, magnify singular voices over communal conversations, and flatten distinctive needs. Prophetic, international articulation of the crisis at hand is significant work, and it is work that Bhagwan and others continue. But proximity to local communities is necessary for such prophetic witness. "The prophetic voice of the church cannot be separated from the pastoral voice of the church," Bhagwan notes—which means that, at times, climate advocacy looks very mundane. It looks like communities remembering how to distribute resources so that no one has an excess while others lack (Exod 16:18). It looks like leaders investing in regional relationships rather than chasing international recognition. It looks like ecumenical cooperation. And it looks like accountability, honoring the cries of far-flung places through grassroots collaborations that reframe creation through the lens of belonging: "I choose you, and you choose me." The New Exodus vision is not the work of a single prophet. It is the work of connected communities. An individual is not able to sail a *drua* alone.

But letting go of control is about more than inclusion. It is also about a willingness to change. When place is viewed with a covenantal lens, it not only invites a broader connection to others. It invites human

---

Presbyterian Church (USA), the *Uto ni Yalo* Trust, and the Feminist Coalition for Climate Justice collaborated with the Pacific Council of Churches on the project.

[153] Jennings notes the ongoing influence of John Locke's distinction between nature and the self in fueling this perspective: Kenan Institute for Ethics, "Facing the Anthropocene."

[154] Kenan Institute for Ethics, "Facing the Anthropocene."

communities to an "affectable"[155] future. The New Exodus envisions place as more than a destination, commodity, or proof of legitimacy. Place has its own story of *becoming*, its own needs, and its own arc of relationship to God and the world. As such, it forms human communities to be "on the way" people,[156] people rooted in relationship, who have not yet arrived.

And this is, perhaps, why a theological lens is so critical to the New Exodus' model of place. Covenantal wayfaring in the climate crisis is slow, formational work—tacking together to catch the wind. Grief and loss are real, and the future is unclear. Banivanua and Bhagwan are circumspect when describing the New Exodus' future. They wonder if their "Connexional [*sic*] Plan" was too complex, if they should have encouraged more local Bible studies, or if the experience of sailing on the ocean made their core metaphor inaccessible to villages living in the nation's interior.[157] In local communities, the seed of the New Exodus vision is crowded alongside an increased "striving . . . for the commodified, urban life."[158] And at the church's head office, attempts to streamline or modernize the vision have threatened the metaphor's particular decolonial specificity. The goals of the New Exodus are now described as "three Es": "education, evangelism, and the economy,"[159] a three-part simplification that works to retain the original plan's holistic understanding of faith. Even as he voices support for the leaders who have followed him, Banivanua worries that significant things have been lost in that reduction. "Where is the ecumenism?" he wonders.[160] And where, for that matter, is the environment?

The impact of sermons can seem so ephemeral. Students in my preaching classes in the United States regularly confess their skepticism that sermons can bring about meaningful societal change. The stubbornness of congregations who hear God's Word and refuse to respond is a familiar grief to prophets and pastors alike, stretching over millennia (Ezek 2:6–7; 3:5–7). It is, perhaps, why I was so arrested during my three years in Fiji

155 I discuss Denise Ferreira da Silva's use of the term "affectable others" to describe Western constructions of racialized others in this book's introduction; see her *Toward a Global Idea of Race* (Minneapolis: University of Minnesota Press, 2007), xv.

156 Referencing Bhagwan's earlier description.

157 Bhagwan wonders how the New Exodus vision might have been more concretely communicated in local contexts in both his September 28, 2021, and January 7, 2023, interviews.

158 Rubow and Bird, "Eco-theological Responses," 165.

159 Banivanua, personal interview, January 2, 2023.

160 Banivanua, personal interview, January 2, 2023.

by Banivanua's preaching, the New Exodus' vision of a decolonial church, and its navigation of biblical narratives for the sake of sustainable climate justice. The New Exodus leaders believed that preaching mattered in the work of climate advocacy. They believed in the power of Spirit-led, communal discernment and the ability of the biblical text to reframe the most urgent contemporary questions. To put a more precise point on it, they believed that God was a significant player in the work of environmental activism and the creation of a decolonial church. To include God in conversations about place allowed these church leaders to embrace relationship and shoulder responsibility, even as they relinquished control.

In 2019, I asked Bhagwan if there was a resonant testimony he heard from local Fijian pastors who were facing the increasingly fierce storms, dying reefs, and relocations of climate catastrophe. He answered straightaway: "God will not allow us to die."[161] Like many claims of faith, the sentence can seem like wishful thinking. There is evidence to the contrary. But tellingly, this sermonic testimony is not a prediction about individual lives. It is an affirmation about divine commitment to a communal "us." The fierce resolve with which Bhagwan speaks the sentence makes it sound more like protest than guarantee. It calls on God to be consistent with who God has claimed to be. A covenanting, wayfaring God. A "wheel within a wheel" God (Ezek 1:16). A God who raises dead communities to life and calls them to account. And a God who continues to create places of communion.

[161] Bhagwan, personal online interview, September 28, 2021.

5

Place and Covenant

Reframing Exodus in Ezekiel's Temple Vision

The New Exodus sermons of the Fijian Methodist leadership underscore a critical point for preachers. Covenantal wayfaring isn't only done through waves. It is also done through proclamation. One way to understand the New Exodus project is to see it as a collection of sermons that reclaim an Indigenous approach to place when that community's land is under threat. But it is more than that. It is also a collection of sermons that reclaim the place of preaching in the midst of climate catastrophe. When Banivanua asserts that the "struggle over the interpretation" of the biblical tradition in light of Indigenous culture is "crucial to political and social life,"[1] he is insisting that biblical interpretation and proclamation have significance for the material realities of his nation and the larger world.

It is a claim that may seem overly precious or abstract. Worse, the claim may seem dangerously misguided. When the ecologies of a community are breaking apart on both a local and a global scale, of what use are interpretive shifts in the reading of millennia-old narratives? I have argued through much of this book that white, Western homiletic education has regularly reduced conversations about place to the realms of hermeneutic philosophy, historical criticism, or persuasive rhetoric. In such streams of thought, place becomes a puzzle to be solved or a power to be commodified. The questions of displaced communities—questions of *sovereignty*, *belonging*, and *loss*—are sidestepped in favor of questions about the nature of meaning, the basis of one's interpretative control, or the scope of one's prescriptive influence. Eve Tuck and K. Wayne Yang's fierce reminder that "decolonization is no metaphor"[2] forecloses such

1 Tevita Nawadra Banivanua, "Staying the Course in the Winds of Change," keynote address to Davuilevu Theological College graduates, Baker Memorial Church, Nausori, Fiji, November 24, 2017.

2 Eve Tuck and K. Wayne Yang, "Decolonization Is Not a Metaphor," *Decolonization: Indigeneity, Education and Society* 1, no. 1 (2012): 1–40 [29].

reductions. Decolonization is not a philosophy or an interpretive method. It is a practice that centers costly conversations about reparations, restitution, and the redistribution of land to those who have been forcibly dislocated throughout colonial history.

But what if these conversations are precisely those convened by Scripture's testimony?

The New Exodus of the Methodist Church in Fiji (MCIF) insists that decolonial biblical interpretation does more than reframe the biblical past. It presses for material justice around climate policy and land distribution in the present. Moreover, and this is key, it makes this claim in ways that do not conflate places separated by time and space, universalizing prescriptive sound bites into transcendent blueprints. Rather, the MCIF's New Exodus draws on Scripture's profound theologies of place—theologies crystalized in crises of displacement—to assert place's relationality, formative character, and sacramental telos. It then asks what such covenantal approaches to place require of communities standing on very different ground.

I have suggested that the most difficult questions to answer in the climate crisis are not technological in nature. They are relational. What is one's responsibility to the places of others given the borders of our understanding and authority? How does one embody covenantal connection in ways that honor difference? Is it possible to belong to each other in life-giving ways, given the death-dealing histories that have shaped our world? When Banivanua argues that Scripture's interpretation is "crucial to political and social life," he does not only mean that biblical preaching has material policy implications for particular places. He also means that these broader relational questions—questions, finally, about *place* itself—lie at the heart of interpretive work.

The second chapter of this book used Ezekiel's critiques of Babylonian place-making to point toward unexpected sites of resistance in Fijian Methodist sermons. Ezekiel illumines questions of sovereignty, belonging, and loss that continue to resonate in displaced communities responding to the climate crisis today, disrupting neocolonial idealizations of place's universality, insularity, and constructed character. This chapter makes an inverse move. It wonders how the covenantal theology of place articulated by the Fijian New Exodus leadership might illumine what is at stake in Ezekiel's own reframing of the exodus story—particularly through the book's closing temple vision (chs. 40–48). How does the relationship between place and covenant shape these final chapters, their function, and their material implications? There is an intentional tacking back and forth in my analysis, allowing biblical text and then contemporary context to catch the Spirit's wind. The goal is not simply to understand Ezekiel better—though I will argue that the MCIF's New Exodus approach to place brings critical aspects of Ezekiel's theology into focus.[3] The goal is to better own where we contemporary readers are standing in Ezekiel's story and in the present day. In the face of apocalyptic environmental loss, how does this text orient us, convict us, and call us to change?

I have argued that displaced communities are critical facilitators in biblical conversations around climate justice. For all their particularity, displaced communities carry a wisdom about place's importance and its loss. They know something about navigating toward land unseen in stormy seas. But in addition to these broad resonances, the MCIF's New Exodus shares similarities with Ezekiel that are more precise. When grappling with the crisis of displacement, Fijian church leaders and the biblical prophet reclaim and reframe the Exodus narratives in ways that call to mind God's work of creation. Both call their communities to repentance, even as they critique imperial ideology. And both share a hard-earned distrust toward false promises. I will argue that from these theological and biblical grounds, they share one thing more: a covenantal commitment to place that is *relational*, *formational*, and *sacramental*—with concrete implications for a world coming undone.

Such commitments inform more than our understanding of Scripture's content and call. They also inform one's posture toward Scripture itself. In the chapter's final move, I will underscore several implications that these displaced theologies of place hold for the work of biblical interpretation. This reflection on interpretive work should not come as a surprise. The place of the biblical text has been visible on this project's horizon from the start. It is, after all, a book about preaching. I will contend that if preachers shaped by colonial paradigms mean to commit to creation's community in a season of environmental crisis, they must reconceive their relationship to the place of Scripture—and Scripture's relation to the place of preaching.

The Place of Ezekiel 40–48: Reframing Exodus and Remembering Covenant

When Category 5 Cyclone Yasa hit the village of Yaro in December 2020, the ferocity of the storm crumbled the majority of the village's homes and

[3] David Stark argues that paying closer attention to the sermons of those impacted by colonial oppression is a significant intervention in mitigating interpretative strategies that dismiss or deny difference in the biblical text: "Whiteness and the So-Called Problem of Preaching the Old Testament," *Theology Today* 79, no. 2 (2022): 115–26.

tore apart the concrete sanctuary that stood at its center. Yaro is one of only three villages on Kia Island, an island that requires an hour's boat ride off the north coast of Vanua Levu to reach. The community has no ports or paved roads. It is off the grid. When I visited the island three years later to interview the pastor assigned to the Kia Circuit, many of the demolished homes were still under construction. Building supplies are difficult to obtain and transport to this remote location—much less purchase.

Nevertheless, the day I arrived in Yaro, the village was singing. A small team from the Fijian military had made a long-delayed trip to the island to assist the village in mixing concrete for the structural supports of a new sanctuary. The village men had joined them in a bucket line, passing loads of wet cement from hand to hand in rhythm. Fijian hymns filled the air with layered harmony. When the men took their break following the work of the morning, they sat cross-legged on mats and prayed. The church was far from finished. The storm had destroyed its walls and shifted its foundation. It stood open to the sky. But the men gave thanks for cement pillars emerging from the rubble and joined their voices in song again. Rebuilding the Yaro sanctuary was more than a construction project. It proclaimed God's commitment to ruptured ground and gave communal testimony to the sacredness of Yaro's *place*.

If the opening chapter of Ezekiel is not enough to convince readers of the significance of place and displacement in the book's theology, the book's final nine chapters do the job.[4] "In the twenty-fifth year of our exile," the prophet explains, ". . . in the fourteenth year after the fall of the city . . . the Lord . . . brought me *there*" (Ezek 40:1, emphasis added).[5] In a

[4] Though there has been significant scholarly discussion about whether Ezek 40–48 is a later addition to the Ezekiel material, Ellen Davis argues that the book's canonical shape has a theological cohesiveness that is convincing evidence for a more holistic interpretation. "The vision of a restored temple is essential to the theological shape of the book," particularly in relation to the return of the divine presence (*Opening Israel's Scriptures* [Oxford: Oxford University Press, 2019], 308). Even if redactions have expanded the text, Paul Joyce argues for a sixth-century date for Ezek 40–48 as a whole. "If the Second Temple had already been inaugurated, . . . the account would surely have been accommodated to the new reality." There is also, in other words, a contextual cohesiveness in evidence ("Temple and Worship in Ezekiel 40–48," in *Temple and Worship in Biblical Israel*, ed. John Day [New York: T&T Clark, 2005], 145–63 [147]).

[5] The word "there," in the context of this sentence, most plainly refers to the site of Jerusalem's destruction. However, scholars like Stephen L. Cook find it significant that Jerusalem is not mentioned in Ezekiel's vision. "In Ezekiel's vision of an ideal redemption, the traditionally 'royal' city of Jerusalem will *not* regain its former status" (*Ezekiel 38–48* [New Haven: Yale University Press, 2018], 125, emphasis original).

vision, Ezekiel stands on a high mountain in the land of Israel overlooking what appears to be the city, though the geography and layout are distinct from the Jerusalem of Ezekiel's memory.[6] Accompanied by a man "whose appearance shone like bronze" (v. 3), the prophet is taken on a carefully orchestrated tour of a temple rebuilt outside the city walls and instructed to declare all that he sees (v. 4). The extended description includes measurements, geographic orientations, ritual instruction, and a promise of God's return to this place of worship. It is by far Ezekiel's longest visionary account—and it is strange. Scholars have been uncertain how to interpret passages that seem, at once, bathed in both dreamlike symbolism and fastidious architectural detail. What is this passage's function? Is it a devotion for prayer,[7] a blueprint for a building project,[8] Zadokite propaganda,[9] or something more like lament? One can almost hear the footsteps of Ezekiel and his spiritual guide echoing through the temple's empty chambers. For all its golden-hued hope, the temple seems haunted by silence, reminding readers of prayers and persons that are no more.

It is also haunted by the account of the Exodus narrative.[10] For many Western preachers, movie images of Red Sea crossings loom large enough

6 The fact that the temple is not included within the city walls would be a significant change in the landscape. But given the reallocation of land described in ch. 48, it also seems the site of the city has shifted from the site of the original Jerusalem. Rather than connect the "very high mountain" to an actual geographical feature, Jon Douglas Levenson connects it typologically with Mount Sinai: *Theology of the Program of Restoration of Ezekiel 40–48* (Missoula, Mont.: Scholars Press, 1976), 37–49.

7 Susan Niditch, "Ezekiel 40–48 in a Visionary Context," *Catholic Biblical Quarterly* 48 (1986): 208–24. See also Cook's description of the visionary account as "one of continuous ascent . . . into God's presence" (*Ezekiel 38–48*, 132).

8 Such an approach is often paired with scholarship that stratifies the redaction history of these chapters, assuming that "realistic details are a sign of secondary elaboration of an original vision" (Joyce, "Temple and Worship," 147).

9 For a description of the "ideological considerations and political aims" of Ezekiel's Zadokite group, see Stephen L. Cook and Corrine L. Patton, "Hierarchical Thinking and Theology in Ezekiel's Book," introduction to *Ezekiel's Hierarchical World: Wrestling with a Tiered Reality*, ed. Stephen L. Cook and Corrine L. Patton (Atlanta: SBL, 2004), 1–26 [14–17].

10 A diverse range of scholars have noted this connection. See, for example, Levenson, *Theology of the Program of Restoration*; Walther Zimmerli, *Ezekiel*, vol. 1, trans. Ronald E. Clements (Philadelphia: Fortress, 1979); and Daniel I. Block, *The Book of Ezekiel: Chapters 1–24* (Grand Rapids: Eerdmans, 1997). Discussions of whether Exodus serves as a referent for Ezekiel or Ezekiel serves as a referent for Exodus are ongoing. While scholars like John Lust argue that Ezekiel influences the composition of Exodus (see "Exodus 6,2–8 and Ezekiel," in *Studies in the Book of Exodus: Redaction,*

in the cultural vernacular that they eclipse the majority of Exodus' text. The gifting of the Ten Commandments on Mount Sinai may make an appearance in Sunday's proclamation, but large swaths of Exodus are remembered hazily, if at all—particularly those sections devoted to the descriptions of cultic ritual, priestly duties, and the construction of the tabernacle (e.g., Exod 25–31, 35–40). These are hard texts to preach. But such descriptions are clearly significant for Ezekiel. The debates about which particular texts and which oral histories influenced the writings of the prophet and his school are ongoing and inconclusive. But the dialogical relationship between the Exodus narrative and Ezekiel is apparent in even rudimentary analyses of their structures.[11] Ezekiel 40–48 brings home the point. Like Moses, who also experienced God's glory and received explicit, detailed visions of communal worship from a mountaintop, Ezekiel declares the significance of place, sanctuary, and ritual practices for his displaced community.[12] And similar to the village of Yaro, more than a building project is at stake. Ezekiel's temple vision is a proclamation of God's commitment to ruptured, uncertain—even haunted—ground.

The New Exodus of the Fijian Methodist Church rearticulates covenantal theologies of place in its readings of the biblical text and through its ecclesial reimagining of church as ocean-faring *drua*. Ezekiel's final nine chapters do something similar, reframing exodus traditions to center conversations of covenant and creation in their affirmation of place's importance. Through Ezekiel's descriptions of a visionary temple, a river that flows through the surrounding land, and a newly allotted city "for ordinary use" (48:15), place's covenantal significance is underscored. Place's relational, formational, and sacramental character is not a theo-

---

Reception, Interpretation, ed. Marc Vervenne [Leuven: Leuven University Press, 1996], 209–24), Moshe Greenberg, Michael Fishbane, and others argue that the influence flows the other direction. The latter is the orientation to which I tend. Rebecca Idestrom helpfully maps these different perspectives in "Echoes of the Book of Exodus in Ezekiel," *Journal for the Study of the Old Testament* 33, no. 4 (2009): 489–510 [506–7].

11 In Joseph Blenkinsopp's view, the close relation between Exod 25–31 and Ezek 40–48 "can hardly be coincidental": *Ezekiel* (Louisville: Westminster John Knox, 1990), 200. Daniel I. Block also maps parallels between Exod 3–4 and Ezek 33; Exod 5–13 and Ezek 34–37; and Exod 14–15 and Ezek 38–39: *The Book of Ezekiel: Chapters 25–48* (Grand Rapids: Eerdmans, 1998), 498–99.

12 Idestrom follows H. McKeating in describing the significant similarities between Moses and Ezekiel in the biblical narrative: "Echoes of the Book of Exodus," 493–96; H. McKeating, "Ezekiel the 'Prophet Like Moses'?" *Journal for the Study of the Old Testament* 61 (1994): 97–109. There are also interesting inversions. Note God's promise to Moses to be "with your mouth" (Exod 4:12), in contrast to God's promise to Ezekiel that the prophet's tongue would "cling to the roof of [his] mouth" (Ezek 3:26).

logical abstraction for Ezekiel. These characteristics have political and economic implications for his displaced community.[13] Indeed, Ezekiel's covenantal theology of place centers the very questions that have been echoing through this book: questions of *sovereignty*, *belonging*, and *loss*. In so doing, Ezekiel does more than reframe exodus. He reframes the work of justice and the role of proclamation in the face of climate catastrophe.

Place and Movement: Navigating a Relational Covenant

At first reading, Ezekiel's visionary temple seems diametrically opposed to the Fijian image of the *drua*. This is no moving tabernacle, carried through the Pentateuchal wilderness. Ezekiel's temple foundations are stationary, its porticos and gates orientated toward particular, cardinal points (Ezek 40:6, 20, 24, 27, 32, 44ff.). It stands solid and still, evoking Bhagwan's worry of "permanence" rather than "mobility" in the life of faith.[14] The vision seems in juxtaposition to the book's opening descriptions of stormy displacement and a mobile God—wheels within wheels countered by stone pillars and carefully aligned courtyards. But Ezekiel does not allow his readers so simple a binary. YHWH's glory, in all of its windy radiance, returns to the temple and fills it (43:1–7),[15] giving the temple a fixed and fluid quality. The God who moves is not in contradiction with the God who stays. To underscore the point, Stephen Cook argues that God's promise to "dwell among" God's people (43:7) carries a different connotation than that of a king who "inhabits" and "settles down" in a palace. There is, in the verb "to dwell," an allusion to the God who dwelt with God's people in their wilderness wandering (Exod 25:8). This God is a "*tremendum:* that which will always remain unsettled, unnerving, and unsafe."[16]

But Ezekiel's final vision includes another, less likely witness to place's movement. It is not quite accurate to say that, apart from the movement of YHWH's glory, the prophet, and his guide, everything in the temple is silent and immobile. Ezekiel is shown an unexpected stream of water

13 Whether these implications were materially enacted is, of course, a separate question.

14 As discussed in chapter 4; James Bhagwan, personal online interview, September 28, 2021.

15 Mary Mills describes Ezekiel's image of the temple site as "twinned" with that of the mobile throne chariot, rather than standing in contradiction, in *Urban Imagination in Biblical Prophecy* (London: T&T Clark, 2012), 81. It should be noted that the MCIF's New Exodus also makes use of "twinned," dialogical images. The MCIF's *drua* is paired with a twelve-pillared "Connexional [*sic*] Plan," alluding to the twelve pillars that Moses places at the base of Mount Sinai (Exod 24:4).

16 Cook, *Ezekiel 38–48*, 183.

flowing "from below the threshold of the temple toward the east" (Ezek 47:1).[17] The image seems out of place given Ezekiel's strong emphases on enclosure, controlled access to the holy, and sacred hierarchy.[18] In the context of a vision preoccupied with the delineation of space, the stream introduces a permeability to the temple structure that is provocative. There are, of course, numerous antecedents that connect water with YHWH's temple in the biblical tradition.[19] The Psalms describe God's Zion temple as both a "fountain of life" (Ps 36:9) as well as a place where God sits "enthroned over the flood" of chaos (Ps 29:10). Stephen Cook sees Ezekiel's river as echoing both of these images, reminding readers of the temple's necessary role in the preservation of ecosystems far beyond its walls—and those ecosystems' vulnerability.[20] Ezekiel's temple does not describe places as impervious to change; instead, it describes a fluid, present-tense relationality as necessary for place's continuity and thriving. God's investment in this relationship is underscored by the prophet. YHWH's restorative power is not locked away behind temple walls. It actively flows into increasingly quotidian spaces of agriculture and commerce, creating a living relationship between everyday realities and the sacred.

Place's relationality is precisely why Ezekiel's vision insists on the protection and maintenance of holy space—a theme that can prick uncomfortably in the ears of those who would rather see distinctions between holy and ordinary spaces dissolved. For displaced communities, the relationality of place does not only connote blessing; it also connotes place's susceptibility to threat.[21] Contemporary Oceanic environmental activists

[17] Significantly, the river flows out of the temple in the direction that the glory of the Lord has just entered (Ezek 43:2).

[18] Cook and Patton, "Hierarchical Thinking," 1–26. In spite of this emphasis on enclosure, it is telling that the description of the rebuilt city describes gates rather than walls (Ezek 48:30–34). For Julia Rhyder, the vision's emphasis on gates connotes "a complex renegotiation of royal, priestly, and communal roles within the post-catastrophe community," in "Gates and Entrances in Ezekiel 40–48: The Social Utopia and the Temple Vision," *Vetus Testamentum* 73 (2023): 752–65 [754].

[19] There are also numerous Near Eastern parallels of sacred rivers flowing from temple structures. See Cook, *Ezekiel 38–48*, 262–63.

[20] Cook, *Ezekiel 38–48*, 268.

[21] Natalie Mylonas reads Ezek 16 as gendered description of place's loss of purity, countered by Ezek 40–48's hyper-attention to boundaries and particularity: *Jerusalem as Contested Space in Ezekiel: Exilic Encounters with Emotion, Space, and Identity Politics* (New York: T&T Clark, 2023).

might point to the Pacific's ugly history of nuclear testing as an example, underlining why place's particularity and purity are bright concerns. In his reflections on the significance of baptism in Oceanic contexts, Jione Havea notes that images of water continue to be ambiguous in places where "contaminated remnants of atomic testing" impact the Pacific.[22] Inhabitants living near the domes where dangerous waste is stored call them "tombs." Havea notes that, given the size of these concrete holding containers and their distance from the sea, "a strong wind can push waves up to and over the tomb," allowing salt water to leak in—and leak out. Sea-level rise only exacerbates the problem.[23] Such present-day realities make preoccupations with place's enclosure, particularity, and purity an issue of survival.[24]

The boldness of Ezekiel's vision, however, is that it not only resists the contaminating threats of a relational world through careful hierarchies of responsibility. It also insists that place's relational vulnerability is integral to its restoration. The temple river flows from its source with miraculous healing, impacting the harvests of fruit trees (47:12), the healing properties of their foliage (v. 12), and the abundance of fish (v. 9). It is on the banks of this river that other humans are first glimpsed in Ezekiel's vision, fishing and spreading nets (v. 10). The passage even attends to the details of salt production in a freshwater ecosystem (v. 11). Though the vision is "not attainable through merely human effort," Ellen Davis notes its "most basic practical concern" for the economic and agricultural welfare of the world. The river makes possible the flourishing of land, humans, and nonhuman creatures together, "in intricate systems of interaction."[25] This robust relationality is what finally characterizes the place of Ezekiel's temple, an ecosystem that simultaneously attends to both continuity and change.

The New Exodus preachers in the Methodist Church in Fiji used the fixed fluidity of the *drua* to describe the ways that place can move, even as it retains its particularity. The land's living relationship to all it sustains requires such movement. Significantly, they did not apply this principle to the place of the *vanua* alone. They also applied it to their readings of Scripture, reimagining exodus texts that had previously been used to exclude

22 Jione Havea, "Holy Baptism," *Liturgy* 34, no. 2 (2019): 3–11.

23 Havea, "Holy Baptism," 4.

24 Havea carefully negotiates the relationship between the purity of water and the holiness of the baptismal ritual, concluding, "Theologies of baptism do not hold water until the troubled waters in Pasifika and beyond are taken into account" (10).

25 Ellen Davis, *Biblical Prophecy: Perspectives for Christian Theology, Discipleship, and Ministry* (Louisville: Westminster John Knox, 2014), 126.

ethnic-religious others in ways that stressed covenantal commitment across difference.[26]

Ezekiel's temple vision does its own reimagining of exodus narratives. There is both resonance and dissonance in their comparison. Risa Levitt Kohn notes that Ezekiel is not "working furiously to preserve . . . traditions for posterity. He convolutes the past, turning history on its head." In Ezekiel, Exodus' "land of promise becomes the land of exile (20:38)."[27] Kohn references Ezekiel 20 in this argument, noting how Ezekiel's retelling of the exodus story postpones the culminating promise of land and covenantal intimacy with God. Wilderness wandering in the "land of Egypt" (20:36) does not culminate in a joyful Jordan crossing, nor does it end in faithful adherence to the "bond of the covenant" (20:37). Idolatry and child sacrifice postpone such intimacy, requiring a second exodus through "the wilderness of the peoples" (20:35). Ezekiel's reframing has a strikingly different narrative shape than the history described in Exodus itself.[28] Similar to the New Exodus of the MCIF, Ezekiel's new exodus calls for more than a return to or repetition of the past. He questions the exodus traditions he engages, "comments on them, and ultimately reformulates them."[29] To use the language of chapter 4, there is a covenantal wayfaring in his analysis, taking stock of these narratives' navigational trajectories and the shifting ground of his community. Wojciech Pikor argues that part of what is at stake for Ezekiel is the telos of the land itself, significant not for its own sake, but as a site of covenantal relation with God. To enter the land and not adhere to the covenantal relationship between divine, human, and nonhuman creation is to misunderstand the land's function and sacral worth. It is,

[26] Banivanua describes how the New Exodus differs from the Fijian ethnonationalist use of exodus imagery in chapter 4.

[27] Risa Levitt Kohn, *A New Heart and a New Soul: Ezekiel, Exile, and Torah* (Sheffield: Sheffield Academic Press, 2002), 111.

[28] Perhaps the most stark example of this reframing is Ezekiel's repurposing of YHWH's recognition formula (i.e., "You shall know I am the Lord"), which is used repeatedly in the Exodus material to describe the purpose of Egyptian plagues (Exod 7:17; 14:4, 8) and is regularly addressed to Pharaoh himself. Ezekiel now addresses this formula to his own "stubborn" community (Ezek 3:7), assigning them an ambivalently fluid role in the exodus narrative. See John Evans, *You Shall Know That I Am Yahweh: An Inner-Biblical Interpretation of Ezekiel's Recognition Formula* (University Park, Pa.: Eisenbrauns, 2019).

[29] Kohn, *New Heart*, 111.

in Ezekiel's understanding, equivalent to not having entered the land of promise in the first place.[30]

In this view, the land of the promise is not simply a destination point; it is a place of covenantal fidelity. This does not mean that the particularity of land is unimportant, but rather that living relationships with land require lived practices of commitment. Practitioners negotiate change and continuity as they move through time and space. One significant omission in Ezekiel's temple is the ark of the tabernacle—God's mobile dwelling place. It is an omission that suggests an end to the nomadic wandering of God's people. They are now at home.[31] What it does not suggest is that "home" and "movement" are contradictory categories. Movement has run through Ezekiel's vision from the start, and not simply through the witness of the miraculous river. The temple itself is now located outside the city walls, and the physical location of the destroyed city has shifted—its relocation implied by the new boundaries of land distribution in Ezekiel 48.[32] This place is not the place it was before. There is a cost to Ezekiel's emphasis on place's relational claim, and the cost is that place changes. It moves. It is not diffuse and all-encompassing. It is vulnerable to the forces of life and death, and it cannot be taken for granted. Neither can it be amassed. The fullness of place's promise is no geographical coordinate; it is revealed in practices of covenantal faithfulness.

Such practices require more than place's movement. They require the movement of God's people in acts of ongoing, committed relation, a movement to which the city's myriad gates and entrances testify (48:30–34). More fundamentally, such relational acts require the moving, restoring life of God. Ezekiel would not be present in this visionary place were it not for

30 Pikor argues, "It is not at all obvious that the first exodus ended in success, even if he speaks in 20:28 of YHWH's bringing Israel 'into the land that he swore to give them.' The prophet concludes his account of the new exodus in 20:42 the same way . . . to show that the Israelites have yet to enter the land and that they are still on the way there" (*The Land of Israel in the Book of Ezekiel* [New York: T&T Clark, 2018], 74–75).

31 *Pace* Ezek 43:7.

32 Margaret Odell notes how the tribal land portions indicate that the city Ezekiel observes is in a different location than that of the original Jerusalem. It now lies outside the holy portion of land that marks the tribal allocations' geographical center (*Ezekiel* [Macon, Ga.: Smyth and Helwys, 2005], 523). Cook describes the "tentative language" that Ezekiel uses in relation to this city, "which appears to rule out both Jerusalem and any historically literal example" (*Ezekiel 38–48*, 125–26). See also Levenson's discussion of the typological location of Ezekiel's vision (*Theology of the Program of Restoration*, 37–49).

God's moving hand (40:1). From this vantage point, Ezekiel's temple needs no ark of the tabernacle because it has itself become an icon testifying to God's covenantal bond. The divine movement required for covenantal relationship may be quieter and less explicit than the movement of winged chariots—but it is necessary. Rootedness implies responsiveness.[33]

Like the MCIF's New Exodus' theology of *vanua*, Ezekiel's theology of place insists that negotiations of continuity and change are necessary in living relationships, be those relationships with persons, with places, or with the divine. These negotiations are glimpsed in his vision's harmonic imagery of temple, chariot, river, and land—but also in Ezekiel's moving relationship with the theological and narrative traditions of his people. Ezekiel's navigation of change and continuity undergirds his interpretation of his community's exodus traditions, moving the place of tradition and text into covenantal relation with the place of exile.

Place and Time: Embracing a Formational Covenant

For the Fijian Methodist New Exodus, the movement of place—its trajectory and its becoming—begs the question of place's relation to time. For those who have endured the erasures of colonial history, do the hopes of the community lie in the recreation of a pre-colonial place, or does place's hope lie on the shores of some unknown future? What, in other words, is the New Exodus' destination? For Ezekiel, the temple vision also raises fraught, time-related questions. Is this a vision of a) some eschatological utopia beyond time's horizon, b) a timeless, theological allegory, or c) a vision meant to impact the material choices of a community in a recognizable world?[34] Negotiating change and continuity makes little sense if place's telos and function are undetermined, which is why scholars like Kalinda Rose Stevenson foreground the question of function in Ezekiel's visionary descriptions. What is the urgency to which these texts respond—and what do they hope to do?[35] It is a

33 The fluidity between the temple sanctuary and God's mobile presence is alluded to in Ezek 11:16: "I have been a sanctuary to them for a little while in the countries where they have gone." See Joyce's helpful discussion of various constructive interpretations for the tabernacle's omission for Ezekiel's theology in "Temple and Worship," 150–51.

34 Stephen L. Cook's introduction to *Prophecy and Apocalypticism: The Postexilic Social Setting* (Minneapolis: Fortress, 1995) outlines some of the differences between proto-apocalyptic and apocalyptic genres, noting the sea changes in how scholars have interpreted the relation between Ezekiel's temple vision and time's horizon.

35 Kalinda Rose Stevenson, *The Vision of Transformation: The Territorial Rhetoric of Ezekiel 40–48* (Atlanta: SBL, 1996).

rhetor's question. But more specifically, it is a preacher's question. In the time-bound event of the sermon, the implications of a given text for a particular, emplaced community can only be discerned in light of a preacher's orientation toward history and her understanding of preaching's place in time's flow.

Stevenson quickly dismisses two possibilities, the first being that Ezekiel's temple is "a construction blueprint for the post-exilic temple."[36] There is enough left out of these descriptions, for all of their measurements, to argue against the vision functioning as a surveyor's map. Unlike Exodus' instruction to build the tabernacle (Exod 25:1), Ezekiel is not called to construct this temple, nor does he instruct his people to do so. He is called to declare all that he sees as he walks through it, suggesting something more akin to proclamation than an architectural plan. But Stevenson also dismisses the possibility that the temple is "only a literary creation, a symbol for the universal presence of God."[37] There is more at stake in this vision than "spiritual realities."[38] Indeed, the dichotomy between the spiritual and the material is undermined by the vision's own description of holiness' flow into everyday life.[39] For Stevenson, there are constructive implications for this revisioning of temple and place. Because societies are formed by space and place, "to change the way a society is organized in space is to change the society."[40] Ezekiel's revisioning of place addresses the profound questions of exilic loss and the "spatial abuse" of imperial

[36] For a writer who takes this prospect seriously, see Ronald E. Clements, *Ezekiel* (Louisville: Westminster John Knox, 1996).

[37] Stevenson lifts up Steven S. Tuell's assertion that the temple represents a "present, on-going reality" free from "historical and geographic limitations" as an example of this perspective: "The Temple Vision of Ezekiel 40–48: A Program of Restoration?" *Proceedings of the Eastern Great Lakes Biblical Society* 2 (1982): 96–103.

[38] Though Daniel I. Block acknowledges that the restoration of Ezekiel's people would have been inconceivable for the prophet without the restoration of land, he distances himself from Stevenson's "territorial" interpretation of the book, arguing for an "ideational" reading that prioritizes "spiritual" rather than "physical" realities (*Beyond the River Chebar: Studies in Kingship and Eschatology in the Book of Ezekiel* [Eugene, Ore.: Cascade, 2013], 171–72).

[39] Cook problematizes this dichotomy between the spiritual and concrete further, noting the ways that literary symbols themselves can serve powerful political, material functions. He notes the long history of "utopian literature" challenging the status quo in revolutionary ways—a theme I will return to later in this chapter. Thomas More's *Utopia*, for example, "was not futuristic or eschatological, but an engaging literary critique of contemporaneous England" (*Ezekiel 38–48*, 5–6).

[40] Stevenson, *Vision of Transformation*, 151.

policies, but it also makes possible a future that has not yet come to pass: "a post-exilic Israel . . . organized around a temple, without a human king."[41]

This book affirms much of Stevenson's point. Place has social significance, shaping the political and economic possibilities of a given culture. It is a point I will turn to later in this chapter. But I would broaden her canvas. I think more is going on in Ezekiel's final chapters than a pitch for anti-monarchical theocracy. Perhaps it is the *iTaukei* dialectic between "time in front" and "time behind"[42] resonating in my interpretive ear, but I cannot ignore the ways that Ezekiel's vision of an alternate future is rooted in his recontextualizing of an ancient past.[43] Place's political and economic implications grow out of a fundamental reclaiming of place's role in shaping what it means to be human. Similar to Terence Fretheim's insistence that Exodus is crucially concerned with God's act of creation,[44] Ezekiel's reframing of the exodus account does deep theological work around what it means to be a creature in a world created by God, embedding those insights into his vision of Israel's restoration.

The Edenic references in Ezekiel's life-sustaining river and healing trees are vibrant, even if these images have transformed in their retelling. A single tree of life is now a stand of trees on the banks of a river (Ezek 47:12). Fruit is provided each month, a fecundity that surpasses the vegetal cycles of Eden's flora—and their leaves bring healing.[45] Stephen Cook sees Ezekiel 47's verdant imagery foreshadowed in Ezekiel 34–36, which describes an intimacy between God, humans, and nonhuman creation in a "garden-world where God might step out from the temple's safety, . . . enjoying the forests and trees (Ezek 34:25)."[46]

[41] Stevenson notes, "It is impossible to know how much this changed society was the result of Persian imperial practice and how much the result of . . . priestly writing, but the radical change in the social structure imagined by this vision actually occurred. There was a new Israel, a new temple, and no king" (*Vision of Transformation*, 153).

[42] Referencing Akanisi Sobusobu Tarabe's literal translation of the *iTaukei* words for "future" and "past," as discussed in chapter 4, n. 40.

[43] Ellen Davis references Ezekiel's "archival mind" in *Swallowing the Scroll: Textuality and the Dynamics of Discourse in Ezekiel's Prophecy* (Sheffield: Almond Press, 1989), 84.

[44] Terence Fretheim argues that in Exodus "redemption is in the service of a creational end": *God and World in the Old Testament: A Relational Theology of Creation* (Nashville: Abingdon, 2005), 125.

[45] Pikor compares and contrasts Eden and Ezekiel's temple vision (*Land of Israel*, 189).

[46] Stephen L. Cook, "Burgeoning Holiness: Fecundity Let Loose in Ezekiel 34–36," in *Ezekiel: Current Debates and Future Directions*, ed. William A. Tooman and Penelope Barter (Tübingen: Mohr Siebeck, 2017), 345–59 [357].

Creation imagery is, in fact, prolific enough that one wonders what is at stake in the arguments of scholars who maintain that "creation traditions are not important for Ezekiel's theological argument."[47] It is certainly not God's character as a creator that is in question. Ezekiel prophesies God's radical "recreation" of God's people (36:26).[48] David Peterson worries, however, that the language of "creation traditions" implies continuity between God's "radical" re-creating work and the emplaced stasis of Zion theology. He sees rupture in God's new creation rather than a cyclical return to preexilic past. His argument presses against John T. Strong's emphasis on the continuities between Ezekiel's vision of Israel's restoration and Zionist theologies prioritizing Jerusalem and the temple.[49] It is finally re-creation's relation to time that is at stake. Does God's new creation necessitate a radical break within history—or is God's new creation the fulfillment of a process already underway?

What strikes me about both of these options is their strong emphasis on fulfillment and finality—rather than process and formation. Banivanua warns that a church that sees itself as fully formed, rather than part of God's ongoing formational, re-creating work, is a danger to itself and others. The New Exodus understanding of place is more than cyclical; it has a trajectory. But it is also "voyaging."[50] Its goal is more than arrival. In light of place's relational, covenantal function, its goal is the formation of a people. This is one of the reasons why the well-being of land and the organization of place are so crucial to Ezekiel. Place reveals who we are, even as it forms in us the connections necessary for relational commitment. And place's organization can malform faithfulness as deftly as it can shape it.

Regardless of Ezekiel's visionary distance from Babylon, he and his people are still physically located near the banks of the Chebar Canal when these chapters are recorded. It is difficult to imagine two more dissimilar streams of water. The Babylonian irrigation ditch has been dug by forced labor in an attempt to ration and control its water's flow.[51] Ezekiel

[47] David L. Petersen, "Creation and Hierarchy in Ezekiel: Methodological Perspectives and Theological Prospects," in Cook and Patton, *Ezekiel's Hierarchical World*, 169–78 [175].

[48] Petersen, "Creation and Hierarchy," 175–76.

[49] The outlines of Strong's disagreement with Petersen can be seen in "Cosmic Re-Creation and Ezekiel's Vocabulary," in Tooman and Barter, *Ezekiel*, 245–84.

[50] James Bhagwan, personal interview, Suva, January 7, 2023.

[51] Rebecca Copeland describes the ecological damage caused by Babylonian irrigation practices, particularly in relation to soil salination, in "'Their Leaves Shall Be for Healing': Ecological Trauma and Recovery in Ezekiel 47:1–12," *Biblical Theology Bulletin* 49, no. 4 (2019): 214–22.

47's visionary stream grows deep and wide through a miraculous overturning of the possible. Rationing and control are not the point. Exiles sit by the banks of the first (3:15); trees that echo the promises of Scripture line the banks of the second. The rivers are juxtaposed, even as they call each other to mind. One wonders, in fact, if Ezekiel's vision has miraculously transfigured those exiles into the psalmist's faithful "trees planted by streams of water" (Ps 1:3).

Ezekiel wades into the water bodily, and the text attends to the stages of his near immersion (Ezek 47:3–5). Significantly, this is no river crossing. Given all of the other allusions to the exodus journey throughout this passage, it is striking that the passage makes a point of the river's impassability. This is not a re-narration of the Red Sea journey, made possible by a pathway of dry ground through the waters. Neither does it describe the Jordan River, traversed to enter promise. This is a river that "could not be crossed" (47:5). Katrina Schaafsma sees the river as a conduit of embodied healing that "opens up space for trauma recovery," rather than as reenacting these territorial initiations.[52] The temple vision is not simply a picture of restoration; it is also a return to the site of Ezekiel's most painful vision: the leaving of God's glory from the Jerusalem sanctuary (11:23). Pain and hope swirl together in this imagery. Following Ezekiel's inaugural vision at the book's opening, he sits in speechless shock (3:15) on the banks of the Chebar River for seven days.[53] Rivers have not been easy places for the prophet. But Ezekiel's bodily encounter with the river of chapter 47 allows for water's re-narration. Former places of loss and grief are not the end of his story; nor are they the end of God's people's story.

These layers of meaning call to mind the harmonies of lament and hope that Luke Powery hallows in his study of the African American spiritual "Dem Dry Bones" and Ezekiel 37.[54] Similar chords resonate between the song "Wade in the Water," also an African American spiritual, and Ezekiel 47. The spiritual connects the wading of God's children with biblical images of Red Sea and Jordan crossings (Exod 14:21–22; Josh 3:14–17) and New Testament references to the "troubled," healing waters of the

[52] Katrina Schaafsma, "Follow the Body: How Play Therapy Makes Sense of Ezekiel's River (47:1–7)," unpublished paper presented to the Theological Perspectives on the Book of Ezekiel section, Society of Biblical Literature Annual Meeting, Boston, 2017, 2.

[53] Schaafsma, "Follow the Body," 10.

[54] Luke Powery, *Dem Dry Bones: Preaching, Death, and Hope* (Minneapolis: Fortress, 2012).

pool of Bethesda (John 5:7). But it also aches with the painful wisdom of previously enslaved fugitives who reminded each other that "wading in water" is a method of avoiding a slave-catcher's dogs.[55] In the song's fierce intertextuality, water becomes a site of freedom and healing, even as it refuses to forget traumatic escapes, and in the words of Tracy K. Smith, the "miraculous many gone."[56] Ezekiel's wading into God's sacred river testifies to a different way of relating to water than the conquest-dependent models of Babylon's agricultural enterprise.[57] But it also actualizes the undoing of the larger project of Babylonian hegemony through its intentional remembering and reframing of a painful history. The passage confronts imperial amnesia and determinism, two of an oppressive regime's most potent weapons. An empire that can erase the past will find it far easier to instill resignation in its subjects.

But more than pain is remembered. Ezekiel's entire body is relearning the core testimony of the Hebrew creation story: he was created in covenantal connection to life. This is formational work, unlearning the lessons of place as possession and remembering place as covenantal partner. Genesis 1 is rarely given credit for its interrelated portrait of the world and the agency it assigns to nonhuman creation.[58] But Ezekiel's tree-lined river underscores the point. These trees "bear" fruit and the river "swarms" with living things, as in Genesis, and the earth "brings forth" life and the waters "swarm" (Ezek 47:9; Gen 1:20).[59] Significantly, Ezekiel's word choice for the trees' "bearing" of fruit (Ezek 47:12) is unusual.

[55] Jacob Uiti, "Behind the Meaning of the Classic Gospel Song, 'Wade in the Water,'" *American Songwriter: The Craft of Music*, February 23, 2023, https://americansongwriter.com/behind-the-meaning-of-the-classic-gospel-song-wade-in-the-water/.

[56] Tracy K. Smith, "Wade in the Water," in *Wade in the Water* (Minneapolis: Graywolf Press, 2018), 15–16.

[57] Thorkild Jacobsen and Robert M. Adams argue that the irrigation practices of ancient civilizations like Babylon were ultimately unsustainable, leading to soil salinization: "Salt and Silt in Ancient Mesopotamian Agriculture," *Science* 128, no. 3334 (1958): 1251–58, quoted in Copeland, "Their Leaves Shall Be for Healing," 187.

[58] An exception is an Oceanic women's Bible study on Genesis 1 that I attended in 2016, sponsored by the Christian Talanoa Network that emphasized these very characteristics. See Jerusha Matsen Neal, "Decolonizing 'Dominion': Oceanic Methodist Women Read Genesis 1," in *Decoloniality and Methodism: A Commitment to Ecological Justice*, ed. Upolu Lumā Vaai (Oxford: Oxford Institute of Methodist Studies Publications, forthcoming 2024).

[59] This echo of Genesis' "swarming" waters is also seen in Exod 1:7. The "swarming" abundance of God's people was precisely the thing that Pharaoh found threatening.

The verb בכר (*b-k-r*) is only used once elsewhere in the Hebrew Bible, denoting childbearing (Lev 27:26). To connect such a verb with the trees' labor emphasizes the birth imagery throughout the passage.[60] Ezekiel doesn't simply pass through these waters. He is immersed and buoyed by the memory of a world created as a carefully woven community—a community in which he participates. This participation is not just symbolic; it is surprisingly concrete. Ezekiel's very center of gravity is impacted as he wades into the water of creation's flourishing. He is more than an observer; he is a part of creation's flow.

When Banivanua talks about what it would mean to undo the impact of colonization on the mindset and spiritual practices of the church, he also draws on the imagery of birth. Wilderness formation is more than an unlearning of colonial objectifications and violence. It is a "dusty womb" where God's people remember who they are.[61] For the leaders of the New Exodus, the exodus narratives speak to the Fijian people's history as Indigenous navigators, bound in fidelity to a relational world. A similar creative act occurs in Ezekiel's river immersion. Place reminds him who he is in God's creation story—not a cog in a colonial wheel, but a person who belongs. He is not cut off from place; he lives in relation to it. He is even carried by it. This insight has implications for the Israelite people. When Ezekiel "[declares] all that [he sees] to the house of Israel" (Ezek 40:4), he is proclaiming more than a reiteration of the past or a reimagining of the future. He proclaims a formational reorientation to the present through a reclaimed covenantal ontology.[62] The bone-rattling resurrection of Ezekiel 37 may seem a long distance from Ezekiel 47's currents. But these texts provide a paired vision of new creation, holding rupture alongside

60 The connection between trees and birth also resonates with the burying of the umbilical cord under a tree in *iTaukei* culture. See also Fulata Lusungu Moyo, "'Ukugqiba inkaba'—Burying the Umbilical Cord: An African Indigenous Ecofeminist Perspective on Incarnation," in *Planetary Solidarity: Global Women's Voices on Christian Doctrine and Climate Justice*, ed. Grace Ji-Sun Kim and Hilda P. Koster (Minneapolis: Fortress, 2017), 179–92.

61 Tevita Nawadra Banivanua, "Kairos, God's Plan, and Wilderness Formation in the New Exodus," sermon preached at Davuilevu Theological College, Nausori, Fiji, November 2016.

62 Marvin A. Sweeney notes that, for all of Ezekiel's theocentricity, the book does more than give witness to Ezekiel's testimony of God. It also gives witness to God's creation and preparation of Ezekiel: "The Ezekiel That G-d Creates," in *The God Ezekiel Creates*, ed. Paul Joyce and Dalit Rom-Shiloni (London: T&T Clark, 2015), 150–61.

formational flow in describing how God's miraculous gift of the people's "new heart" (36:26) is given.[63]

Creation theologies that are overly invested in pristine pasts or utopian futures have been leveraged by the colonial project for centuries.[64] Richard Davis describes the dangerous eschatologies of settler-colonial theology in particular.[65] The desire to fix, change, and rearrange the world—even in an attempt to undo the sins of the past—continues to define hope on its own terms and place as an object to control. But the embrace of creation in Ezekiel 47 reveals a humbler vision. Place is not the resolution to or the origin of God's story. It is the site of God's ongoing formation of a people remembering who they are and navigating where they are headed in time's current.

Place and Divine Presence: Discerning a Sacramental[66] Covenant

Place's relational, formative character speaks to the significance of its role in a covenantally connected world, a key resonance between the theology of place found in the Fijian New Exodus and the prophet's revision of the exodus in his temple vision. It is a vision that emphasizes creation's "dynamic interconnectivity . . . [where] life within every corner of the lattice is precious, to be uplifted and ennobled."[67] But place's covenantal significance does not only impact and illumine creaturely relationships. In Ezekiel, place mediates creation's relationship with God. Cook uses the language of "real presence" in his description of God's filling of the

63 As different as they appear, the two chapters are more than complements. They also amplify each other's emphasis on interconnection. They resonate. The tendons and sinews that reconnect Ezek 37's bones foreshadow a restored web of creation in Ezek 47.

64 See, for example, Susan Bruce, "Utopian Justifications: More's 'Utopia,' Settler Colonialism, and Contemporary Ecocritical Concerns," *College Literature* 42, no. 1 (2015): 23–43.

65 Richard A. Davis, "Decolonial Settler Theology: A *Pakeha* Theological Proposition," public lecture, Wesley House Cambridge, January 26, 2023, https://www.youtube.com/watch?v=kzWNYjHANpY.

66 I am guided by Pikor's broad, nonsectarian understanding of this term in describing Ezekiel's understanding of the land's role in covenantal encounter. For Pikor, the term connotes the "sacralizing and revitalizing role of the land in Israel's relationship with YHWH," a mediatory place of "communion" that is part of YHWH's plan for self-manifestation in the world (*Land of Israel*, 113–14). See also Zimmerli, *Ezekiel*, 1:261.

67 Stephen L. Cook, "Ezekiel's Recovery of Premonarchic, Tribal Israel," in Tooman and Barter, *Ezekiel*, 360–76 [365].

temple,[68] which makes place more than covenant's guarantee. It is, more accurately, a site of covenantal encounter. For Ezekiel, the goal of covenantal fidelity is not land. The goal is knowledge of and participation in the divine presence, mediated through the place of God's promise. For scholars like Pikor, this sacramental function gives Israel's land a "theophanic character." It is not an object to possess, but rather "a key element in Israel's relationship with YHWH."[69] More specifically, it is a key element in Ezekiel's understanding of the "covenant of peace" (Ezek 34:25; 37:27) that includes "not only . . . the good that the land and the people can partake of, but also God's presence among [God's] people."[70]

I am reticent to use the word *sacramental* in relation to Ezekiel's theology. The word is anachronistic and carries with it centuries of obscure debates on the mechanisms and manners of mediatory signs. But the word *sacred* is not performative enough for the tensive, tender relationship between God and place in Ezekiel. The land is more than a sacred symbol—or even a sacred object unto itself. It is place of "communion."[71] This relational quality underscores the crisis of exile for those who took Israel's right to the land and God's localized presence in the Jerusalem temple for granted. Ezekiel's mobile vision of God's glory (ch. 1) and God's removal of that glory from the temple (chs. 8–11) are chilling reminders that true relationship entails freedom—the freedom to stay and freedom to go. And yet, in spite of God's freedom from material space, Ezekiel emphasizes God's faithful commitment to Israel's place.[72] The book of Ezekiel carefully balances the "transcendent and the territorial" in its depiction of divine relation to land and temple.[73] In the book's final chapters,

[68] Cook, "Ezekiel's Recovery of Premonarchic, Tribal Israel," 365.

[69] Pikor, *Land of Israel*, 3.

[70] Pikor, *Land of Israel*, 230. The multivalent meaning of *shalom* in this covenantal context echoes the multivalence of Ilaitia S. Tuwere's description of the *iTaukei* concept of *sautu* (i.e., peace, well-being, spiritual intimacy, robust relations) as described in his *Vanua: A Fijian Theology of Place* (Suva: Institute of Pacific Studies Press, 2001), 107, 130, 147.

[71] Pikor, *Land of Israel*, 114.

[72] God's glory may have followed the people into exile, but Ezekiel's fierce protest against Edom's claiming of Israel's mountains is based on the fact that "the LORD was there" (Ezek 35:10). This is no empty land. In a meaningful way, God's presence does not abandon the place of God's promise.

[73] Cook, *Ezekiel 38–48*, 159. Elsewhere, Cook suggests that "the being of glory may entail something akin to a 'hypostatic union' . . . paradoxically, both fully localized and transcendent," in "Ezekiel's God Incarnate!" in *The God Ezekiel Creates*, ed. Paul Joyce and Dalit Rom-Shiloni (London: T&T Clark, 2015), 107–31 [141]. Brittany Wilson

God dwells among God's people, permeating the land and revealing God's character and intention through it.[74] Place functions as a site of revelatory relation between God and God's creation.

It is worth pausing to reflect on the radicalness of that statement for many theological streams in Western thought. Ezekiel stands at a great distance from contractual or transactional understandings of covenant that objectify nonhuman creation and place. He is also far from schools that would use the language of "covenantal theology" as a "code word" to indicate a bifurcation between theologies that center place and theologies that center time.[75] Finally, he has pressed past the romantic idea that covenant is about an experience of intimacy alone—as if place and its covenantal ties were only about sentiment and connection. In Ezekiel's framing of the exodus' goal, the promised land is a place where covenantal relationships grow from and mediate a revelation of God. This mediation is not exclusive, of course. Place does not domesticate divine freedom. Place does, however, become a means through which YHWH calls God's people to covenantal responsibility and repentance, just as it is a place where covenantal responsibility is made manifest.

The river in Ezekiel 47 is not Godself; the text makes clear that the temple is God's dwelling place, and God is described as inhabiting the temple in concrete ways.[76] But it is also more than metaphor.[77] The river

---

traces this "localized and transcendent" quality of divinity in the Hebrew scriptures into New Testament descriptions of divine embodiment in *The Embodied God: Seeing the Divine in Luke-Acts and the Early Church* (Oxford: Oxford University Press, 2021).

74 Pikor argues that God's sanctity is manifested "in the land," giving it a "sacral character" (*Land of Israel*, 231).

75 Walter Brueggemann describes how the language of "covenantal theology" was leveraged by twentieth-century biblical scholars to codify an "either/or" dichotomy between theologies that emphasize place and those that emphasize time, in "The Loss and Recovery of Creation in Old Testament Theology," *Theology Today* 53, no. 2 (1996): 177–90 [184]. In other words, "covenantal theology" became shorthand for theologies that emphasized God's "redemption history" over against theologies grounded in creation—a distinction deconstructed by Ezekiel's vision of creation's redemption and covenantal role.

76 For example, Ezek 43:7 gives a strikingly anthropomorphic description of God's "feet" within the halls of the temple. Cook notes that such embodied, localized imagery is characteristic of Zadokite spirituality (*Ezekiel 38–48*, 186–87), even as Joyce makes clear that YHWH is not "shut up in God's house" like a bird in a cage: "Temple and Worship," 155.

77 Pikor uses the term "metaphor" to describe Ezekiel's river, with YHWH as the "metaphorized element" (*Land of Israel*, 189). But I find the term insufficient. The

speaks to God's deep desire for the world, and it manifests those desires through its bearing of God's holiness in and through the land. There are ethical implications to that manifestation. It is significant that Ezekiel's text does not end with a visionary river and trees of healing. Immediately following the prophet's watery immersion, the text describes redistributions of land and reallocations of power (47:13–48:35). The book ends with concrete relinquishments of autonomy in the embrace of a collective dependence.[78] The river may be a source of healing, but it is also a rebuttal of imperial logics of control and imperial models of dominion. Similar to Mount Sinai's fiery cloud (Exod 19:18–23), the river embodies a sobering call to repentance and commitment—a recurring characteristic of God's revelation. "Revelation" is, after all, the English translation of the Greek *apokalypsis*, in all its world-upturning exposure. One does not experience divine communion without also experiencing an unveiling of Godself and God's unmasking the world's creation-enslaving powers. One does not experience divine communion without God's unveiling of where one stands in relation to God's covenantal call.

Ezekiel and the New Exodus of the Fijian Methodist Church disrupt many binaries in their proclamation. These displaced prophets navigate between continuity and change, memory and hope, and freedom and fidelity, reclaiming relational understandings of place, narrative, creation, and God. For many in the Western theological academy, such covenantal wayfaring disrupts a different binary still—a contemporary binary between *place* and *apocalypse* in framings of God's self-manifestation in the world. Theologies that center emplaced revelation regularly stress God's immanence in institutions, in ritual, in communal traditions, and in cyclical continuities. They celebrate the gift of rootedness, even in seasons of dislocation.[79] Noting the hybridity of places and practices, Sally A.

---

passage describes more of an actualized solidarity between YHWH's holiness and the river. The river's water causes life "wherever [it] goes" (Ezek 47:9).

78 While some scholars have understood Ezek 40–48 as Zadokite propaganda (see, for example, Ezek 44:15–16), Cook notes the significant restrictions that these chapters place on priestly power. The restriction on priests possessing land is "one of several pieces of evidence against a hypothesis of sectarian self-promotion" and suggests instead an interdependent "balancing of power between the priests, the Levites, the ruling chieftain and the lay tribes" (*Ezekiel 38–48*, 222).

79 For example, John Inge's influential reflection on "holy places" emphasizes the continuities between place and revelation, finding "roots" even in pilgrimage's transitory passages (*A Christian Theology of Place* [New York: Routledge, 2003], 100). Aden Cotterill wonders if Inge has fully accounted for the discontinuities

Brown notes that one danger of emplaced theologies is that "what strikes [one] as 'obviously Christian' may have more to do with unconscious assumptions about what is appropriate in the eyes of a dominant class, ethnicity or cultural group."[80] They can, in other words, become insular. On the other hand, theologies that center apocalyptic revelation (certain contemporary strands of neoorthodoxy, for example, or pneumatologies that efface the significance of materiality) emphasize God's radical otherness, the divine inbreaking of history, and God's call to God's people to leave the places they know behind. While associated with visions of cataclysm at the world's end,[81] apocalyptic understandings of revelation impact theology's parsing of the present in significant ways. Peter Leithart notes, "Apocalyptic theology disturbs settled opinion, resists fixities of all sorts, and demands existential response. . . . It attacks dogma, often seen as an illegitimate attempt to tame a God too lively to be confined in categories, a God who . . . breaks in to shatter every pathetic, little intellectual and cultural structure we feebly erect."[82] Such a framing can mute place's significance—or even set it in opposition to God's self-revelation. Biblical studies and theology have worked to bring these distinct streams into

---

between place and pilgrimage, asserting the distinct insights of dislocation in the spiritual life: "The Dialectics of Place and Pastoral Practices," *Ecclesiology* 19, no. 2 (2023): 186–204 [198–99]. Stanley Hauerwas extends Inge's reflections to call pastors and priests to the "slow, hard, rewarding" work of remaining in a place: *War and the American Difference: Theological Reflections on Violence and National Identity* (Grand Rapids: Baker, 2011), 162–63.

80 Sally A. Brown, *Sunday's Sermon for Monday's World: Preaching to Shape Daring Witness* (Grand Rapids: Eerdmans, 2020), 17. Oliver O'Donovan and Joan Lockwood O'Donovan note another concern. While celebrating place's importance for theology, they worry about uncritical conflations of place and theology in relation to political speech. It is possible for the church to appropriate "prevailing local forms of idolatrous and worldly communication" in service to emplaced commitments: *Bonds of Imperfection: Christian Politics, Past and Present* (Grand Rapids: Eerdmans, 2004), 291.

81 This emphasis is seen in popular cultural vernaculars, but also through the precise, scholarly definitions of apocalypticism as a biblical genre which "disclos[es] a transcendent reality which is both temporal, insofar as it envisages eschatological salvation, and spatial, insofar as it involves another supernatural world" (John J. Collins, *The Apocalyptic Imagination: An Introduction to Jewish Apocalyptic Literature*, 2nd ed. [Grand Rapids: Eerdmans, 1998], 5). This transcendent reality broadly testifies to "the imminent end of the ongoing struggle between good and evil in history" (see John J. Collins, ed., *The Origins of Apocalypticism in Judaism and Christianity*, vol. 1 of *Encyclopedia of Apocalypticism*, ed. Bernard McGinn, John J. Collins, and Stephen J. Stein [New York: Continuum, 1998], vii).

82 Peter Leithart, *Revelation 1–11* (London: T&T Clark, 2018), 59.

generative dialectic, but attempts at integration often remain at the level of abstraction.[83] It is in the crucible of practical theology where theoretical dialectics reveal their loyalties. Binaries often crystallize and harden in the homiletic classroom.

These two approaches to revelation have led to distinctly different practical theological models in relation to preaching. One attends to God's indwelling of the place of God's people, embracing the cadence of their cultural dialect and rooting down into the soil of their concerns. The other proclaims God's transcendent call, disrupting all territorial domestications. Despite homileticians who have worked to press beyond these reductions,[84] chapter 1 describes how difficult it has been for the white, mainline homiletic guild in North America to hold these approaches

83 Brueggemann describes, for example, the dialectic in twentieth-century conversations between "natural theology" and theologies that emphasize a "recital of YHWH's transformative activities": "Loss and Recovery," 189. Susan Eastman describes the reframing of apocalyptic theology in the work of J. Louis Martyn through the "incarnational logic implicit in Paul's message of the cross": "Apocalypse and Incarnation: The Participatory Logic of Paul's Gospel," in *Oneself in Another: Participation and Personhood in Pauline Theology* (Eugene, Ore.: Cascade, 2023), 3–20 [4]. Wesley Hill underscores Martyn's significance in reorienting apocalyptic theology's engagement of the world through the place of the cross in "Apocalyptic Theology," in *The New Cambridge Companion to Christian Doctrine*, ed. Michael Allen (Cambridge: Cambridge University Press, 2023), 315–29. These integrations and recategorizations of place and apocalypse are significant—but can be challenging to translate into preaching's embodied practices. R. David Nelson's discussion of the interdependence of apocalyptic interruptions of faith (*theologia crucis*) and mature theological thought (*theologia viatorum*) acknowledges that this dialectic has often been engaged as an issue of "epistemology" rather than "temperament" and concrete practice: "The Word of the Cross and Christian Theology: Paul's Theological Temperament for Today," *Theology Today* 75, no. 1 (2018): 64–76 [72].

84 Various homileticians have employed distinct strategies in their resistance to this binary. For example, James F. Kay's concept of "stereoscopic imagination" stresses a preacher's Spirit-gifted ability to interpret place through a cruciform apocalypse: "The Word of the Cross at the Turn of the Ages," *Interpretation* 53, no. 1 (1999): 44–56 [51]. Charles L. Campbell notes how form itself can disrupt expectation in his Christology of "grotesque hope": *The Scandal of the Gospel: Preaching and the Grotesque* (Louisville: Westminster John Knox, 2021). See also Donyelle McCray's work on "genre-bending": "Playing in Church: Insights from the Boundaries of the Sermon Genre," *Liturgy* 36, no. 2 (2021): 11–17; Lisa Thompson's negotiation of alterity and community in *Ingenuity: Preaching as an Outsider* (Nashville: Abingdon, 2018); and my own work on the relation between embodiment and pneumatology in Luke-Acts, *The Overshadowed Preacher: Mary, the Spirit, and the Labor of Proclamation* (Grand Rapids: Eerdmans, 2020).

in generative tension. Preaching practices often vacillate between the romanticizing of place and its dismissal. I do not attempt to trace that difficulty to a root cause, but I have noted how this vacillation serves the long-standing, rarely acknowledged investments of settler colonialism, asserting "native" connections to place at the same time that one asserts universal claims on the places of others. For the homiletic guild in which I participate, disrupting the binary between *place* and *apocalypse* would require a costly accounting of these investments and the bitter histories that haunt them—a critical first step toward more honest practices and pedagogies. Given the environmental catastrophes that are on their way, it would also press the guild toward a more costly reckoning with the contemporary climate crisis and more active solidarity with displaced communities that are even now proclaiming God's Word in places of loss.

If the witnesses of the displaced communities collected in this volume are to be believed, *place* and *apocalypse* are not as opposed as certain Western theological models have suggested. In their proclamation, apocalyptic crises of place have led to vivid revelations of what place was created to be: a site of covenantal encounter and transformation. Creation becomes a part of God's redemptive promise, mediating divine communion and witnessing to divine holiness. When God's presence returns to Ezekiel's visionary temple, the earth itself shines with God's glory (Ezek 43:2). Such encounters do not dismiss the emplaced value of communal practices or institutional structures. And yet they call forth resistance to reductive idolatries—be they within the community or without—that treat place as an object to exploit and control. More than this, they call for an active discernment of what righteousness in relation to place requires.[85]

Choctaw elder and retired Episcopal bishop Steven Charleston explains the relationship between emplaced history and apocalyptic revelation in his reflections on the North American annihilation of Indigenous nations. Charleston takes the systematic genocide of Indigenous nations with utmost seriousness—an "end of the world" for entire people groups. But he argues that in the midst of this historical catastrophe, a different sort of apocalypse emerged—an apocalypse of spiritual insight. "Apocalypse

[85] For other examples of how ancient apocalypses engaged place, see Liv Lied, *The Other Lands of Israel: Imaginations of the Land in 2 Baruch* (Leiden: Brill, 2008), and Hindy Najman, *Losing the Temple and Recovering the Future* (New York: Cambridge University Press, 2013). Both describe textured relationships between apocalyptic theologies and the investments of displaced communities in an emplaced future.

as revelation," he explains. "An uncovering . . . a vision."[86] In Charleston's view, such revelations are not disconnected from place. They engage an Indigenous "spiritual tradition as deep and rich as any culture on earth." They speak as those "immersed in tradition even as they talk about how that tradition will need to change to meet new apocalyptic challenges."[87] For Charleston, the apocalypse of the climate crisis makes these negotiations more crucial than ever. His description of the woven quality of place and apocalypse is reminiscent of Banivanua's frequent calls for spiritual discernment. For both leaders, place reveals and mediates God's presence, even as it testifies to a moving, relational God who calls the community toward an unknown future.[88] What if, both suggest, the most radical resistance one can mount in the face of environmental catastrophe and imperial aggression is a radical recommitment to place's sacramental function as a site of divine encounter? What might such a recommitment reveal—and how might it reorder the world?

Artist Cannupa Hanska Luger grew up on the Standing Rock Sioux Reservation. In the midst of the Standing Rock community's resistance to the Dakota Access Pipeline, Luger used art to create an apocalypse (i.e., a revealing) of that place and an unveiling of the forces that threatened it. The Mirror Shield Project[89] involved community members in making simple rectangles of wood and covering them with reflective Mylar. At its most basic level, the wood shielded Water Protectors from the rubber bullets of militarized law enforcement officers.[90] But the shields could also

86 Steven Charleston, *We Survived the End of the World: Lessons from Native America on Apocalypse and Hope* (Minneapolis: Broadleaf Books, 2023), 8.

87 Charleston, *We Survived the End of the World*, 10.

88 Tevita Nawadra Banivanua places spiritual discernment at the center of his address at the 2016 MCIF Annual Conference following the devastation of Cyclone Winston. "God flows into our existence when we least expect it," he preaches, ". . . [making] inexplicable intersections with our daily walks." One must be able to "know the difference between the voice of God, the voice of our own thoughts, the noise of the world and the voice of temptation" ("Signposts of the New Exodus," keynote address, Centenary Church, Suva, August 15, 2016).

89 Cannupa Hanska Luger, "The Mirror Shield Project (2016)," accessed October 15, 2023, https://www.cannupahanska.com/social-engagement/mirror-shield-project.

90 Barbara R. Rossing vividly describes how Standing Rock's Water Protectors (their preferred designation) were met with "militarized police violence and water cannons from privately hired security thugs": "Waters Cry Out: Water Protectors, Watershed Justice, and the Voice of the Waters," in *Decolonizing Ecotheology: Indigenous and Subaltern Challenges*, ed. S. Lily Mendoza and George Zachariah (Eugene, Ore.: Wipf and Stock, 2022), 39–57 [50].

be used in communal performances of solidarity. Rory Erler Wakemup filmed the Oceti Sakowin camp Protectors from above as they held the mirrored rectangles over their heads and moved in a spiral, imitating a dancing water serpent. The community shone like the water of the Missouri River that they were protecting.[91] The performance claimed the protestors' relational solidarity with nonhuman creation, grounding their resistance in a communal vision of life.

The mirror shields had another purpose as well. Inspired by a photograph of Ukrainian women holding up mirrors to the faces of riot police,[92] Luger wanted to create an artifact that could be held up to the faces of the heavily armed officers tasked with removing the Protectors from the site, so that the officers might see themselves. In the language that Charles Campbell uses to describe preaching as a model of opposition to the powers of an oppressive world, these mirrors not only "envisioned" an interconnected creation; they also "exposed" the long, multilayered violence at the heart of a project that put the water rights of Indigenous people at risk.[93] As objects incorporated into a communal dance, the mirror shields named place's significance. In confrontations with militarized law-enforcement officials, the mirror shields unveiled a pointed question: *where are you standing in this struggle?* Whom were these officers serving—and at what cost?

God's indwelling presence in Ezekiel's temple vision does something similar. God's holiness gifts a vision of a connected world, rooted in core memories of creation and the formative promise of creation being made new. But it also exposes approaches to place that threaten that vision, calling for concrete changes in how geographic, economic, and political structures are organized. As my discussion below will show, Babylonian

91 Cannupa Hanska Luger and Rory Erler Wakemup, "Water Serpent," video edited by Sheila Regan, Oceti Sakowin Camp, December 21, 2016, https://www.youtube.com/watch?v=78epQ8GspaQ.

92 Karla Cloete, "To Protect and Protest: Artist Cannupa Hanska Luger and the Mirror Shield Project," Arts Help, August 28, 2022, https://www.artshelp.com/mirror-shield-project/. See Dylan C. Robertson, "Kiev Protestors Hold Mirrors to Police—Literally—as New Confrontation Looms," *Christian Science Monitor*, January 17, 2014, https://www.csmonitor.com/World/Europe/2014/0117/Kiev-protesters-hold-mirrors-to-police-literally-as-new-confrontation-looms.

93 Charles L. Campbell, *The Word Before the Powers: An Ethic of Preaching* (Louisville: Westminster John Knox, 2002). Chapter 6 (pp. 105–27) deals explicitly with "exposing" and "envisioning" as strategies for confronting the powers through preaching.

hubris is implicated, but also the monarchical abuses of preexilic Judah. These two moves, envisioning and exposing, are dependent on each other. Ezekiel's vision serves as a mirror that helps his community better see their external oppressors. But it also allows them to see themselves, even when they would rather not.

In contexts where visions of divine restoration are regularly used to mute human responsibility for the environment, Ezekiel 40–48 can seem a dangerously beautiful siren song, lulling to sleep a world on fire. Scholars like Faafetai Aiava note the frustration of environmental activists interested in concrete legislative action with those who stress the "spiritual dimension of ecology."[94] From Aiava's Samoan perspective, however, the spiritual and the social are "a wedded pair." Naming creation's covenantal relation and its role in mediating the sacred is not "a return to . . . mysticism." It lays the groundwork for an ethic that supports a "life-world where everything is interconnected."[95] The second half of this chapter will connect Ezekiel's ontology of place—and its relational, formational, and sacramental character—with ethical implications of those affirmations. For Ezekiel, emplaced covenantal commitment is a site of radical exposure, unveiling false hopes and false righteousness. It is a site of overturned empires and upended expectations, where God's people discover a new creation they could not construct on their own. Significantly, it is also a place where Ezekiel returns to the concrete ethical questions of the displaced communities described in this project: questions of *sovereignty*, *belonging*, and *loss*.

Sovereignty and the Sharing of Responsibility

The significance of the carefully recorded date of Ezekiel's temple vision—"at the beginning of the year, on the tenth of the month" (Ezek 40:1)—has presented a puzzle for scholars. If the beginning of the year is measured by the Babylonian calendar, the date may indicate the season of Passover, a time for remembering God's restoration of the people and recommitting to God's covenant.[96] Moshe Greenberg also suggests the Day of Atonement as an option, emphasizing its call to solemn repentance.[97] Both arguments have theological coherence. Kalinda Rose Stevenson, however, suggests

[94] Faafetai Aiava, "Eleele Interrupts the Eden Wedding: From Mother Earth to Mistress," in Mendoza and Zachariah, *Decolonizing Ecotheology*, 93–107 [104].

[95] Aiava, "Eleele Interrupts the Eden Wedding," 104.

[96] Blenkinsopp, *Ezekiel*, 200.

[97] Moshe Greenberg, "The Design and Themes of Ezekiel's Program of Restoration," *Interpretation* 38, no. 2 (1984): 181–208 [190].

a different possibility. She notes that the Babylonian New Year celebrates a very specific festival: the annual enthronement of the Babylonian god, Marduk. In the *akitu* festival, Marduk enters the city and processes to the temple of his residence, entering the sanctuary on the "tenth day of the month."[98] The resonances between this annual ritual and YHWH's return to Ezekiel's visionary temple (43:1–7) are clear for Stevenson—but so is one significant difference. In the Babylonian celebration, a human king leads the divinity into the temple. But in Ezekiel's reframing, "there is no human king. . . . [Israel's] god comes alone."[99] Similar to how Genesis 1 draws on and distinguishes itself from the Babylonian creation myth in order to critique imperial ideology, Stevenson sees Ezekiel 43:1–7 as an assertion of YHWH's superiority over the Babylonian deity. More than this, she understands the passage as a broader critique of the preexilic Judean monarchy.

Given Ezekiel's privileged position as a Zadokite priest in his nation's preexilic order, this critique of monarchical power seems strange.[100] Ezekiel likely benefited from the monarchical model. The reestablishment of the monarchy seems the clearest way for a defeated, exiled nation to reassert its sovereignty among the nations. But Ezekiel's vision of restoration does not transfer Babylonian models of power to Israel's land; nor does it reinstate the political structure operative when the city of Jerusalem fell. Instead, it reimagines political sovereignty in light of YHWH's kingship—and in so doing, decenters prior hierarchies of governance, creating a "dynamic interdependency" between the center and peripheries of power.[101] Cook argues that this political vision is grounded in local responsibility for land, local commitment to kinship ties, and "covenantal

[98] Stevenson, *Vision of Transformation*, 52.

[99] Stevenson, *Vision of Transformation*, 53.

[100] Andrew Mein emphasizes the "privileged class status" of the Zadokite priesthood prior to exile (*Ezekiel and the Ethics of Exile* [Oxford: Oxford University Press, 2001]). It is possible that Ezekiel's critique of the monarchy is also a negotiation of Persian policies that allowed religious institutions to be restored with significant autonomy, but not royal houses. My thanks to Dr. Laura Lieber for this insight. Leo Perdue and Warren Carter describe how Persian kings projected the image of religious tolerance to prevent revolt, in *Israel and Empire: A Postcolonial History of Israel and Early Judaism*, ed. Coleman Baker (London: Bloomsbury, 2015), 109ff.

[101] Cook, "Ezekiel's Recovery of Premonarchic, Tribal Israel," 365. I agree with the general trajectory of Cook's argument that Ezekiel is wanting to "push back the clock" (360) to premonarchic, tribal models of social organization, but think his language of "repristination" (364) is overstated. Ezekiel continues to innovate, even as he reclaims the relevance of former models of governance and land tenure.

ethics" that counter the tendencies for property and power to be "concentrated in the hands of a few."[102]

Ezekiel has made his disdain for the "princes of Israel" clear. Ezekiel 22 calls out these leaders for bloodshed, bribery, and the extortion of the "alien" (vv. 6–8). He also speaks of their willingness to "[destroy] lives to get dishonest gain" (v. 27), suggesting unjust administrations of the death penalty. Ezekiel 45:9 indicates that the monarchy has evicted God's people from their land through violent means, and 43:7–9 suggests a kind of idolatrous worship of Israel's kings after their death. Cook adds "the imposition of military service and forced labor" to the list, noting the ways that "respect and empowerment do not easily flow down to the outer layers of society in a centralized monarchic state."[103] Such models of sovereignty—models based on the unchecked power of a human ruler—are particularly ill-equipped to foster care for the land. In Cook's view, "monarchic systems . . . tend to divest land of its sacral character, focusing on militarizing state capitals and fortifying royal cities."[104] Such emphases regularly lead to the exploitation of human and nonhuman creation alike.

In establishing YHWH as sovereign, Ezekiel changes the ways that the sovereignty of God's people and God's land are understood. The power of the ruler is limited, and responsibility for the land is localized and shared. The ruler of Ezekiel's visionary city is not named "king" but *nasi* (prince or chieftain), and while that ruler does have a role to play, the role is clearly bounded. The ruler's land is not adjacent to the temple; nor does the ruler preside over a state cult.[105] The fields adjacent to the city belong to "the workers of the city . . . from all the tribes of Israel" (48:17–19) rather than serving as the ruler's private domain. Crucially, the ruler is no longer able to "violate covenantal guarantees of land" through force.[106] Evictions and land grabs are prohibited (45:8–9; 46:18).

But Ezekiel's model of human sovereignty in relation to a divine sovereign does more than limit the *nasi*'s power. It also redistributes land in

[102] Cook, "Ezekiel's Recovery of Premonarchic, Tribal Israel," 362. Cook and Corrine L. Patton note that, for all of Ezekiel's emphases on hierarchies of purity, the prophet "problematizes all notions of hierarchy in light of a divine sovereign" ("Hierarchical Thinking," 19).

[103] Cook, "Ezekiel's Recovery of Premonarchic, Tribal Israel," 362, 367.

[104] Cook, "Ezekiel's Recovery of Premonarchic, Tribal Israel," 367.

[105] Zimmerli notes that the temple is purged of all royal references (*Ezekiel*, 2:418).

[106] Walter Brueggemann, *Interpretation and Obedience: From Faithful Reading to Faithful Living* (Minneapolis: Fortress, 1991), 252.

ways that emphasize a shared responsibility and shared voice for the community. The land divisions described in 47:13–48:35 are, in one reading, "artificial" and idealized.[107] In conversation with Joshua 13–19, Ezekiel's land allotments are equally designated among each of the twelve tribes, regardless of the tribe's size or status. Though never enacted, these boundaries are materially significant in their critique of preexilic practices of land hoarding. In redistributing equal portions of land to each tribe, there is an economic and ecological modeling of one of the core principles of God's covenantal commitment to the exiled community: "the worth and holiness of all God's people."[108] Shared political decision-making and shared care for the land fleshes out the relationality in Ezekiel's theology of place, creating postures of interdependence between land and community that thwart the "massive flow of goods and services from the many to the few."[109] For those with ears to hear, Ezekiel's vision of shared sovereignty raises pointed questions about inequitable distributions of place and power in today's climate-justice movement.

Fijian Methodist pastors regularly insist on the significance of social models of interdependency in addressing environmental threat—even when these models are embodied at a village level. Rev. Samuela Delai uses the traditional *iTaukei* social structure in a tribe (*yavusa*) as a way to describe the dependent connections between human and nonhuman creation. "There are seven clans of *matagali* [in a tribe]," he explains, each with a unique, traditional role. Whether chiefs (*turaga*), chief-makers (*sauturaga*), heralds (*matanivanua*), priests (*bete*), warriors (*bati*), fishermen (*gonedau*), or tradesmen (*mataisau*), they all have a function. Each depends on the other, and none are expendable. Similarly, he argues, creation is interconnected, with each of God's creatures playing a significant function. "The problem of the world today is that humanity considers themselves superior to other creations, causing abuse and breaching the systems that connect us," he states. In case I've missed the point, he breaks it down: "[Humanity] is as equally important as trees."[110] Delai grounds

107 Brueggemann, *Interpretation and Obedience*, 253.

108 Cook and Patton, "Hierarchical Thinking," 16. This worthiness has political implications. Ezekiel's land distribution reintegrates the northern and southern kingdoms, in spite of their history of national fracture.

109 Gerhard Lenski particularly notes the role of this priesthood in mitigating abuses and excesses in this flow: *Power and Privilege: A Theory of Social Stratification* (New York: McGraw-Hill, 1966), 266.

110 Samuela Delai, personal email correspondence, June 29, 2023.

his theological argument in the relationality of the Trinity, but he might have also grounded it in Ezekiel's attention to how the restoration of land and restoration of social relationships connect. As a pastor serving an Indigenous community, he moves fluidly from a relational theological claim to the linking of relational social structures with relational models between human and nonhuman creation. Sovereignty, in this case, means something different than atomized power. It implies the gift of responsibility in community.

Such models of shared sovereignty are increasingly significant given the realities of Fijian village relocation. When the village of Cogea contacted the Fiji Council of Social Services (FCOSS), it was the community's sovereignty that was at stake. They wanted the village to have voice, responsibility, and decision-making agency in their relocation process. And they had reason to be concerned. The government at the time had taken a heavy hand in the relocations of other communities, with less than successful results.[111] One village had been quickly rebuilt in the outdated manner of housing barracks, with kitchen spaces excluded from the building plans. Vani Catanasiga, the FCOSS executive director, notes that there were no women in that planning process. Catanasiga also describes another government-sponsored relocation that moved a village onto land susceptible to landslides, apparently ignorant of the findings of the technical survey data.[112] It was not long before a second natural disaster hit the new site, requiring the village to move again. Still another village was split in two by government efforts at rebuilding, half the community being relocated away from their neighbors.[113] Cogea wondered what it would mean for the government to move "off-center"[114] in the crucial decisions

[111] FCOSS executive director Vani Catanasiga suggests that the Bainimarama government of the time—though publicly supportive of climate resilience efforts on the international stage—was often more interested in photo opportunities than community engagement (personal interview, Suva, January 11, 2023).

[112] Catanasiga names the example of Tukuraki in the Ba province as an example of how the government's standard plans for home construction were not adequate to the concerns raised by geo-tech surveys of the "waterlogged" building site (personal interview, January 11, 2023). In later correspondence, she reiterates, "The landslide occurred within meters of the newly constructed houses of relocated families" (November 22, 2023).

[113] A former minister to this area of Kadavu, Rev. Netani Radravu, noted that this was a significant factor in the relocation's failure.

[114] Echoing Cook's description of Ezekiel's attempt to displace monarchical structures of power "off-center"; cf. Cook, "Ezekiel's Recovery of Premonarchic, Tribal Israel," 360.

it needed to make about land, lumber, and communal resources. Was it possible for financial, scientific, and political stakeholders to share their expertise, while allowing Indigenous decision-making processes to lead? Such an interdependent model of decision-making would necessarily include all genders and ages, and it would take time to come to fruition. But in the words of the village's youth leader, Semisi Nabiri Curuki, such communal processes also had the potential to "heal the heart"[115] after land's loss. Through the advocacy of local social-service organizations like FCOSS and funding agencies that have recognized the importance of local ownership in relocation processes, Cogea's community-centered relocation is currently underway. The process attempts to maintain the village's sovereignty by investing in relationships and shared accountability, even as the land under their feet shifts.[116]

Centering Indigenous voices in climate decision-making is more than a local issue. The concern drove the protests that marked COP26's global environmental summit,[117] and it continues to challenge climate negotiations.[118] Shared responsibility in a global crisis and respect for the interdependence of local land claims are not easy tasks, particularly given colonialism's long silencing and displacing of Indigenous communities. Such histories require more than an amplification of Indigenous input. And they certainly require more than a withdrawal from Indigenous conversations. They press those currently at the center of climate decision-making processes to move "off-center" and listen, even as they remain critically engaged.

Cogea's youth leader had a fierce response when asked what advice he would give other villages facing relocation. "I have *no* advice," he

115 Semisi Nabiri Curuki, personal interview, Cogea, Fiji, July 6, 2023.

116 Catanasiga describes the "deliberate . . . three-phase planning structure that accessed both governmental structures and the traditional structure of decision making—the Bose ni Vanua." She notes that community engagement has led to significant conversations around "the relevance and appropriateness of the infrastructure, gender insensitivities, access to traditional sites, . . . [and] sources of livelihoods" (personal interview, January 11, 2023). The relocation is currently in process through the funding agency, Brot für die Welt (Bread for the World), which hopes the village's experience can provide a pathway for future community-engaged relocations.

117 Kate Nakamura, "COP26: Five Indigenous Activists and Movements You Should Know About," Global Citizen, November 9, 2021, https://www.globalcitizen.org/en/content/cop26-indigenous-activists-organizations-movements/.

118 Nina Lakhani, "'Shameful' UN Silencing Indigenous Voices, Say Banned COP27 Activists," *Guardian*, November 18, 2022, https://www.theguardian.com/environment/2022/nov/18/shameful-un-silencing-indigenous-voices-say-banned-cop27-activists.

said, speaking in as strong a tone as I'd heard him use. He was tired of others who did not know his community's situation leading the way, he explained, and he would not play that role for others. "Trust the community," he told me. "Trust the community."[119] His exhortation is profound, even as it is complex. Communities do not always have answers. They can make poor decisions, as others can—and their processes are not always just. Local models of decision-making can exclude the insights of women, for example, as easily as governmental models.[120] But ignoring a community's insights, overriding members' responsibilities and relationships to each other and the place where they live, will not lead to covenantal wholeness. When the heart's healing is at stake, humble, honest, expectant relationships are necessary.

In significant ways, Ezekiel's redistribution of land concretizes the political, economic, and social implications of a sovereignty rooted in shared responsibility and relationship.[121] When I ask Marika Bale how the environmental movement might better model such shared responsibility, he is similarly concrete. "If wealth was distributed fairly instead of bigger nations getting more—that might change the picture. Redistribution of wealth would create new pockets of voice." And then he laughs. "I don't know how [the world] will accept it!"[122] It seems as "artificial" and idealistic a suggestion in relation to our contemporary landscape as it did in Ezekiel's own—and perhaps as impossible, without a higher Sovereign.

Belonging and the Allotting of Land to the Stranger

In his description of how biblical texts have been leveraged for colonial conquest in the Americas, Oscar García-Johnson underscores the dangers that Joshua 13–22 have held for Indigenous peoples. He

[119] Curuki, personal interview, July 6, 2023.

[120] Ezekiel could be accused of such. Ezekiel 13:17–19, for example, expresses disdain for "daughters of your people . . . who prophesy."

[121] In other ways, Ezekiel leaves the concrete significance of shared sovereignty underexplored. Ezekiel does not, for example, flesh out the implications of land distribution for women or describe their role in shared responsibility for the land. Gender equity is not part of this prophet's agenda. Regardless, Ezekiel's themes of shared responsibility and shared power can be extended in ways relevant to that work. In Fiji, shared responsibility across gender relationships is a subject of increasing urgency—particularly in relation to climate resilience. See Priyatma Singh, Tammy Tabe, and Tess Martin, "The Role of Women in Community Resilience to Climate Change: A Case Study of an Indigenous Fijian Community," *Women's Studies International Forum* 90 (2022): 1–13.

[122] Baleidelabaci "Bale" Tamanikaigavo, personal online interview, January 25, 2022.

describes the divisions and reorganizations of the land in these chapters as a "theological cartography," providing historical justification for the "acquisition practices [of] those in control of both maps and canons."[123] Robert Allen Warrior's fierce naming of the ways that the displacement and dehumanization of Canaanite communities in the biblical text fueled the settler-colonial imagination and the genocide of Native Americans leads him to describe the texts as irretrievable for generative theology. He acknowledges these texts' usefulness in excavating the roots of the settler-colonial project, but he doubts if they can be trusted in the interpretive hands of local communities.[124] In parts of the world that have known the conquests of empire, Exodus and Joshua have been extensively critiqued as texts leveraged to support the "dispossession, enslavement, and elimination"[125] of first peoples.

It is part of what makes the Fijian Methodist Church's New Exodus a vital conversation partner in biblical framings of place and displacement. It has an unusual vantage point in these conversations. A predominately Indigenous community with a living memory of colonial rule and ethnonational violence has turned to the exodus narratives, not to justify imperial histories, but to unravel and rewrite them.[126] Their strategy has been to focus on different moments in the exodus account—wilderness moments that speak to the creation of a community marked by covenantal relationships of faithfulness to God, to place, and to each other. But it

[123] Oscar García-Johnson, "Faith Seeking for Land: A Theology of the Landless," in *Theologies of Land: Contested Land, Spatial Justice, and Identity*, ed. K. K. Yeo and Gene L. Green (Eugene, Ore.: Cascade, 2021), 38–68.

[124] Warrior critiques "catechetical, narrative, and canonical" approaches to Scripture. But he also worries about the biblical interpretation of "Christian base communities [like the] Solentiname peasants." "No matter what we do," Warrior states, "the conquest narratives will remain": "A Native American Perspective: Canaanites, Cowboys, and Indians," in *Voices from the Margin: Interpreting the Bible in the Third World*, ed. R. S. Sugirtharajah, 25th anniversary ed. (Maryknoll, N.Y.: Orbis Books, 2016), 283–90 [288].

[125] Hulisani Ramantswama's "Negotiation and/or Conquest of the Land: Reading the Land of Promise Motif in the Hexateuch through Decolonial Lenses," in Yeo and Green, *Theologies of Land*, 98–131 [114]. In contrast, see Kenyatta R. Gilbert's discussion of the well-documented significance of the Exodus narrative in African American preaching, *Exodus Preaching: Crafting Sermons about Justice and Hope* (Nashville: Abingdon, 2018).

[126] While the sermons of Banivanua address unique concerns in the Fijian context, the use of Exodus narratives for regional liberation does have a long pan-Pacific history. See for example, Suliana Siwatibau and David Williams, *A Call to a New Exodus: An Anti-Nuclear Primer for Pacific People* (Suva, Fiji: Pacific Council of Churches, 1982).

has also meant reconceiving of that biblical narrative's trajectory. When Banivanua refuses to appropriate the "conquest" at the end of Exodus' account to the Fijian context, he is boldly reimagining a new end to Exodus' story and a new framing of God's emplaced promise.[127]

In Ezekiel 47:22–23, the prophet does something similar. In a "stunning" declaration,[128] Ezekiel insists that "aliens who reside among you and have begotten children among you" (*gerim*) should receive an allotment of land as "an inheritance." "They shall be to you as citizens of Israel," the voice of the Lord proclaims (v. 22). It is a very different vision of the promised land than Joshua describes (Josh 13:1–7). Ezekiel has previously shown an attentive eye toward the vulnerability of resident aliens, condemning former political leaders for their "[extortion of] the alien without redress" (Ezek 22:7, 29). It is a concern that Cook understands to be rooted in Ezekiel's commitment to the Holiness School, an interpretation of covenantal law that "understands that foreigners who settle in God's land do not have kin and tribe to protect them and so come under the protection of YHWH's covenant and Israel" (see, for example, Lev 19:33–34).[129] But the extension of land to foreigners in the legally binding, "inalienable" nature of an inheritance is something else altogether.[130] It is an "absolute novelty"[131] in the Hebrew Bible and has been the subject of significant debate. Cook notes that "classic, premodern Jewish interpretation wrestled hard to comes to terms with the radical inclusivity of Ezek. 47."[132]

One of the questions raised for modern scholars is the definition of *gerim* in Ezekiel's text. Ezekiel's call for *gerim* to put away their idols (Ezek 14:7) has led some to argue that this provision was not meant for

127 Tevita Nawadra Banivanua, personal interview, Suva, January 2, 2023.

128 Walter Brueggemann, *The Land: Place as Gift, Promise, and Challenge in Biblical Faith*, 2nd ed. (Minneapolis: Fortress, 2002), 134.

129 Cook, *Ezekiel 38–48*, 280.

130 Stevenson says that "the inalienability of inherited land" is a crucial issue for Ezekiel (*Vision of Transformation*, 87). The concept of *nahalah* (נחלה) in Ezek 47:22 does not mean "owned" in the contemporary sense. The word implies land held in trust for future generations. See Benjamin Gordon, *Land and Temple: Sacred Real Estate and the Second Temple Priesthood* (Munich: De Gruyter, 2020).

131 Pikor, *Land of Israel*, 220. Leviticus 19:33–34 and Num 15:15 may be relevant precursors to Ezek 47:22–23, but neither describes land inheritance—a key innovation.

132 Cook, *Ezekiel 38–48*, 280. Significantly, one of the justifications of this passage in the interpretative tradition is the *gerim*'s "suffering alongside the exiles." In other words, a common grief and experience of loss was understood to tie the community together.

uncircumcised foreigners, but for proselytes only.[133] Though even this argument, grounded in the changing circumstances, religious commitments, and communal investments of resident aliens over time, works to undermine static categories of insider and outsider.[134] Persian-era Judean society was socially complex. All sorts of people had been displaced and traumatized by imperial force. The historical fluidity of *gerim* as a category bears witness to this.[135] As has been argued, communities that have valued land and lost land foreground questions of *belonging* in their common life, and, if Ezekiel is any example, those negotiations are more complex than bloodlines, bank accounts, or even biblical narratives indicate. They are discerned on the ground, in the context of actual differences, common commitments, and present need. For Ezekiel, such discernment leads to a bold extension of Holiness School thought—perhaps growing out of his own displaced experience. In Brett's words, this privileged, native (*'ezrach*) status (Ezek 47:22) "should not be invoked in order to deny the land rights of others who live within Israel, particularly when they have raised children within the community."[136] For Brett, the expansiveness of this "counter-imperial imaginary" is that it allows for freedom "to engage with strangers in the same social space."[137] Brett, then, would not dilute Ezekiel's affirmation "of equality before the law" with a "religious test."[138] Cook shares this general view, even as he differentiates Ezekiel from the "modern liberal value of human autonomy."[139] Much of Cook's scholarship has taken to task biblical interpreters that describe

133 Ranier Albertz describes how the meaning of *gerim* changes over time in "From Aliens to Proselytes: Non-priestly and Priestly Legislation Concerning Strangers," in *The Foreigner and the Law: Perspectives from the Hebrew Bible and the Ancient Near East*, ed. Reinhard Achenbach, Ranier Albertz, and Jakob Wöhrle (Wiesbaden: Harrassowitz, 2011), 53–69.

134 Pikor accessibly summarizes scholarship focusing on the shifting meaning, social assimilation, and legal concerns relating to *gerim* (*Land of Israel*, 220–22).

135 For a rich account of the many ways that the biblical text relates to the category of the "stranger," see Adriane Leveen, *Biblical Narratives of Israelites and Their Neighbors: Strangers at the Gate* (New York: Routledge, 2017).

136 Mark Brett, *Political Trauma and Healing: Biblical Ethics for a Postcolonial World* (Grand Rapids: Eerdmans, 2016).

137 Brett, *Political Trauma and Healing*, 109.

138 Brett, *Political Trauma and Healing*, 109. Brett does note that this does not mean that "strangers had no responsibility for maintaining the sanctity of the land" given the "stringent regulations in the Holiness Code that bind all those who live in the land (e.g., Lev. 20:2–5) . . . [but that] some room is left for the negotiation of identities."

139 Cook and Patton, "Hierarchical Thinking," 16.

Ezekiel's commitment to place as a "narrow, exclusivist and parochial . . . national agenda,"[140] often grounded in what he believes to be a misreading of Ezekiel 44:6–9.[141] In contrast, Cook insists, Ezekiel 47 pushes Israel to "fulfill its covenantal commitment to protect the resident aliens in the broadest manner possible."[142]

I do not need Ezekiel to align with Western models of liberal autonomy or human rights in order for his allocation of land to the stranger to bring me up short. His fierce emphases on purity and protocol, alongside his deep investment in the survival of his community's particular witness, underscore the costliness of Ezekiel 47's allocation of land to strangers—be they proselytes or no. I would expect that someone displaced by foreigners would bear a deep-seated anger and fear toward the outsider, but Ezekiel's displacement presses him toward a broader recognition of holiness' requirements. The Holy One cannot establish a covenant of peace in a place where the displaced have no home. It is an affirmation as difficult and simple as it sounds. Tellingly, the anger and fear toward outsiders I expect to see in Ezekiel is more in evidence among those in my own United States. The settler-colonial desire to exclude the resident alien from land and citizenship is one of the reasons that García-Johnson connects the contemporary "global migration crisis" to more than climate disasters. He sees the wound of forced displacement connected to "a five-hundred-year problem spawned by global designs built with great political, economic imagination and legitimized by westernizing theologies of land."[143] I would add westernized biblical readings to this list—i.e., readings which continue to interpret place as a possession, rather than a living relation.

Theologies of relational correspondence between people, land, and God can sediment into unassailable logics of control, particularly when these categories are objectified, quarantined, or used to justify imperial expansion. It is part of why conversations about this "relational triangle"

[140] Cook, *Ezekiel 38–48*, 270.

[141] Cook's argument about Ezek 44 is worth reading in full (*Ezekiel 38–48*, 216–18). He interprets "sanctuary" (*miqdas*) in a narrow sense—denoting temple areas that are not only restricted to resident aliens, but to "all native Israelites outside of Aaron's lineage" (*Ezekiel 38–48*, 280–81). In summary, while Ezekiel constructs plenty of hierarchical strata into his temple vision, access to these strata are linked with communally designated roles and responsibilities rather than xenophobic categories.

[142] Cook, *Ezekiel 38–48*, 280.

[143] García-Johnson, "Faith Seeking for Land," 39.

are "polyvalent"[144]—even among communities united in their commitment to decoloniality. Construals of the relationship between God, land, and people have underwritten colonial conquests. This relationship has also been colonialism's first casualty, denying the sacral connections between Indigenous communities and the places that form them. One key distinguisher between colonial uses of the relational triangle and many Indigenous uses of the same are the distinct theologies of place that undergird each. In colonial models, where place is understood as possession, relationships between God, land, and people become rigid, ideological assertions. When place is understood as relational, formational, and sacramental, this relational triangle takes on a different quality. It becomes a living, responsive commitment, negotiated in time. For Ezekiel, this requires a different articulation of the land's relation to religious others than the one described in Joshua's conquest (Josh 13:1–7). And it has meant something similar for Fiji.

The Fijian *iTaukei* community is presently negotiating the question of the Fijian land's relation to outsiders. How does one frame belonging in ways that do not objectify and commodify land or diminish the bonds of responsibility between people and place? On the other hand, given the displacements of colonial history and climate catastrophe, how can a community establish right relationship with creation and their neighbor if the displaced among them have no home? I have described elsewhere a radically inclusive attempt to address the belonging of Indo-Fijian *girmitiya* descendants through *iTaukei* protocol on July 9, 2016.[145] On this day, the high chief of Rewa, Gone Marama Bala Roko Tui Dreketi Ro Teimumu Kepa, claimed Indo-Fijians as part of the Rewa *vanua*. The decision was based on an event over a hundred years prior, when the village of Noco had saved the lives of hundreds of Indian indentured laborers endangered by a storm when their British transport ship went down. Those lost were buried on Noco land—buried, in other words, in the Noco *vanua*.[146] That

144 K. K. Yeo summarizes the "polyvalent" perspectives regarding this "relational triangle" presented in a panel discussion on decolonizing theologies of land in "Theologies of Land: Contested Land, Spatial Justice, and Covenantal Identity," in Yeo and Green, *Theologies of Land*, 132–56.

145 Jerusha Matsen Neal, "Oceanic (Fijian)," in *The Bible in Folklore Worldwide*, vol. 2, *A Handbook of Biblical Reception in Folklores of Africa, Asia, Oceania, and the Americas*, ed. Eric Ziolkowski (Berlin: De Gruyter, 2023), 183–97.

146 Sashi Kiran, "Home, Where They Belong," *Fiji Times*, July 9, 2016, https://www.fijitimes.com.fj/home-where-they-belong/.

stand of graves, which still exists, testifies to a shared suffering, a shared resistance, and a shared history. And it was a sufficient call for the high chief of Rewa to imagine new possibilities of belonging, even as the implications of that new relationship continue to be discerned. Significantly, when the ceremony of tribal inclusion was complete, the chief of Noco turned to a representative from the Fijian Methodist Church who was on hand to bless the proceedings. "This is the *Lako Yani Vou*," he stated. "This is the New Exodus."[147] Together, the community was reframing place in the context of covenant and re-narrating the exodus story.

The shared suffering and shared displacements of the climate crisis are also inviting new negotiations of belonging. An *iTaukei* woman sheltering at a local school following a cyclone noted the presence of her Indo-Fijian neighbor and would later reflect, "There was so little that divided us." Though of course, there was also much—centuries of imperial categories of personhood and place designed to divide, isolate, and control. It is why, for Bhagwan, the New Exodus is not only a question of navigation. Nor is it simply a description of where the *drua* of the New Exodus is heading. It is also a reimagining of "who may join this journey as crew or passenger."[148] Negotiating belonging is a central challenge of covenantal commitment in a season of climate crisis, as it is a central challenge in the formation of a decolonized church. Banivanua minces no words. To navigate to a place of justice, "the *Lako Yani Vou* must carry everyone."[149]

Loss and the Apocalypse of Place

The question of hope is fraught in Ezekiel's temple vision. On first reading, it seems a satisfying narrative resolution to the crisis of exile—a full-circle denouement of restoration that is concrete, lush, and detailed enough to counter the bloodshed and horror of exilic loss. But, like a painting that covers over a prior portrait, the temple vision never fully obscures that pain. The shadows of another world and the lines of other histories

[147] James Bhagwan, "Gratitude," Girmitiya welcome service sermon, Nabudrau village, Rewa, Fiji, October 9, 2016.

[148] James Bhagwan, "Sailing to the Island of Hope: A Wesleyan Ethical Framework for 21st Century Fiji" (unpublished master of theology thesis, Methodist Theological University, Seoul, 2013), 99.

[149] Banivanua, "Signposts of the New Exodus." The scope of Banivanua's rhetoric reaches beyond Fiji's borders with this statement, explicitly naming the "abuse and injustice from parts of our Body of Christ in West Papua and [cries for] freedom . . . in Maohi Nui – French Polynesia."

bleed through. Rather than resolving the exile, the vision vibrates with the knowledge that, in Edward Said's words, the world is "contingent" and "homes are always provisional."[150] There are certain losses that cannot be restored. Even in Ezekiel's commitment to the promise of a land suffused with "holiness radiated from an indwelling God, . . . [a God] profoundly present,"[151] there is the ache of absence.

Steven Tuell notes that, in this restored landscape where the temple lies outside of the city gates, "YHWH's presence is no longer identified unambiguously with the city"[152] and its people. The apocalyptic crisis of place's destruction has meant that even the category of presence is reframed. He suggests that the text of Ezekiel's vision itself becomes a new "medium of divine presence" in a world where the physical temple has been destroyed.[153] Contrary to Cook's insistence on the profound presence of YHWH throughout the vision, Tuell insists that "Ezekiel is the prophet of the absence of God"[154]—or at least the prophet who most radically reconsiders the meaning of divine presence from the ground up.[155] Stevenson agrees. The questions Ezekiel asks about place in the temple vision are not merely anthropological; they are theological. In their affirmation of place's mediation of presence, unvoiced questions emerge. *Where is God to be found . . . now?*[156] This is the cost of theologies of place that take seriously the provisionality of the world, the covenantal requirements of belonging, and the freedom of a living God. This is no "romantic ecological

150 Edward Said, "The Mind of Winter: Reflections on Life in Exile," *Harpers Magazine*, September 1984, 49–55, quoted in Daniel Smith-Christopher, *The Religion of the Landless: The Social Context of the Babylonian Exile* (Eugene, Ore.: Wipf and Stock, 2015), 206.

151 Cook, "Ezekiel's Recovery of Premonarchic, Tribal Israel," 365.

152 Steven S. Tuell, "Divine Presence and Absence in Ezekiel's Prophecy," in *The Book of Ezekiel: Theological and Anthropological Perspectives*, ed. Margaret Odell and John Strong (Atlanta: SBL, 2000), 97–116 [104].

153 Tuell, "Divine Presence and Absence," 109.

154 Tuell, "Divine Presence and Absence," 106.

155 Jonathan Z. Smith's classic reflection on Ezekiel's temple vision also makes this point, arguing that Ezekiel transmutes God's presence and even the place of the temple itself into structures of social hierarchy: *To Take Place: Toward Theory in Ritual* (Chicago: University of Chicago Press, 1987), 73.

156 Stevenson ends her reflections on Ezekiel's "territorial rhetoric" with the fundamental theological question at stake for this exiled people. Alongside questions of "Who are we?" and "Where is home?" is the question, "Where is God?" (*Vision of Transformation*, 165).

worldview"[157] that shuts its eyes to the continuing devastations of empire and real losses of land. Even as Ezekiel walks temple pathways and wades in holy rivers in visionary ecstasy, the reader can never completely forget the physical location of his exiled body.

The final line of the temple vision makes sure of it. The last verse in Ezekiel pronounces the name of the restored city, and given YHWH's commitment to dwell with God's people (43:7), one might expect that name to be "The Lord Is Here." But that is not the city's name. It is, instead, "The Lord Is There" (48:35). Cook's reading of that name is that it gestures to the temple complex—as in, the Lord is over there, up the hill, inside the temple walls.[158] But I find Adriane Leveen's reading more convincing.

Leveen notes how Ezekiel's bodily tour of the temple invites readers to participate in his visionary encounter with place. She describes the many textual references to his body, his movement, and his vantage point and how these markers orient us in this restored world.[159] We begin to believe we are with Ezekiel, walking through the city and its surroundings. Alongside the prophet, we can see it, measure it, wade into it. But with the words, "The Lord Is There," the vision becomes a hoped-for reality, unfulfilled. More than this, YHWH's presence becomes located on that far-away, uncertain horizon. The city's name is no comfort; it is an expulsion. We are thrown out of the illusory, present-tense idyll and back to the banks of the Chebar. Indeed, in naming YHWH's location, the prophet's dislocation is underscored. In the vision's final verse, Ezekiel points "there," to a landscape and Lord impossibly far from his present experience. Despite all of the vision's place-making, Leveen explains, Ezekiel "is still in exile—desperately wanting to leave there (Babylon) and to arrive there (Jerusalem), but . . . doubtful of success." The intimate "here" of God's presence and the fulfillment of place's promise are "again deferred."[160]

The embedded lament in the promise of Ezekiel's restored city was apparent to me when I visited the village of Cogea in the summer of

157 Mendoza and Zachariah, introduction to *Decolonizing Ecotheology*, 2.

158 Cook, "Ezekiel's Recovery of Premonarchic, Tribal Israel," 369. Others suggest that the name echoes Ezekiel's protest against Edom's population of Israel's mountains by underscoring that "the Lord was there" and continues to be (Ezek 35:10). See Pikor's discussion in *Land of Israel*, 108, n. 83, and also Moshe Greenberg, *Ezekiel 21–37* (New York: Doubleday, 1997), 715. This multiplicity of opinion underscores the unexpectedness of the name's construction.

159 Adriane Leveen, "Returning the Body to Its Place: Ezekiel's Tour of the Temple," *Harvard Theological Review* 105, no. 4 (2012), 385–401; see particularly 394–98.

160 Leveen, "Returning the Body to Its Place," 401.

2023.[161] It had been two and a half years since the majority of the village had been washed away, and the community's relocation process was ongoing. A new building site had been named, funding had been secured, and conversations about the placement of the village's new sanctuary had finished. I thought my conversations with villagers would be focused on hope-filled plans and purposeful logistics. What I found, instead, were fierce currents of grief. I do not say this to undermine the community's commitment to relocation or their determination to solve problems and move forward. I say it to underscore that foreign funding and inclusive protocols—as necessary as they have been—have not compensated for the loss of the community's land. Even as they imagined the new site and God's presence "there," questions about God's presence "here" remained.

I sat in a morning session with around twenty women from the community, thinking we would speak of gender inclusivity and the significance of female leadership in relocation's decision-making—a commitment that has marked Cogea's process. But that is not the story they wished me to hear. They wanted to tell, over and over, the story of the 2021 flood. "It took away everything we had worked for." "I had a one-month-old baby." "The water was a waterfall." "I pushed to high ground in the dark." "We lost everything." "I had only the clothes on my back." The pain was still raw. There were the expected testimonies to God's goodness in the midst of the loss. The fact that no one in the village had died was remarkable, and the women said so. "*Masu* (prayer), sharing, and working together" has helped, several said. But the heaviness in the room could not be denied. I asked one woman where she had seen God's hand over the past two years. She did not answer. She looked out the window in silence, and her tears streamed. For all of God's presence in the "there" of future restoration, she struggled to find language for the God who was "here" in the *vanua* where she had raised her family—the *vanua* that she would soon leave.

When our conversation ended, the Fijian pastor who had been assisting with translation offered to pray a prayer for those gathered,[162] and then he asked if we could sing a hymn: "Loloma i Jisu sa dau Voliti Au"[163]—the

161 My deep appreciation to the leadership of Cogea for their hospitality and engagement. I am particularly grateful for the Tuirara Levu Wainunu, Suliasi Curuki, for the welcome provided by his family.

162 My great thanks to Rev. Esala Nacuva's leadership as our *matanivanua* (community spokesperson) and translator during this visit.

163 "Loloma Jisu sa dau Voliti Au, #65," in *Ai vola ni sere ni Lotu Wesele e Viti* (Suva: Methodist Church in Fiji, 2003). The hymn's first line is translated, "The love of Jesus surrounds me."

Fijian Methodist version of "It Is Well with My Soul."[164] The women sang, in four-part harmony, and when they finished, a hushed peace shadowed the room. One can only speak tentatively of God's presence after so many tears. The peace in that place was hard-fought and unexplained. Multiple members of the village had used the language of "trauma" to describe their pain during our visit. Such language has become increasingly familiar in village contexts through trainings conducted by the Bible Society of the South Pacific, the Methodist Church in Fiji, and the University of the South Pacific.[165] These women knew something of grief that is traversed and revisited, reframed and shared. The community did not take this work lightly or conceive of healing in linear ways. And yet, sitting on mats together in one of the village's handful of remaining homes, singing the chorus' call, "Cegu tu, vei Jisu, tikoga" (Rest in Jesus always), there was more than absence in the room. This place of memory and loss became a site of God's unveiling.

Psalm 77 speaks with a similar tentative hope regarding divine presence. The psalm begins in existential crisis, naming the experience of God's absence even for those who are faithful in religious practice and prayer (vv. 7–10). It then imagines a God present in thunder and storm, a God whose presence destroys even as it creates (vv. 16–18). But in the psalm's final verses, its imagery shifts again. God's location is with God's people, leading them through Exodus waters like a shepherd leads a flock—"[though divine] footprints were unseen" (vv. 19–20). Unseen footprints were precisely what these Cogea women had as testimony. And yet, in their prayer, worship, and shared grief, God's countenance shifted from One "changed" (v. 10) and unrecognizable, to a God of whirlwind and storm, creating in the midst of catastrophe. And finally, God was revealed as shepherd, accompanying the community through mighty waters—present, even when hidden.

When I read the name of Ezekiel's visionary city, I do not read it as a signifier on a map, marking the spot of treasure.[166] I read it as proclamation.

[164] Horatio Spafford, "It Is Well with My Soul, #377," in *The United Methodist Hymnal* (Nashville: United Methodist Publishing House, 1989).

[165] For an example of *iTaukei* training materials used by local faith communities to respond to trauma, see Harriet Hill et al., *Na I Wali Ni Mavoa Se Tadoka Ni Yalo*, trans. Apenisa Lewantoro (Suva: Magi Magi, 2013).

[166] In spite of Corrine Carvalho's beautiful use of this metaphor in "Finding a Treasure Map: Sacred Space in the Old Testament," in *Touching the Altar: The Old Testament for Christian Worship*, ed. Carol M. Bechtel (Grand Rapids: Eerdmans, 2008), 123–51.

My argument has been arcing toward a performatively weighty truth. Place *preaches*—for better or worse, in contradiction with the preacher or in complement. And when the name of Ezekiel's city is read sermonically, its trajectory changes. It does not name itself. The city speaks a reflexive blessing over a broken world, arms outstretched—a blessing particularly relevant for communities standing in places similar to Ezekiel's own. For those in crucibles of climate disaster, the city's name is a sermon that unveils arguably the most compelling revelation of Ezekiel's temple vision. Indeed, coming full circle, the city proclaims to the prophet the word he received at the book's start: *The Lord is there, in the place of exile where you stand. The Lord is not quarantined in a realized vision of future glory. But there, Ezekiel, with you.*

Like a double-edged sword—and like a word of faithful proclamation—the city's name does two things at once. More precisely, the meaning of the city's name changes depending on who is speaking and *where that speaker stands*. When spoken by Ezekiel, the city's name locates God's presence in a future beyond the community's control. This deferral has concrete, ethical significance. It resists technological solutions as sufficient to the challenge of place's healing, even as it resists theologies of place that strip it from its sociopolitical context. Indigenous and decolonizing climate activists have critiqued both of these postures within contemporary, mainstream ecotheologies and climate movements.[167] Ezekiel's theology of place remembers and responds to his exiled reality, and in so doing he refuses reductive framings of hope. For those who know place's worth and place's work, naïvete about the challenge ahead is dangerous—as are simplistic resolutions of loss. In the temple vision, even visions of restoration hold grief.

But when the city's name is proclaimed back to the prophet by the city itself—and back to so many others who weep on the haunted ground of displacement's loss—the name of God's dwelling reveals something more. Haunted ground will be a site of communion and covenantal reckoning. God will encounter God's people even there. Even here. This is place's proclamatory "apocalypse"—its revelatory testimony. And in it, human preachers find their task in the climate crisis: to witness to the costly

[167] In discussing the dangers of prioritizing technological solutions at the expense of more costly conversations about power and sovereignty, Mendoza and Zachariah note, "Environmental problems cannot be solved by reproducing structures of colonialism and capitalism" (*Decolonizing Ecotheology*, 6). See also García-Johnson's discussion of "salvation by the colonial economy and civilizing Christianization" ("Faith Seeking for Land," 49ff.).

repentance that covenantal relationships with place require and to the God who steps onto the muddy banks of place's catastrophe to unveil a divine heart breaking.

Covenantal Wayfaring in the Pulpit

In the Christian tradition, Scripture is a place that mediates God's revelatory call. Like Luger's mirror shields, it envisions a covenantally connected world, while also exposing the apocalypse of imperial conquest, our creaturely limits, and our dependence on God's creation. Scripture reveals where its readers stand in relation to God's call and Godself—like a double-edged sword discerning the heart (Heb 4:12). Or at least, it can.

Biblical interpretations marked by the postures and expectations of settler colonialism obscure that witness—appropriating others' places as one's own or constructing rigid borders between oneself and another. Such interpretations can dull Scripture's sociopolitical edge, assume a hegemonic authority, or exoticize and dismiss Scripture's claims. "I'm sorry to say," Banivanua tells me wryly, "that the West changed the meaning of the Bible somehow. . . . It twisted biblical theological definitions to . . . make the world a 'better place.' A better place *for whom*? This is the big question."[168] For Banivanua, methods of biblical interpretation are also a mirror, revealing a reader's assumptions about her place in the world—and her place's "affectability" in relation to others. Such assumptions are amplified in the preaching event.

Sermons can hide behind historical-critical objectifications or imagine themselves the gauzy sites of individual, existential insight. They can purport to teach a univocal grammar, rather than engage a living vernacular. Such approaches discourage responsiveness and relationality between the place of the interpreter and the place of the text. Whether through an assumed universality of place or a carefully preserved insularity, these interpretive postures suggest that the biblical text can be quarantined and controlled, rather than navigated and encountered. And yet, regardless of sermons that pretend otherwise, the place of preaching is not settled. It is a place of covenantal wayfaring—fixed and fluid, as deep as any ocean. Bridging memory and eschaton, grounded in the call of a covenanting God, preaching does more than teach a covenantal theology *of* place. It enacts a covenantal relation *to* place—the place of the biblical text and the place of the world God

[168] Banivanua, personal interview, January 2, 2023. This observation is part of Banivanua's larger critique of Western hermeneutic strategies that have undermined the innate value of nonhuman creation.

loves—from the particular ground of the pulpit. Given this relationality, the sermon is accountable to the testimonies of displaced communities. It cannot pivot away from decolonization's material consequences, especially in a world coming undone. It must steer into that gathering storm.

A student tentatively approached me after a lecture I had given on climate-justice preaching that had included a photograph of the "Mirror Shield Project." He had seen those shields used by other Water Protectors resisting a different pipeline near the Mississippi River—the Line 3 Pipeline—and had joined the movement. He had been walking down to the river on a snowy day, alongside several Protectors holding mirrors. Construction equipment ripped the ground around them. "Law enforcement agents learned that they would lose the media narrative if they fired bullets," he said. "So that day, I remember an official quoting Scripture at us. 'Submit to the governing authorities' (Rom 13:1) was the verse."[169] In response, the shields reflected their mirrored witness, changing the trajectory of the weaponized words, and placing them in relation to the river flowing behind the official's back. With the river as their backdrop, the words unveiled the place where the official stood and raised critical questions about who exactly its "governing authorities" might be. One use of Scripture dissociated it from place and time—while the other reflected the relationship between Scripture and place, formed a community through practices of resistance, and unveiled an apocalypse of human disregard. The Protectors are people of diverse faiths, and at times, no faith—but on this day, they gave witness to the presence of a seemingly absent God, a Presence hidden to those indifferent to place's unveiling. They gave witness to a God intent on revealing where God's children stood in creation's struggle. "I wish there had been a preacher there," this student said to me. *Perhaps*, I thought, *there had been.*

I began this chapter with Banivanua's claim that the "struggle over the interpretation" of biblical traditions in light of Indigenous culture is "crucial to political and social life."[170] I have argued that, for both Ezekiel and the MCIF, covenantal framings of the exodus narratives hold material implications relevant for communities displaced by the climate crisis. Covenantal framings of place prioritize key questions these communities are asking regarding sovereignty, belonging, and loss, questions that are neither easy

[169] I am grateful to this student for sharing his story and honor his decision to remain anonymous in order to foreground the communal Anishinaabe leadership of the Water Protectors' resistance to the Line 3 Pipeline.

[170] Banivanua, "Staying the Course."

nor risk free. Prescriptive biblical work requires dangerous discernments. One cannot conflate places across time and space. We stand in relation to these narratives, sometimes closer and sometimes farther away, but always in ways that are provisional and time bound. Determining "where one stands" in the biblical text is hard. We often stand in multiple locations, and those locations shift in relation to a shifting world. Indigenous communities are not immune from these shifts. Carolyn Sharp is right. Colonized and displaced communities can be co-opted by the hermeneutic strategies of their oppressors.[171] Fiji's fraught history navigating exodus texts testifies to this. In the words of Sally A. Brown, faithful witness is a "constant negotiation" that doesn't "leave one's 'world' behind at the sanctuary door."[172] Such negotiations do not yield one-size-fits-all solutions. A navigational star might direct one community east and another in an opposite direction depending on the starting location of each. Our journeys are particular.

Given this complexity, the temptation to sentimentalize or spiritualize the text is strong, as is the temptation to analyze the book as a historical document or ethnographic curiosity. Such methods make Scripture easier to control. But there is a simpler method. If control and autonomy are one's goals, one is better served by ignoring Scripture altogether. When I speak with seminary students in Western contexts about climate-justice preaching, they regularly express ambivalence about the use of Scripture in their sermons—at least initially. They fear the Bible has little to add to this urgent contemporary conversation. Worse, they fear this text is a hostile witness, reinscribing models of domination resigned to place's destruction. They have seen it used as such. I am grateful for these students' reserve. Heaven knows, preachers can be far too quick in making prescriptive pronouncements that disregard the contextual limits of their understanding. Navigating Scripture with an eye to one's limits, however, is different than disengagement. It would be self-servingly convenient if, just at the moment when Western theological education determined that biblical texts were not meant to aid the ends of empire, it also decided the text had little material value at all. For a settled theological academy and a settled church only beginning to recognize the resonances between its colonial legacy and the legacy of Babylon, such dismissals silence the

171 Carolyn J. Sharp, "Sites of Conflict: Textual Engagements of Dislocation and Diaspora in the Hebrew Bible," in *Interpreting Exile: Displacement and Deportation in Biblical and Modern Contexts*, ed. Brad E. Kelle, Frank Ames, and Jacob Wright (Atlanta: SBL, 2011), 365–76 [365]

172 Brown, *Sunday's Sermon for Monday's World*, 27.

biblical material at precisely the moment its witness calls us to account. Something more difficult is needed: reading the biblical text in ways vulnerable to reflexive revelation. Such reading moves us beyond analysis of where others stand and challenges us to take responsibility for the covenants we have broken and the covenants that may yet be restored. This does not mean that preachers silently acquiesce to biblical texts that have become the tools of empire. Acquiescence is disengagement by another name. Instead, it reads Scripture with the expectation that communities and texts from other places have a covenantal claim on one's choices and relinquishments. It means reading Scripture in affectable relation to such communities.

The journey of this book, *Holy Ground*, attempts to describe the difference such covenantal commitments might make in climate-justice preaching. When place's covenantal character is allowed to influence the place of biblical interpretation, questions of power, responsibility, locality, and commitment are foregrounded—and practical implications faced, for all of their difficult discernment. Covenantal framings of place chasten preaching's sovereignty, relativize its belonging, and loosen its interpretive control. Such framings help preachers and congregants acknowledge together where they stand in the world and in Scripture's pages—and discern, with greater clarity and will, what they are called to do.

For the Western practical theological guild, covenantal framings of place charge the theological academy to embrace sustained relationships with displaced communities that impact the academy's conversations, power structures, and practices. Such relationships are costly and time-consuming, requiring granular interpretative work, shared leadership, honest conflict, and sacrificial solidarity. They also require new relational methodologies that attend to place's particularity, even as they embrace a common responsibility.[173]

[173] See, for example, Cresantia Frances Koya-Vaka'uta, "Rethinking Research as Relational Space in the Pacific," in *Relational Hermeneutics: Decolonising the Mindset and the Pacific Itulagi*, ed. Upolu Lumā Vaai and Aisake Casimira (Suva: University of the South Pacific Press, 2017), 65–84. Timothy Eberhart's attention to "how the seats are arranged" in global relationships has relevance for the theological academy: "To be faithful to the God who is triune means, first of all, critiquing any arrangement of the public household in which economic wealth and political power are concentrated in the hands of an elite few" ("Sustaining the Planetary Household: Methodist Contributions to an Ecological Economy," in *Methodist Revolutions: Evangelical Engagements of Church and World*, ed. Joerg Rieger and Upolu Lumā Vaai [Nashville: Wesley's Foundery Books, 2021], 117–48 [138–39]).

For preachers in communities that are only now coming to terms with the environmental crisis, covenantal framings of place center the concerns and biblical interpretations of displaced communities when discerning a congregation's own covenantal call. They ask preachers to consider the relationship between the place of the pulpit and places of loss in the work of climate justice. I will reflect on the hope and challenge of communal discernment in responding sermonically to the climate crisis in the chapter that follows. In the wake of environmental disasters impacting our connected world, such discernments are increasingly necessary—and they are increasingly close to home. Displaced communities do not only exist "there," in other hemispheres. They are here.

In the reckoning of relationship, God is waiting to be revealed.

6
Leaves of Healing
Preaching from a Place of Apocalypse

Roughly a hundred years ago in Durham, North Carolina, the U.S. city where I currently reside, the federal government decided to assist certain communities with their mortgages, but only in neighborhoods deemed "safe" investments. Given the segregated, racialized realities of the U.S. South, these neighborhoods were disproportionately, if not exclusively, white.[1] The impacts of this practice—now referred to as "redlining"—persist to this day, even when statutes have changed and protections for communities of color have ostensibly been put in place.[2] Redlining's history is particularly evident in the shape and scope of Durham's tree canopy. A 2016 study compiled by Duke's Nicholas School of the Environment outlined the tree-planting initiatives championed by the University's Forestry School during those fiercely segregated decades.[3] Not surprisingly, trees were planted in the same communities as those receiving assistance from the federal government. The city's investment in a green future was carefully contained in white neighborhoods. When one takes a picture of

[1] Steve Maas notes how the Home Owners Loan Corporation (HOLC) maps reinscribed inequities already present between black and white neighborhoods arising from decades of unequal treatment: "Searching for the Origins of Redlining Black Neighborhoods," *National Bureau of Economic Research Digest* 2 (February 2021), https://www.nber.org/digest/202102/searching-origins-redlining-black-neighborhoods. The HOLC maps are part of the records of the FHLBB (RG195) at the U.S. National Archives and Records Administration.

[2] Matthew Desmond describes the persistence of housing inequities across racial lines in *Poverty by America* (New York: Penguin Random House, 2023), 165–71.

[3] Charlie Shelton-Ormond and Frank Stasio, "Planting Inequality: How Durham's Trees Show a History of Marginalization," WUNC, July 4, 2017, https://www.wunc.org/environment/2017-07-04/planting-inequality-how-durhams-trees-show-a-history-of-marginalization. See also *Duke Today* staff, "What Durham's Tree Canopy Tells Us about Its Racial and Class Divides," *Duke Today*, July 2, 2021, https://today.duke.edu/2021/07/what-durhams-tree-canopy-tells-us-about-its-racial-and-class-divides.

Durham from the air at present, one sees green where white communities lived in the 1930s—and asphalt elsewhere. Further research has correlated the presence of tree canopies with lower rates of asthma, increased exercise, and greater mental health. It also correlates the absence of trees with higher heat indexes, higher heart-attack rates among the elderly, and—given the impact that heat can have on flared tempers—greater incidents of gun violence.[4]

In short, not all trees lead to place's healing. When the planting of trees takes no account of the inequities already present in a location, divisions are exacerbated, with life-and-death consequences. It is a lesson that the environmental movement, haunted by colonialism's centering and universalizing of Western concerns, is still learning. This book has argued that preaching in the climate crisis cannot ignore these colonial specters, nor can it ignore the forced displacements that colonialism's loyalties to accumulation and white supremacy have caused. These displacements continue—and they are increasingly linked to climate disaster and environmental exploitation.

In 2014, the Lutheran World Federation met in Medan, Indonesia, organized around the theme "Like a Tree Planted by Streams of Water" (Ps 1:3). But Bishop Willem T. P. Simarmata chose a different Scripture passage for his opening sermon. The image was the same, but it was rooted in different soil. Preaching on Jeremiah 17:7–8, Bishop Simarmata wondered what it might mean to reflect on trees planted by water when one's community is facing dislocation and exile. He questioned the significance and viability of such an image when faced with the massive destruction of Indonesian rain forests "that have caused the disappearance of streams of water from many lands."[5] Ironically, these forests are being stripped, and the communities that live in them displaced, because of the planting of oil palm trees—trees used in the production of biofuel. In response to economic incentives privileging renewable sources of energy, a key environmental concern, multinational companies have radically expanded their

[4] Ruoxue Chen, "Master's Project Investigates Health and Environmental Disparities in Durham, N.C.," Duke Nicholas School of the Environment, April 8, 2022, https://nicholas.duke.edu/news/masters-project-investigates-health-and-environmental-disparities-durham-nc. See also Lindsey Key, "Medical Student-Led Initiative Plants Change for Environmental Justice in Durham," *Magnify*, Duke School of Medicine, January 12, 2022, https://medschool.duke.edu/stories/medical-student-led-initiative-plants-change-environmental-justice-durham.

[5] Terri Miller, "LWF Council Meeting in Medan Indonesia Opens with Traditional Worship," Lutheran World Federation, June 12, 2014, https://lutheranworld.org/blog/lwf-council-meeting-medan-indonesia-opened-traditional-worship.

palm oil plantations, with disastrous environmental and human cost. The problem is not with palm oil itself. Palm oil can be sustainably grown on smaller-scale farms.[6] The problem is the exclusion of local communities from the decision-making processes regarding the places that sustain their lives. The problem is their forced displacement, displacements that are increasingly violent, to produce palm oil on a massive scale.[7] Not all trees lead to place's healing.

The claims of this book are rooted in the conviction that the voices of displaced communities, particularly those with deep connections to land and a living memory of land's loss, are crucial conversation partners in developing practical theologies of place marked by covenantal commitment. I have argued that covenantal approaches to place honor place's relationality, its formative character, and its sacramental mediation of revelation in ways that crucially reframe the climate crisis. That reframing is not only significant for places like Fiji. It calls for a vital course correction—or in biblical parlance, repentance—in places where settler colonialism holds sway. Place's covenantal reframing has concrete implications for Western economic policies, climate priorities, and even the theological pedagogies of its academies. It also has implications for Western preachers. Covenantal relationships with place impact the ways that Scripture is interpreted and God's call to faithful discipleship is understood. Not every tree leads to place's healing, and neither does every sermon preached in response to the climate crisis. If a preacher's call to creation's preservation reinforces the preservation of a languishing empire, it will do more harm than good. Not unlike the prophets that Ezekiel disdained, such preaching peddles in false hope and false gods. For sermons to serve as leaves of healing (Ezek 47:12), a costlier revelation is required.

At the close of the last chapter, I charged climate-crisis preachers to read Scripture as if texts and communities from other places have a covenantal claim on their choices and commitments. This chapter will expand

6 The Roundtable on Sustainable Palm Oil, established in 2004, sets environmental and social criteria to minimize the negative impact of palm oil production on the environment, wildlife, and local communities: https://rspo.org/who-we-are/ (accessed November 17, 2023). See also World Wildlife Federation, "8 Things to Know about Palm Oil," accessed November 17, 2023, https://www.wwf.org.uk/updates/8-things-know-about-palm-oil.

7 For just one of many stories of increasing violence, see Pailin Wedel, "'When We Lost the Forest, We Lost Everything': Oil Palm Plantations and Rights Violations in Indonesia," Human Rights Watch, September 22, 2019, https://www.hrw.org/report/2019/09/23/when-we-lost-forest-we-lost-everything/oil-palm-plantations-and-rights-violations.

on that exhortation, risking practical theology's prescriptive move. Prescriptive work is never easy and is often where practical theology goes wrong. Throughout this project, I have criticized practical theologies that assume univocal relevance.[8] But I have also critiqued insular disengagements and objectifications that ignore place's connected character. In a season of environmental upheaval, description alone is irresponsible. Acknowledged or no, we are tethered together. My interlocutors in this project did not gift their sermons for the purpose of disinterested analysis. They carry the conviction-filled hope that it is possible to learn from those in other places—and respond. Fijian preachers are not alone in this hope. Numerous climate scientists carry it as well. Scientific experts are increasingly asking some version of the question I asked in my introduction. *Can communities rooted in different soils work together, grow together, and learn from each other in ways that are life-giving?*[9] Environmentalists are, in other words, pressing toward more robust, responsive methodologies of *place*. They understand that the climate crisis demands it. One of this book's broader goals is to frame the significance of this question for the practical theological guild and point to the preaching of displaced communities as a significant site of the question's discernment.

That question has personal significance. One of the reasons that this book has taken so long for me to write is that I myself have struggled to describe how the apocalypse of Fiji's place intersects the place I live and teach in the U.S. South. I am quite certain that Taniela Balenaikorodawa's terrible truth ("you are part of our family"; "you are destroying our homes")[10] is not limited to the Fijian context. In the following pages, I will

8 See particularly my discussion of place in chapter 1 and the specters haunting North American biblical studies in chapter 2.

9 When confronted with conflicting biophysical and cultural constructions of land, scientists are asking how to negotiate "philosophical tensions and controversies that act to reinforce particular discourses and render languages outside specific limited 'fields' unintelligible" (Deirdre Wilcock, Gary Brierley, and Richard Howitt, "Ethnogeomorphology," *Progress in Physical Geography* 37, no. 5 [2013]: 573–600 [574]). Following a description of the ontological differences between European and Aboriginal conceptions of earth and water, for example, Siri Veland argues that our "lack of ability to act on climate change underscores the urgency to transform our way of thinking, seeing, and doing." Veland proposes a model of "ontological conjunctures—places of 'networked dialogue among multiple ontologies,'" in "Transcending Ontological Schisms in Relationships with Water, Air, and Ice," *Weather, Climate, and Society* 9, no. 3 (2017): 607–19 [616].

10 As described in the preface.

discuss multiple forms of forced displacements—some the result of rising sea levels, some the result of divestment and racialized urban planning, and some the result of white supremacist violence. The preaching of Fijian Methodists has helped me see a common thread running throughout these particular devastations: the unraveling of covenantal commitment between place, community, and God.

This chapter will name how the covenantal approaches to place affirmed by the Fijian Methodist preachers in this study and the biblical prophet Ezekiel reframe practical theological research, the priorities of climate advocacy, and the act of preaching in the face of environmental catastrophe. In the chapter's first section, I outline how covenantal approaches to place call practical theological researchers to responsive accountability in relational methodologies of listening. Such postures are needful in the theological academy, but they are also relevant for preachers—practical theologians on the front lines of place's negotiation and threat. In the chapter's second section, I argue that place's covenantal character reorients Western climate advocacy around the theological and material concerns of displaced communities. Foregrounding these concerns shifts the trajectories and touchstones of climate-justice preaching. The chapter's third section speaks to the function and value of preaching itself in the face of unthinkable loss. Place's covenantal character calls preachers to recommit to the critical role that sermons play in revealing who God's people are called to be and where they stand in the coming storm. Preaching matters in the climate crisis far more than many believe, but not for the reasons often assumed. Because biblical preaching engages theologies born out of experiences of displacement and exile, it is poised to address the environmental crisis' most searing questions. And through the sermon's covenantal wayfaring, tacking between past and present in the deep water of the world's need, preaching forms a congregation in dependence: dependence on a covenanting God and on the creation of which we are a part. Preaching becomes a place of divine encounter, relinquished control, and reclaimed relationship—unveiling the place of apocalypse as holy ground.

Relational Accountability in Research: Listening Attentively and "Affectably"

In 2006, Ilaitia S. Tuwere presented a paper at the University of the South Pacific on "Navigating Knowledge" in light of Indigenous understandings of place, particularly focusing on the theological significance of the *iTaukei were* (yam garden). The paper describes the importance of "bodily knowledge," relationality, and interdisciplinary research in the "disruption of Western epistemological sovereignty."[11] I have argued that such disruptions are a crucial precursor to climate justice. Amitav Ghosh cites a disturbing quote from the director of Survival International, an advocacy group for the rights of tribal people. The director states that over decades of confrontations with governments and corporations about their abuse of tribal peoples' rights, "none have been as duplicitous as big conservation NGOs."[12] The director's frustration is not focused on the work of conservation, per se. It censures a universalizing approach to climate preservation that ignores the communities most affected. As García-Johnson describes, the West has a dangerous history of "othering" and objectifying land and Indigenous communities in attempts to save them. Such attempts have repeatedly devolved into exploitation.[13]

Practical theology will need new relational methodologies if it is serious about the disruption Tuwere describes—methodologies that attend to place's particularity, while embracing a shared responsibility. Having more voices in climate-justice conversations is critical, but it is not enough. Diverse perspectives do not necessarily lead to equitable distributions of power and influence when methods and research processes are shaped by deep-seated Western investments in authorization. This is one of the reasons my engagement with the environmental crisis has underlined the ethical difficulties of cross-contextual learning, even as it emphasized its importance. The ethics of place impact Fiji's shoreline and coral reefs, as well as the research assumptions of the theological academy. Modeling reflexive collaboration and interdisciplinary engagement is work that Western practical theologians have begun.[14] But Amiria Salmond presses

[11] Ilaitia S. Tuwere, "At Home: A *Were-Kalou* Response to Epistemology," paper presented at Navigating Knowledge: Pacific Epistemologies Conference, University of the South Pacific, Suva, July 6, 2006.

[12] Amitav Ghosh, *The Nutmeg's Curse: Parables for a Planet in Crisis* (Chicago: University of Chicago Press, 2021), 226.

[13] Oscar García-Johnson, "Faith Seeking for Land: A Theology of the Landless," in *Theologies of Land: Contested Land, Spatial Justice, and Identity*, ed. K. K. Yeo and Gene L. Green (Eugene, Ore.: Cascade, 2021), 36–68. García-Johnson describes Western colonialism as an "othering epistemic machine" (55).

[14] See, for example, recurring themes of "reflexivity" and the significance of "process" in the "seven theses about research in practical theology" described in Zöe Bennet et al., *An Introduction to Research in Practical Theology* (New York: Routledge, 2018), 28–32. Generative interdisciplinary conversations between anthropology, ethnography, and theology are happening in this regard. See Jeanette Matthews and Matt

those engaged in cross-contextual research toward encounters of greater "vulnerability and uncertainty." Difference, in her view, is neither discovered nor "solved before leaving the academy. . . . Difference is something that is done."[15] It is worked out in real time, through reflective engagement in living relationships.

Marika Bale and I were speaking of the challenges involved in this work as we drove back from an in-person visit we had made to one of the Fijian preachers in this study. Methods and processes are hard to change, not only because they require bold reframings of what counts as knowledge, but also because they are enfleshed in on-the-ground practices and postures. Ilaitia S. Tuwere's call to relationality and embodied knowledge is pivotal—but it doesn't describe *how* such work is done. Bale had given that some thought. "It is about listening," he told me. And then, he broke open that seemingly simple word in light of four distinct qualities of the Fijian practice of *veirogorogoci* (listening to one another).[16]

The significance of Bale's insights for academic research in North America's practical theological guilds cannot be overstated. Through concrete practices, grounded in *iTaukei* values and lived commitments, Bale describes how difference is "done"—encountered, attended, and responded to—in Fijian performances of communal listening. In his forthcoming doctoral thesis, Bale uses these practices to shape the ethical criteria that ground his historical research method.[17] I point scholars to that fuller

---

Tomlinson, "Conversations between Theology, Anthropology, and History," introduction to "Anthropology, Theology, and History in Conversation," special issue, *St. Mark's Review* 244 (June 2018): 1–8. Christian Scharen describes several of theological ethnography's usefully disruptive methodological contributions to the study of religious practices in "Theological Ethnography and World Christianity: A Response," *Journal of World Christianity* 10, no. 1 (2020): 109–19.

15 Amiria Salmond, "Re-building Ships at Sea: Ontological Innovation in Action," in *Environmental Transformations and Cultural Responses: Ontologies, Discourses, and Practices in Oceania*, ed. Eveline Dürr and Arno Pascht (New York: Palgrave Macmillan, 2017), 215–26 [223].

16 The significance of listening in scholarly work is also underscored by African American feminist Jacqueline Jones Royster in her groundbreaking essay, "When the First Voice You Hear Is Not Your Own (1996)," in *Landmark Essays on Rhetoric and Feminism, 1973–2000*, ed. Cheryl Glenn and Andrea Lunsford (New York: Routledge, 2015), 122–31.

17 The working title of Marika Baleidelabaci "Bale" Tamanikaigavo's doctoral dissertation is "Silences and Representations: A Postcolonial-Indigenous Historiographical Exploration of the Histories of Christianity in Oceania Using the *Mata ni Katuba* Concept" (Pacific Theological College, Suva, forthcoming 2025). I am grateful for Bale's

work; what I share here is only a thumbnail sketch of a larger project. But Bale has given me permission to share this sketch, as the qualities of deep listening that it describes are regularly absent in Western scholarship of "other" places. Relational justice is undergirded by such listening—as is climate-justice preaching. Centering displaced voices in the climate crisis requires more than careful descriptions. It requires postures of embodied attention and affectability. Bale's description of Fijian listening provides a starting place for such reorientations.

Vakarorogo ena vakanomodi (Listening to Silence)

When ethnographer Matt Tomlinson was first learning the *iTaukei* language prior to his fieldwork, his Fijian instructor gave him a valuable lesson. She told him that he would find himself at kava-drinking sessions where there would be long periods of quiet.[18] He "should not rush to fill the silence with words. . . . Silence [is] not a bad or awkward thing." Tomlinson then describes how hard this advice was to follow for a researcher anxious to connect, learn, and question. He had needed multiple lessons about "needing to shut up sometimes."[19] The gift of silence, however, became plain to him as he got better at the skill. He describes the "warm, liquid silence" of late-night kava sessions, where, finally, someone would ask him to "talanoa mada!" (Tell a story; make conversation!). Silence had been the necessary precursor to deeper, more substantive engagement.[20]

When Bale speaks of "listening to silence," he suggests that it is possible to "find meaning in and behind silences."[21] In Bale's historical work, that has meant paying attention to who is included in conversations and who is not. Bale's research focuses on how the voices of Oceanic communities are represented—and at times, excluded—in historical surveys of the region. He is interested in the assumptions and social structures exposed by these omissions. When Rev. Eric Baleloa connects the environmental crisis to domestic violence against women, he is paying attention to other exclusions. In his preaching, Baleloa argues for the recentering of *iTaukei*

---

willingness to share the outlines of his research methodology through the conversation described, but also through personal email correspondence, November 5, 2023.

18 Tomlinson outlines a brief history of the social significance of drinking kava—a mild narcotic made from the *yaqona* root—and the Fijian debates around the practice in "A Consuming Tradition: Kava Drinking in Fiji," *Expedition* 48, no. 3 (2006): 8–17.

19 Matt Tomlinson, *God Is Samoan: Dialogues between Culture and Theology in the Pacific* (Honolulu: University of Hawaii Press, 2020), 29.

20 Tomlinson, *God Is Samoan*, 29.

21 Bale, personal email correspondence, November 5, 2023.

women's voices within the Fijian context itself.[22] To hear what is said and what is not takes time, as does interpreting those silences. Deep understanding rarely comes on a researcher's or preacher's preferred timetable. Time is especially significant if nonverbal forms of communication are being engaged. Sophie Chao and Dion Enari describe the "sensory attunement" necessary to listen to material places. Such attunement requires a "slowing down of thought and movement" to make "space for the voices of others"—including nonhuman others.[23]

"Ice Watch, 2014" is an art installation created by Olafur Eliasson, designed to allow the silence of the Arctic's melting ice to speak into the urban din.[24] Eliasson collects ice blocks that have broken off glaciers and are currently melting in the Icelandic Sea. He moves these blocks to outdoor courtyards in public places like the Tate Museum in London so that communities that do not live near the ocean can experience the urgency of sea-ice melt. Without words, the ice testifies to a warming world—a "liquid silence" of another sort. Visitors are able to touch the ice and witness its disappearance. It is, in many ways, a silent sermon of lament. Paying attention to the presence of such nonverbal communication, and ultimately its absence, awakens participants to voices they would otherwise miss, calling forth a more honest story.

Vakarorogo vata (Listening with Mutuality)

For James Bhagwan, the woven sail of the Fijian church's *drua* is created through careful plaiting of the community's narratives. Each narrative strand is interlaced with others, like the pandanus leaves of a Fijian mat, over and under in turn. Diverse narratives are a "challenging opportunity for renewal and reform, . . . woven into the sail to strengthen it, to enlarge it and enable

22 Eric Baleloa, "What Must I Do to Ensure Blessing upon My *Vanua*: Luke 15:8–10," sermon preached to the Buca Circuit, Fiji, February 2, 2020. In the context of the United States, African American environmental advocates like Heather McTeer Toney have pushed back against "silences" in environmental discourse created by the exclusion of people of color: "Collards Are Just as Good as Kale," in *All We Can Save: Truth, Courage, and Solutions for the Climate Crisis*, ed. Ayana E. Johnson and Katharine Wilkinson (New York: Random House, 2020), 75–83.

23 Sophie Chao and Dion Enari, "Decolonising Climate Change: A Call for Beyond-Human Imaginaries and Knowledge Generation," in "Tropical Imaginaries and the Climate Crisis," special issue, *eTropic* 20, no. 2 (2021): 32–54 [40], http://dx.doi.org/10.25120/etropic.20.2.2021.3796. The authors cite a Marind proverb in making their point: "Stop talking, start listening. Stop thinking, start walking."

24 Olafur Eliasson, "Ice Watch, 2014," images from the Tate Museum Installation, London, 2018, https://olafureliasson.net/artwork/ice-watch-2014/.

it to harness the wind of the Holy Spirit."[25] Bale's description of the Fijian practice of mutual listening or "two-way listening" emphasizes the woven character of such conversations as well. It is not only that each narrative has significance. Bale is interested in the dialogical process of listening. In his words, "the speaker listens too, [just as] the listener also has a chance to speak."[26] *Vakarorogo vata* describes a listening where silent attention is woven together with speech and where distinct narratives become visible in turn. It is a description of listening that honors difference, even as it imagines how the preservation of difference might serve a communal purpose.

For Rev. Samuela Delai, mutual listening opens new possibilities for listening to scientific narratives around the causes, effects, and mitigations of the climate crisis. "We need to integrate science and theology to be able to deal with this ecological crisis," he says. "The problem now is [that] we work in isolation from one another."[27] For Fijian pastors who associate climate science with the negation of theological framings of the world, Delai's call to relationality and mutual listening is provocative. But it is perhaps more provocative for climate scientists in Western contexts to imagine research that takes Indigenous understandings of place and spiritual practice seriously. There are fears that "relational" understandings of landscapes and Indigenous culture can be written off as "quaint or romantic."[28] But climate scientists are increasingly bold about holding Indigenous knowledges and scientific knowledge in conversation, recognizing the importance of "contingency and the mutual constitution of time-space for understanding place."[29] In a recent research visit to the Alaskan North Slope, climate scientists noted the ability of Inuit peoples—often portrayed as vulnerable victims in the climate narrative—to adapt to rapid shoreline retreat and loss of sea ice. "A long history of interacting with a dynamic and changing Earth renders Inuit knowledge key to securing Arctic futures," the researchers conclude. But more than disparate threads of wisdom emerge from such engagement.[30] Climate science, in particular,

[25] James Bhagwan, "Sailing to the Island of Hope: A Wesleyan Ethical Framework for 21st Century Fiji" (unpublished master of theology thesis, Methodist Theological University, Seoul, 2013), 99.

[26] Marika Baleidelabaci "Bale" Tamanikaigavo, personal email correspondence, November 5, 2023.

[27] Samuela Delai, personal email correspondence, June 29, 2023.

[28] Wilcock, Brierley, and Howitt, "Ethnogeomorphology," 592.

[29] Wilcock, Brierley, and Howitt, "Ethnogeomorphology," 581.

[30] Veland, "Transcending Ontological Schism," 615.

has begun to articulate its need for "moral imagination" in its research, "challeng[ing] the assumption that politics can be excluded from scientific method."[31] The hope is that distinct narratives will weave a stronger sail.

Woven narratives do not necessarily imply compatible narratives. Tuck and Yang repeatedly push back against conflation of displaced voices, noting that decolonization will require an "unsettling" recognition of the "uncommonality" of priorities between immigrants, refugees, and various Indigenous communities.[32] Bale's model insists that the particularity of these witnesses is part of what creates a weave useful to communal life. Trusting the dialogical process—particularly in the face of conflict—is key. Holding voices in woven relation, rather than insisting on uniformity, can lead to the emergence of new insight. Such insights can become places where wiser decisions are made and deeper relationships formed. Bhagwan notes that when an ocean voyage is completed, the sail of a *drua* is taken down and used "as a mat for the community to rest on."[33]

Vakarorogo ena vakarokoroko (Listening with Respect)

S. Lily Mendoza cautions that in a world of "unbridled commodification and marketization, . . . Indigenous spirituality quickly becomes one more commodity to mine, market, and manipulate."[34] When unaccompanied by respectful accountability to the places, people, and protocols of Indigenous communities, insights can be too easily appropriated as "footnotes to dominant paradigms."[35] Like listening to silence, such accountability takes time and a hard-earned reticence in pressing for relevant "takeaways." Good intentions, as important as they are, are not enough for meaningful cross-contextual engagement—particularly when "larger historical, economic, and material structures . . . constrain our relations" to

[31] Wilcock, Brierley, and Howitt, "Ethnogeomorphology," 589.

[32] Eve Tuck and K. Wayne Yang, "Decolonization Is Not a Metaphor," *Decolonization: Indigeneity, Education and Society* 1, no. 1 (2012): 1–40 [35–36].

[33] Bhagwan, "Sailing to the Island of Hope," 100.

[34] S. Lily Mendoza, "Transdiasporic Indigeneity and Decolonizing Faith: Recovering Earth Spirituality in a Settler Colonial Context," in *Decolonizing Ecotheology: Indigenous and Subaltern Challenges*, ed. S. Lily Mendoza and George Zachariah (Eugene, Ore.: Wipf and Stock, 2022), 259–80 [274].

[35] Tuck and Yang, "Decolonization Is Not a Metaphor," 22. Vanessa Watts also critiques the use of Indigenous lore as "theoretical jump-offs" for Western critiques: "Indigenous Place-Thought and Agency amongst Humans and Non-Humans (First Woman and Sky Woman Go on a European World Tour!)," *Decolonization: Indigeneity, Education and Society* 2, no. 1 (2013): 20–34.

each other.[36] There are no meaningful translations of cultural knowledge that sidestep the long commitment of relationship.

This is true not only for research that engages Indigenous communities. Susan Dunlap describes the "conversion" she experiences in worshipping alongside a displaced community of a different sort: the unhoused community in Durham, North Carolina. Through her ongoing work as a shelter chaplain, Dunlap has developed relationships of solidarity with persons whose lives have been overturned by poverty, racism, addiction, and Durham's increasing lack of affordable housing. *Shelter Theology: The Religious Lives of People without Homes* is Dunlap's careful description of the theologies of resistance she finds in this community, but it also confesses a growing awareness in herself. "I have and wield the power that I do," she states in the book's conclusion, "because of forms of privilege that have nothing to do with virtuous decisions and good behavior."[37] It is a confession born of yearslong, respectful listening to the stories of Durham's unhoused and their descriptions of the entrenched injustices that limit their agency. There is a kind of "helping" that jumps over such historical, economic, and material constraints, assuming the listener has ready answers to intractable problems. Dunlap asks her readers to slow down and pay attention to the depth of her interlocutors' challenges. She bears witness to their naming of poverty's structural violence.

When Bale reflects on the technological solutions to the problems of the climate crisis offered by outside nations, it is their avoidance of the historical structures of environmental disaster that worries him. It is not that he is against technological solutions, in principle. He is against solutions that obscure the colonial roots of the crisis and dismiss the ongoing impact of that history. To listen respectfully to those being displaced by ecological disaster is to acknowledge the complexity and depth of the problems these communities name. Such listening also acknowledges

36 Mendoza, "Transdiasporic Indigeneity," 276.

37 Susan Dunlap, *Shelter Theology: The Religious Lives of People without Homes* (Minneapolis: Fortress, 2021), 149. I also point to M. Jan Holton's study of salvation narratives in displaced Sudanese communities as an example of respectful, reciprocal listening in pastoral theology: "Imagining Hope and Redemption: A Salvation Narrative among the Displaced in Sudan," in *Interpreting Exile: Displacement and Deportation in Biblical and Modern Contexts*, ed. Brad E. Kelle, Frank Ames, and Jacob Wright (Atlanta: SBL, 2011), 217–34.

their shared resources. Displaced communities are not only sites of need, Bale insists. "We also have . . . solutions."[38]

Vakarorogo vakavinaka (Listening for Right Response)

Bale insists that listening does not happen for its own sake. It listens to respond rightly in the world. Bearing witness to the complexity of problems is significant work, but if practical theology stops short of enacted commitments of solidarity on behalf of the world God loves, it has missed its goal. There is a danger to practical theological methods that emphasize listening—or to homiletic methods that emphasize "voice." In Faafetai Aiava's view, such linguistic formulations can lead "to a false confidence that being 'a voice' is doing our part."[39] In a context where Western churches regularly avoid the subject of climate change altogether, Kathrine Hayhoe's assertion that the most important thing one can do for a planet in crisis is "to talk about it"[40] makes sense. But facilitating hospitable conversation between emplaced and displaced communities—or even between climate advocates and climate deniers—is not enough.

Decolonial justice in the climate movement requires more than conversation. It requires a commitment to covenantal practices of restitution. In her 2022 speech before the World Trade Organization, Prime Minister Mia Mottley of Barbados articulates her argument for debt relief as reparation for the climate disasters caused by rich nations. Austerity measures have become impossible for the small island nation to maintain in the face of the infrastructure damage done by increasingly fierce storms. "The inequity of the world in which we live," she states, ". . . is preserved fundamentally because we've not changed the power structure." Her diagnosis of the problem is precise. "We are more concerned with generating profits than saving people," she says, "[which] is perhaps the greatest

38 Marika Baleidelabaci "Bale" Tamanikaigavo, personal online interview, January 25, 2022.

39 Faafetai Aiava, "Eleele Interrupts the Eden Wedding: From Mother Earth to Mistress," in Mendoza and Zachariah, *Decolonizing Ecology*, 93–107 [103].

40 Katharine Hayhoe, "The Most Important Thing You Can Do to Fight Climate Change: Talk About It," TEDWomen 2018, November 2018, https://www.ted.com/talks/katharine_hayhoe_the_most_important_thing_you_can_do_to_fight_climate_change_talk_about_it?language=en.

condemnation that can be made of our generation."[41] The question of reparations is not only a question of finances. It is also a question of authorization and power. Are Western theological institutions and practical theological guilds willing to have our universal, insular, and constructivist understanding of place disrupted? And what would such a willingness cost us? I wonder if we have even scratched the surface of that question. Listening attentively and "affectably" to displaced communities would allow time-consuming, descriptive work to disrupt our present constructions of the guild's relevance and import. It would place us, instead, in dependent, bounded relationships of responsibility, accountability, and belonging.

Throughout this book, I have suggested that one of the best ways for Western preachers to form congregations in "affectability" in relation to the concerns of contemporary displaced communities is to listen "affectably" to displaced witnesses in the biblical text. I own that Ezekiel is no easy conversation partner in making this point. I intentionally chose a biblical prophet for this project who is often written off as "audacious, confusing, offensive,"[42] and dangerous to the environmental movement to acknowledge the challenges of attentive listening in practical theological method—not to mention affectable listening! Sarah Jobe's call to "read and re-read" such texts in order "to hear them on their own terms" is a place to begin.[43] Such listening does not mean that readers silence the difficult questions such texts raise or justify the violence described in them. When Ezekiel uses graphic descriptions of rape as a metaphor for divine justice, it is not "respectful" to turn away. Such images raise pointed questions about the prophet's theology of God and his justification of sexual violence. But as Carolyn Sharp notes in her wrestling with the conquest texts of Joshua, rejection of a text's violence is a commitment held alongside a refusal to dismiss Scripture's complicated witness.[44] Dismissals foreclose the recognition of ourselves in Scripture's grief-stained story—and

[41] Abrahm Lustgarten, "The Barbados Rebellion: An Island Nation's Fight for Climate Justice," *New York Times Magazine*, July 27, 2022, https://www.nytimes.com/interactive/2022/07/27/magazine/barbados-climate-debt-mia-mottley.html.

[42] While Sarah Jobe does not explicitly name Ezekiel in these categories, she chooses Ezek 17 as the case study she chooses to engage in her chapter on interpreting challenging, regularly dismissed biblical texts: "A Hermeneutics of Trauma: Trauma-Informed Care as Method for Biblical Interpretation," in *A Hermeneutics of Trauma* (Sheffield: Sheffield Phoenix Press, forthcoming 2025).

[43] Jobe, "Hermeneutics of Trauma."

[44] Carolyn J. Sharp, "'Be Strong and Resolute!' Reading Joshua in the Contemporary Church," *Anglican Theological Review* 97, no. 1 (2015): 19–32.

the reckoning that follows such recognition.[45] Such reckonings hold the potential for repentant transformation. And in Sharp's view, they are far more likely when we "read [the Bible] with the Other."[46]

It is my conviction that reading the Bible with *displaced others* holds this transformative potential—particularly when that reading is marked by the kind of listening to one another that Bale describes. It transforms not only a preacher's understanding of Scripture's place. It transforms a preacher's understanding of her own place in the world. More significantly, it transforms her understanding of what *place* is: a relational, formational, and sacramental site of covenantal claim. This book has attempted to read the unraveling ecosystems of the climate crisis through the sermons and emplaced theologies of two distinct communities that have experienced the loss of land firsthand. In noting how Scripture is interpreted in these contexts, particular questions about environmental disaster and forced displacement have emerged—questions distinct from those that many scholars shaped by colonial epistemologies are asking. For Western preachers to listen "affectably" to these conversations is to prioritize the insights and concerns of these displaced communities when engaging the climate crisis.

These communities do not, of course, speak for all displaced or Indigenous communities. I have underscored the particularity of these communities throughout my discussion. I would celebrate greater breadth of engagement, both in relation to the displaced witnesses in the biblical text and in relation to displaced communities today. My next section outlines, however, how these specific conversation partners reframe significant trajectories and touchstones in Western climate-crisis preaching. In prioritizing questions of sovereignty, belonging, and loss, Fijian Methodist sermons and Ezekiel's prophesies make core theological claims that disrupt imperial theologies of place. For preachers whose pulpits are presently insulated from the climate crisis, this reframing reorients environmental sermons—and indeed, environmental advocacy—toward the limits and dependencies of creaturely existence. It highlights the continuing impact of colonial history in our present cataclysm, and in so doing, foregrounds reparative repentance as necessary for a just, sustainable future.

45 Daniel Smith-Christopher calls for "social reflexivity" in reading exilic texts: "Reading Exile Then: Reconsidering the Methodological Debates for Biblical Analysis in Dialogue with Sociological and Literary Analysis," in *By the Irrigation Canals of Babylon: Approaches to the Study of the Exile*, ed. John Ahn and Jill Middlemas (London: T&T Clark, 2012), 139–57 [157].

46 Sharp, "Be Strong and Resolute!" 32.

Reframing the Climate Crisis: Prioritizing Different Questions

Questions of sovereignty, belonging, and loss steer climate sermons toward the material, economic implications of environmental sustainability, reorienting congregations toward concrete acts of restitution in its relationships. But these questions have more than practical value. They have deep theological roots. As a preacher, to ask such questions is to navigate some of the most contested shoals in biblical anthropology. To foreground these questions is to articulate *who we are not*, *who we are*, and *where we stand* in a world under threat. Such reframings press Western proclamation beyond emplaced theologies of control and stubborn legacies of imperial normativity. They also challenge Western expectations regarding climate preaching's primary claims and ends.

Who We Are Not: Neither Divinities nor Norms

"God is God, and we are not." The sentence reads like a Sunday school platitude. It is certainly not a common place to begin conversations about environmental stewardship in eco-friendly, academic settings—and for good reason. The differentiation can reinforce platonic separations between sacred presence and materiality that ground Lynn White's well-worn critique of Western Christianity's ecological disregard.[47] But differentiation does not necessarily lead to siloed binaries; it can lead to relational interconnection and attentive humility.[48] Alternatively, the collapse of Creator and creation into a single category can threaten relationality. Ezekiel and the Fijian Methodist preachers highlighted in this project are suspicious of governments and technologies that claim salvific, divine-like power. Too often, such claims serve the purposes of imperial idolatries. For these leaders, creation's sacred character and humanity's active commitment to the thriving of nonhuman creatures are aided by an appreciation of difference, between creatures themselves and between creatures

[47] Lynn White Jr., "The Historical Roots of Our Ecologic Crisis," *Science* 155, no. 3767 (1967): 1203–7.

[48] Upolu Lumā Vaai's long theological project on Pacific relationality underscores that "communality" and "individuality" are meaningful in and through the other. "Being relational is about an ongoing wrestling together in a harmonious, holistic fashion. It implies both unity and distinctiveness": "Relational Hermeneutics: A Return to the Relationality of the Pacific Itulagi as a Lens for Understanding and Interpreting Life," in *Relational Hermeneutics: Decolonising the Mindset and the Pacific Itulagi*, ed. Upolu Lumā Vaai and Aisake Casimira (Suva: University of the South Pacific Press, 2017), 17–41 [29–30].

and Creator alike. Such an appreciation orients and orders relationships marked by responsible limits, plurivocity, and solidarity.

The words, "You shall know I am the Lord," occur more than fifty times in Ezekiel's text, and for many preachers, they are chilling. Known to biblical scholars as YHWH's "recognition formula,"[49] the words are regularly associated with Exodus' account of the Egyptian plagues, intended to bend Pharaoh's will toward restorative justice through holy—or at least, self-preserving—fear (e.g., Exod 7:5; 7:17; 8:10). Ezekiel repurposes the formula with a different audience in mind: the people of Israel. Though the phrase sometimes follows miraculous restoration (Ezek 36:11), it is most commonly associated with violence and grief (6:1–14; 15:6–8). For Norman Habel, Ezekiel's use of the formula creates a damning portrait of a divinity "with an ego problem [and] an incessant need to have the divine name and identity displayed as the banner associated with every violent deed."[50] But Habel's analysis diminishes what is at stake for Ezekiel in YHWH's self-revelation. Idolatry does more than threaten divine pride; it has fueled the land's destruction. The horror that Ezekiel's people and the land have endured is linked to the false hope created when objects (8:3, 10–11), military stratagems (17:11–18), or even the corpses of dead kings (43:7–9) are hallowed with salvific power apart from the limits and obligations of covenantal relationship. If idolatry is the sin that displaces Israel and pollutes the land, then the self-revelation of God does not necessarily indicate a divine ego problem. It is knowledge necessary to restoration and return.[51]

Ezekiel's fierce rejection of idolatry shares an unexpected resonance with secular diagnoses of our contemporary climate crisis. Writing for a broad audience, Roy Scranton notes the ways that the reality of global climate change intrudes "on our collective fantasies of perpetual growth, constant innovation, and endless energy, just as the reality of individual

[49] For Walther Zimmerli's discussion of this repeated phrase, see *Ezekiel*, trans. Ronald E. Clements, 2 vols. (Philadelphia: Fortress, 1979), 1:37–38.

[50] Norman Habel, "The Silence of the Lands: The Ecojustice Implications of Ezekiel's Judgement Oracles," in *Ezekiel's Hierarchical World: Wrestling with a Tiered Reality*, ed. Stephen L. Cook and Corrine L. Patton (Atlanta: SBL, 2004), 127–40 [136].

[51] Wojciech Pikor connects the practice of idolatry in Ezekiel with the objectification of the land and its subsequent "devastation and desecration." Significantly, the land's objectification also "makes it a space of violence for the economically deprived who lose their . . . property rights and the right to partake in the gift of the land": *The Land of Israel in the Book of Ezekiel* (New York: T&T Clark, 2018), 55.

mortality shocks our casual faith in permanence."[52] In response, he rejects false hopes that ignore a "consciousness of one's own limits" as a creature in a fragile world.[53] Similarly, Lisa Wells uses the language of idolatry specifically in relation to the environment's false saviors—be they charismatic figures, the latest environmentally friendly purchase, or a technological fix. "The cycle of worshipping idols, investing them with our disowned energies, becoming disillusioned, then lunging for the pedestal we placed them on—it's a dead end, numbing in its repetition . . ."[54]

More pointedly, playing god is the sin that lies at the heart of the imperial project. Divinizing cultures, economies, and technological advances does not require religious altars and rituals; it simply requires that one "privilege the norms of a [single] assumptive world."[55] Such reductions continue to place creation's thriving at risk. George Zachariah notes the dangers of benign-sounding generalizations that assert the existence of "one human family and one Earth" when communities share unequal responsibility for environmental loss.[56] For those who "feel on their bodies the effects of empire and dictatorship," the connection between human hubris and environmental degradation "is perfectly comprehensible."[57] For these communities, "God is God, and we are not" is no theological cliché. It is a theology of defiance. It refuses to submit to the imperial methods of control that promise more than they can fulfill. Zachariah notes how, when global negotiations on climate change are controlled by the world's wealthiest nations, these nations often use the crisis "as an opportunity to continue their economic colonization of the Global South through 'disaster capitalism.'"[58] To lay aside such false hope and false gods would mean "system change, not climate change."[59]

[52] Roy Scranton, *Learning to Die in the Anthropocene: Reflections on the End of a Civilization* (San Francisco: City Lights Books, 2015), 22–23.

[53] Scranton, *Learning to Die*, 91.

[54] Lisa Wells, *Believers: Making a Life at the End of the World* (New York: Farrar, Straus and Giroux, 2021), 305.

[55] Courtney Goto, *Taking on Practical Theology* (Leiden: Brill, 2018), 133.

[56] George Zachariah, "Whose *Oikos* Is It Anyway? Towards a *Poromboke* Ecotheology of 'Commoning,'" in Mendoza and Zachariah, *Decolonizing Ecotheology*, 201–18 [203–4].

[57] Christoph Uehlinger, "The Cry of the Earth? Biblical Perspectives on Ecology and Violence," in *Ecology and Poverty: Cry of the Earth, Cry of the Poor*, ed. Virgil Elizondo and Leonardo Boff (London: SCM Press, 1995), 41–54 [47].

[58] Zachariah, "Whose *Oikos* Is It Anyway?" 206.

[59] Zachariah, "Whose *Oikos* Is It Anyway?" 207.

Differentiations between Creator and creation can also provide an unexpected source of communal resilience. Revs. Lynne and Gerald Godette arrived at St. Paul AME Zion Church in Aurora, North Carolina, when Lynne was appointed as pastor in 2012. The devastation of Hurricane Irene was still visible when they arrived, and the church was rebuilt under her leadership. Following Hurricane Florence in 2015, the church flooded again. "Knee-deep flooding. Pews pressed against the pulpit."[60] Rev. Godette recounts the sermon she preached on the Sunday after the second hurricane. The text was Psalm 24, and her theme was simply, "The Earth Is the Lord's."[61]

Such theocentricity might seem precisely the wrong impulse in such a situation, indicating a dangerous resignation, or worse, a divine callousness regarding disaster. One might be grateful that Ezekiel—a prophet who regularly links devastation with divine action—was not preaching that Sunday.[62] But Adriane Leveen insists that, even for Ezekiel, theocentricity is never yoked with fatalism or human passivity.[63] Following Jonathan Levenson, she notes that God's gracious acts of restoration do not absolve the community from the "human striving"[64] that holy living requires. Similarly, for Rev. Gerald Godette, the claim that "the earth is the Lord's" is not meant to diminish human responsibility in the face of its destruction. God's ownership of the land serves as a communal "impetus to help prevent [repeated flooding] from happening." As an example, he notes the importance of restoring the health of saltwater marshes as

60 Rev. Lynne Godette describes the scene in an interview with Avery Davis Lamb, co-executive director of Creation Justice Ministries, "Day 4: Disaster Relief Climate Action," EcoAmerica, April 8, 2022, https://www.youtube.com/watch?v=yU1f9tU2noA.

61 Rev. Godette shared the content of her sermon in a presentation at the Pastoral Care for Climate: Weaving Science and Theology for Justice conference, Duke Marine Lab, Beaufort, N.C., June 15, 2021. "The earth is the Lord's" is a theme that recurs in her descriptions of her proclamation at St. Paul's AME.

62 This link between devastation and the divine hand continues to be the most troubling aspect of Ezekiel's prophetic witness for many—though, as described in chapter 3, proclaiming God's agency in disasters can have surprisingly life-affirming implications for displaced communities. See in particular Taniela Balenaikorodawa's sermon on Gen 1:1–4: "God the Creator, Recreating," Cogea, Fiji, January 29, 2021.

63 Adriane Leveen, "Returning the Body to Its Place: Ezekiel's Tour of the Temple," *Harvard Theological Review* 105, no. 4 (2012): 385–401, esp. 386.

64 Jon Douglas Levenson, *Theology of the Program of Restoration of Ezekiel 40–48* (Missoula, Mont.: Scholars Press, 1976), 45–46.

sea levels rise. "The earth is not ours," he reiterates.[65] His trusting, active stance echoes chapter 3's descriptions of displaced communities in Fiji, where affirmations of God's ownership of creation amplify material practices of responsibility for the environment. Crucially, for Rev. Lynne, God's presence and power are not leveraged to explain the storm's cause.[66] Instead, these divine characteristics encourage the congregation in their call as "stewards of God's creation" and remind them that there is a "God who will bring [them] through."[67] In her sermonic witness, recognition of God allows for the congregation's recognition of itself.

Who We Are: Creatures in Covenantal Relation

I was in the middle of a stimulating week of teaching on the climate crisis and the Bible at a historic summer-vacation community situated on the Great Lakes. In the evening, a participant mentioned how his family was responding to rapid environmental change. "We're selling our family property in Texas," he said, "and we're buying each of our children plots of land around the Great Lakes. Our family is going to be ready when the water crisis comes." The comment left me shaken. Here was a man who needed no convincing that climate change was real; he was planning on it. And yet, his response was frighteningly insular. The conversation reinforced my conviction that persuading congregations to trust the validity of climate science is, finally, not the telos of climate-justice preaching. People can believe that environmental threats are urgent and still act in ways that reinforce existing imbalances of power and wealth. One can read the data and double down on settler-colonial ways of distributing land.

One of the most significant implications of facing *who we are not* in relation to climate disaster is a corresponding affirmation of *who we are*: creatures in covenantal relation to God and to the world we inhabit. Our limits are not meant to cause shrugs of resignation. They are meant to press us into deeper relational commitment. Genesis 1 does not speak directly of covenantal relation between God and creation—but it does describe an intricately connected world. Different borders and different

65 Rev. Gerald Godette, interviewed in "Faithful Resilience: Stories from North Carolina," Creation Justice Ministries, August 26, 2021, https://www.youtube.com/watch?v=7OKCe6lT9Yc.

66 For an accessible guide for preachers responding to natural disasters in their communities, see Jeff Stanfill, *Disastrous Preaching: Preaching in the Aftermath of a Natural Environmental Disaster* (Eugene, Ore.: Wipf and Stock, 2022), 56–60.

67 Rev. Lynne Godette, "Day 4."

roles create an interdependent reality, where earth and sea, human and nonhuman, are necessary to the survival of the other.[68] Genesis 9:9–10 uses the language of covenant explicitly. For those who have not discerned creation's weave, God makes it plain. The divine covenant extends beyond humankind. God's faithfulness stretches to "every living creature" (v. 10). This covenantal connectedness throbs under the grief of displacement in Ezekiel, and it orders his understanding of restoration. The covenantal responsibility between land and people is a recurring motif in Fijian Methodist sermons, as well—a manifestation of the community's commitment to God.

Tevita Banivanua worries that, in response to a globalizing economy, even the Fijian Methodist Church is becoming "individualistic in its outlook." Part of his New Exodus sermonic work was to press the church toward a "wider vision"—to be "a church with wisdom, not just a clever church."[69] For Rev. Tevita Kete, such wisdom means "patching holes in the [water] drum of the *vanua*."[70] When the relational container of creation is damaged, adding more resources will not bring abundance. "There are holes we need to fill," he states. Samuela Delai describes the significance of interpersonal reconciliation in his own work for environmental sustainability in a village in which he ministered. Patching the holes in the relational fabric of the community meant, in this village context, "trying to forgive each other" for offenses long past. For Delai, climate change is related to a lack of "social holiness . . . [indicated by] a relational problem with God and within the community."[71]

Attending to the health of ecosystems through attention to the health of interpersonal relationships may seem misguided, but it is just such work to which Robin Wall Kimmerer calls settler communities who desire deeper relationships with the land on which they live. In describing how "a foreigner" might become "naturalized" in a place, she describes the need to listen and learn from those native to the land. This means learning from human communities, certainly. But for Kimmerer,

68 Faafetai Aiava makes explicit the voice and agency of the earth in his drama based on the Genesis text, "Eleele Interrupts the Eden Wedding." Norman Wirzba has also stressed the connections between the soil and the human creature Adam in Gen 2: "Dramas of Love and Dirt: Soil and the Salvation of the Earth," *The Cresset* 77, no. 3 (2014): 6–15.

69 Tevita Nawadra Banivanua, personal interview, Suva, January 2, 2023.

70 Tevita Kete, personal online interview, January 27, 2022.

71 Samuela Delai, personal online interview, January 23, 2022.

it also means learning from nonhuman creation. To describe how settlers might humbly make a home in a new place, she turns to the *Plantago major* plant—the common plantain. The plant was brought to the North American continent by European settlers, and unlike many non-native species, it did not "take over others' homes and [grow] without regard to limits."[72] Instead, "its strategy was to be useful, to fit into small places, to coexist with others . . . , [and] to heal wounds."[73] For Kimmerer, whose Potawatomi ancestors were displaced from the land surrounding the Great Lakes, this plant's witness to humble interconnection is a model for human relationships. Tellingly, her emphasis on limits does not lead to insularity or rigid categories of self and other. Human limits nourish a relational lifeworld—even for non-Native communities. "Being naturalized to place," she notes, "means [living] as if this is the land that feeds you, as if these are the streams from which you drink. . . . Here you will give your gifts and meet your responsibilities."[74] For Kimmerer, the climate crisis requires more than technological solutions; it requires a willing dependence on others in the world.

I was invited out to dinner by a different resident of that Great Lakes vacation community to talk about how the association might change its relationship to the land on which it was located. Similar to Rev. Delai, the resident had reached the conclusion that ecological justice required a heightened attention to "relational problems" between the community and its neighbors. "A place to start might be allowing Native communities in the area access to the association's private beach," she said—speaking in low tones, as if the suggestion were provocative. *Such a small step*, I thought. And such a significant indicator of the distance between many North American settler communities and the places they reside. Covenantal relationships, particularly those that disrupt postures of entitlement, seem remote possibilities. But this resident had an astute understanding of what the first step in changing such postures would require. The community needed to remember its history, she told me—including those decades where implicit and explicit mechanisms excluded certain racial and religious identities from purchasing land in the area. It would also need to remember the displacements that had made the community's

72 Robin Wall Kimmerer, *Braiding Sweetgrass: Indigenous Wisdom, Scientific Knowledge, and the Teachings of Plants* (Minneapolis: Milkweed Editions, 2015), 213–14.

73 Kimmerer, *Braiding Sweetgrass*, 213–14.

74 Kimmerer, *Braiding Sweetgrass*, 215.

land "available for purchase" in the first place. To embrace covenantal relationality, the community would have to face specters from the past that continue to haunt its present.

Where We Stand: Accountable to Creation and to God

The land where I teach is the homeland of several Native nations, including Tutelo- and Saponi-speaking peoples. Many of these communities were displaced or killed through war, disease, and colonial expansion.[75] As part of an interdisciplinary report on the complexity of Duke University's history, Mumbi Kanyogo tells the story of twelve children from the Eastern Band of the Cherokee Nation who enrolled in Trinity's College's "Cherokee Industrial School" in 1880, at the expense of the federal government. Some of these children "struggled to wear different clothing and refused to eat college food" and were as young as eight.[76] One student escaped and traveled over a hundred miles to return to his home. Kanyogo notes that "despite the fact that North Carolina has the highest Native American population east of the Mississippi River," Native American students and faculty continue to be underrepresented in the community—evidence that the university's work of rebuilding trust is unfinished.[77] The history is bitter, and it intersects other histories of oppression. The past violence and harmful legacy of racialized slavery has also left its mark on the land.

[75] Thanks to Drs. Ryan Emanuel and Malinda Lowery of the Lumbee tribe for their collaboration with Duke's Nicholas School of the Environment (NSOE) in its process of creating a meaningful land acknowledgment, and to Dr. Nicki Cagle, NSOE associate dean of diversity, equity, and inclusion (DEI), for her prioritization of this work: see Nicki Cagle et al., "Land Acknowledgments: An Exploration," accessed December 3, 2023, https://sites.nicholas.duke.edu/diversity/expand-your-knowledge/teaching/land-acknowledgements-an-exploration/.

[76] Mumbi Kanyogo, "Native American Students," in *Activating History for Justice at Duke: A Report by Constructing Memory at Duke*, final report of the Constructing Memory at Duke Bass Connections Project, Durham, N.C., 2017, 72, https://www.activatinghistoryatduke.com/uploads/8/9/2/3/89234082/print_activatinghistoryatduke_final_1up_highres.pdf.

[77] Kanyogo, "Native American Students." Duke University's Native American and Indigenous Student Alliance (NAISA) has recommended several actions for the school to take in order to receive an official acknowledgment from the group, including the establishment of a Native American and Indigenous Studies program, the creation of an Indigenous Cultural Center, the consideration of full-tuition scholarships for Indigenous students, and the hiring of multiple senior Indigenous faculty: see Cagle et al., "Land Acknowledgements."

The *Activating History for Justice at Duke* report notes that Duke Forest includes four sites where it is known that slavery was practiced.[78]

There are plenty of specters to go around.

Some of those specters are solidified in Durham's concrete. Displacements are not just a thing of the past. For much of the first half of the twentieth century, the Hayti neighborhood was the thriving hub of the city's African American economic base. Professors at North Carolina Central University, businessmen working at the African American–owned North Carolina Mutual Life Insurance Company, doctors, and other professionals lived and invested in the community—leveraging Black-owned institutions to build the neighborhood's financial independence.[79] This concentration of Black wealth was, in part, a matter of necessity. Jim Crow–era segregation laws meant that many neighborhoods in Durham were off-limits to these African American families. When the 1960s brought millions of federal dollars to the city for the purpose of "urban renewal," there was hope that the community's resources might be more equitably shared. The result, instead, was a different sort of segregation created by a highway constructed through the heart of the Hayti community. The historic neighborhood has yet to recover.[80]

The displacement caused by the highway's construction decimated the local economy, and the road's environmental impact continues to be felt.[81] Expensive apartment complexes rise on the opposite side of the freeway, while affordable housing developments in the Hayti neighborhood have seen continuous delays. Its landscape is presently marked by telltale signs

[78] Cagle et al., "Land Acknowledgements."

[79] W. E. B. DuBois, writing in 1912, praised Hayti as the thriving model of a self-sustaining "new group economy" for African Americans: "The Upbuilding of Black Durham: The Success of the Negroes and Their Value to a Tolerant and Helpful Southern City," *The World's Work* 23 (January 1912): 334–38 [334], cited in "The Hayti Restorative Justice Report, Spring 2023," Social Justice and Racial Equity Institute, North Carolina Central University, April 15, 2023, 2.

[80] Joel Brown, "Durham's Hayti Community: A Black Business Mecca Crushed by Empty Promises Reckons with Its Past," ABC Eyewitness News, February 22, 2021, https://abc11.com/durham-hayti-community-freeway-history-black/10362504/.

[81] Christina Greene recounts how the Black community's assessment of "urban renewal" changed when "it became clear that the poor would have no voice in its implementation." When the Durham Redevelopment Commission announced plans in 1966 for the construction of the East-West Expressway, "200 predominantly black and low-income families were threatened with eviction. The crisis was aggravated by a waiting list of over 1,800 families for existing public housing": *Our Separate Ways: Women and the Black Freedom Movement in Durham, NC* (Chapel Hill: University of North Carolina Press, 2004), 128. Thank you to Cullen McKenney for pointing me to this resource.

of a struggling economy and unemployment. One might argue that the highway was a necessary evil if one ignores its trajectory. But from aerial photographs, one notices that the road contains a significant curve. It swerves around a historically white downtown congregation—protecting its beautiful stained glass and its pride of place in the community. African American congregations were not given the same courtesy. Multiple churches lost their sanctuaries.[82] One does not understand the place of Durham without reckoning with that curve and its preservation of racialized privilege. No matter one's congregation, such understanding is vital for Durham's preachers. Covenantal relation to the community and the biblical text means discerning one's place within that story.

When grief is entrenched over centuries, layering multiple narratives of betrayal, how does place heal? How is the trust that grounds generative relationship restored? Norman Wirzba comes at this question through a practical theological case study where the wronged party is the land itself.[83] To create a new start in a broken relation with the earth, Wirzba argues, one must start with the difficult work of confessing the "details of where we are" and the broken relationships that got us there. Restored relationship begins by knowing the wrong one has done and owning one's need for repentance.

Wirzba tells the story of Don and Marie Ruzicka, Canadian farmers who invested in industrialized production methods to maximize crop yields until the land, their bodies, and their financial resources were pressed to exhaustion. In 1995, they took stock of the destructive toll. "What would a different way of farming look like," they wondered, "and what would it require of us?" They educated themselves about species diversity, the value of native species, and the dangers of quick-fix chemicals. They also began to own "their complicity in histories that were systematically degrading and destroying the land."[84] This transformation of themselves and their relationship with the land took time. It required the mourning of species

[82] For example, the Brookstown community congregations of West Durham Baptist Church and Mt. Olive AME Zion Church: Eddie Davis, "West Durham Baptist Church—Fifty Years at 1901 Athens Avenue," West Durham Baptist Church, accessed December 3, 2023, https://westdurhambaptist.org/jubilee-athens-avenue/. Joel Brown notes that St. Joseph's AME Church, one of Hayti's oldest original structures, "was once the center of [the community]. . . . Now it's all that's left of old Hayti—steps away from a highway exit" ("Durham's Hayti Community").

[83] Norman Wirzba, "The Soil of Spirituality: What Agrarians Can Teach Us about the Life of Faith," *Laudato Si' Research Institute* 2 (2023): 1–11. The essay is based on a public lecture given at the University of Oxford, March 10, 2022, https://lsri.campion.ox.ac.uk/node/1632.

[84] Wirzba, "Soil of Spirituality," 4.

loss that could not be restored. It required the unlearning of old methods of control and practicing new methods of relational, ecological respect. It required letting the land rest and putting its needs alongside their own. It required that Don and Marie "acknowledge wrong-doing . . . repent, change direction, and practice metanoia."[85] On May 21, 2000, Don heard a birdsong that he describes as among the most beautiful sounds on the prairie: the song of the Western Meadowlark. "He had not heard it since the spring of 1989," Wirzba recounts, "when, for some reason, the birds stopped showing up at his farm. . . . And now, here it was."[86] It felt something like forgiveness, like a new way of being in the world was possible.

Kimmerer insists that nonhuman creation has something to teach humans about relational respect. The time, reeducation, limits, release of control, and consistent prioritizing of others' needs that mark Don and Marie Ruzicka's restorative journey with land are not bad places to begin difficult journeys of restorative justice with the human communities the land sustains. The musculature needed for the one is not so different from the musculature needed for the other. Repentance is necessary for both. Honest acknowledgments of wrongdoing are also needed—which means that, for human communities, conflict and confrontation are a significant part of the healing process. Such acknowledgments, however, are only a start. Restoration requires the slow, consistent process of repairing damage, over years. Place is not simply something to be fixed; it is a site of longing and dependence. For those with ears to hear, it is a God-inspired preacher, forming a community that is simultaneously willing to shoulder responsibility and relinquish control. Each are qualities sorely needed in a relationally tethered world.

Returning land is a significant part of that formative process. Speaking at the 2022 Truth and Transformation Conference at Harvard's Kennedy School, Alvin Warren argues that returning the land to the control of Native peoples is "crucial to the health of our mother earth and potentially our survival as humans on this planet."[87] Having served as the director for the Trust for Public Land's Restoring Tribal and Indigenous Lands program, Warren has long been an advocate for returning traditional tribal lands

85 Wirzba, "Soil of Spirituality," 5.

86 Wirzba, "Soil of Spirituality," 5.

87 Dan Harsha, "Returning Land to Native Nations Is about Righting Historical Wrongs—and Also Averting Future Environmental Disaster," Harvard Kennedy School, Ash Center for Democratic Governance and Innovation, October 27, 2022, https://ash.harvard.edu/articles/returning-land-to-native-nations-is-about-righting-historical-wrongs-and-also-stemming-future-environmental-disaster/.

to Native control—known as the "landback" movement. "Native cultures are profoundly interwoven with the natural environment, the plants, the animals, [and] natural features," Warren explains. Such knowledge is necessary for "reciprocal and healthy relationships" with land—as is the formal holding of the land's title. Pressing his audience beyond collaborative learning with Native communities and toward tribal land-acquisition efforts, Warren urges, "We have to move from a place of panic, anger, guilt or willful ignorance to a place of responsibility, solidarity and equity. . . . Only then will this country begin to get serious about righting its historic wrongs in taking 1.5 billion acres of land from Native people."[88]

When President Andrew Jackson signed the Indian Removal Act in 1830, forcing all Native tribes to move west of the Mississippi, the Wyandotte Nation initially refused. After one of their young tribal leaders was murdered by white hunters, they reconsidered, fearing the dissolution of their communal identity.[89] When they left their home in northern Ohio, the Wyandottes gave three acres of their land to the local Methodist church, in part to honor the work of John Stewart, a missionary and son of slaves who had worked with the Wyandottes to build a church on that land. More urgently, the Wyandotte community asked the settler Methodist community to protect the graves buried there. In 2019, the United Methodist Church regifted the land to the Wyandotte Nation. For Thomas Kemper, the general secretary of the church's global ministries at the time, "giving back the land is . . . an act of repentance for times when Methodists mistreated Native Americans." Noting that Col. John Chivington, a Methodist, was the leader of the 1864 Sand Creek Massacre, Kemper minces no words: "People have been killed and we have been complicit in this."[90]

Such a small step, one may think. Three acres falls decidedly short of Warren's broader call to accountability and meaningful landback policies. And yet, in the climate fight and in the reweaving of a covenantally connected world, small steps matter. The danger of knowing where one stands in the climate crisis is that the ongoing impacts of racialized violence and

88 Harsha, "Returning Land to Native Nations."

89 Mary Stockwell describes the history of the removal of the Ohio's Native American tribes in *The Other Trail of Tears: The Removal of the Ohio Indians* (Yardley, Pa.: Westholme, 2016).

90 Marisa Iati, "The US Once Forced This Native American Tribe to Move. Now They Are Getting Their Land Back," *Washington Post*, September 20, 2019, https://www.washingtonpost.com/religion/2019/09/19/us-once-forced-this-native-american-tribe-move-now-theyre-getting-their-land-back/.

the traumatic theft of colonial histories can overwhelm—just as anxieties of a future one cannot control overwhelm. I listen to Samuela Delai's description of his choice to combat the fierce storms threatening his village through the strengthening of interpersonal relationships in his community, and I press him on this. "*Talatala* [Rev.] Samuela," I ask him, "you know that the broken relationships that are causing these changing weather patterns are not just in your village. They are caused by broken relationships between nations and within governments across the ocean. How does the restoration of relationships in your small place make a difference in such a world?"

He takes a long pause and a deep breath. This is not a new question for him.

"Change happens when we start small," he says. "We must start somewhere. I am a person of faith, and that is what I believe. I must start somewhere."[91]

Reclaiming Preaching's Place in the Storm: Covenantal Wayfaring in Community

Preaching seems a weak intervention in such a world. It is a small act—regardless of congregational size. Sermons are provisional, time bound, and fraught with risk. They can be easily blown off course by winds of conflict or run aground on unseen shoals. And yet this book has argued that sermonic speech is a crucial component in the fight for relational sustainability and climate justice. Leah Schade, who has been at the vanguard of mainline homiletic engagements with the environmental crisis for more than a decade, argues that preaching is "uniquely positioned to present a vision of what is possible when the faith community addresses [the climate disaster] and clearly communicates what it means to live in right relationship with Earth, community, and God."[92] Preaching can change "the kinds of stories we tell about ourselves."[93]

[91] Delai, personal online interview, January 23, 2022.

[92] Leah Schade, "Building the Eco-ethical Ark in the Age of Climate Disruption," in *Rooted and Rising: Voices of Courage in a Time of Climate Crisis*, ed. Leah Schade and Margaret Bullitt-Jonas (New York: Rowman and Littlefield, 2019), 17–22 [18]. The significance of preaching's theological engagement with climate-justice advocacy is the subject of Shade's influential *Creation-Crisis Preaching: Ecology, Theology, and the Pulpit* (St. Louis: Chalice Press, 2015).

[93] Schade, "Building the Eco-ethical Ark," 20. To make her point, Schade draws on the argument of Amitav Ghosh in *The Great Derangement: Climate Change and the Unthinkable* (Chicago: University of Chicago Press, 2016).

Schade is not alone in insisting on preaching's value. Over the past three years, I have heard her sentiment echoed in surprising ways by secular climate scientists from Duke's Nicholas School for the Environment.[94] In annual three-day retreats that bring together natural and social scientists, theologians, and area clergy, the importance of preaching has been underscored even by those who are not persons of faith. These scientists are interested in preaching's capacity to accessibly communicate environmental data in ways that mobilize grassroot communities. But it is preaching's ability to change the "kinds of stories we tell about ourselves" that interests them more. For communities steeped in stories of settler-colonial production and possession, such work is vital.

Communities displaced by imperial aggression have unique insights into these transformed narratives, precisely through their lived experience with place's loss. The sermons from the *iTaukei* pastors described in this volume and the prophesies of the exiled Ezekiel are only two examples of the diverse, under-engaged witnesses present in these communities—but their synergies are significant. Both resist destructive imperial theologies of place still operative in the world and in the contemporary environmental movement. More constructively, both reframe place as a site of covenantal responsibility and divine encounter. Both also believe that proclamation matters to place's restoration—though, again, they are less interested in rhetorical persuasion or accessible translation. In a way similar to Salmond's description of Indigenous environmental activists "rebuilding their ships at sea," proclamation has served as a site of "ontological innovation"[95] for these displaced communities. The proclamation of God's Word to God's people has allowed for bold, communal navigations beyond tired binaries and false choices that reinforce the powers that be. Blown by Spirit-wind and bound to the fixed fluidity of a world in threat, these witnesses have retold fixed and fluid stories about themselves. They have proclaimed a relation attentive to difference, a continuity alive with change, and an agency chastened by dependence. Proclamation has become a place of relational, formational, sacramental encounter.

94 Pastoral Care for Climate: Weaving Science and Theology for Justice was a three-year, interdisciplinary collaboration between Duke Divinity School, Duke's Nicholas School for the Environment, the Duke Marine Lab, and Creation Justice Ministries, Beaufort, N.C., 2022–2024.

95 Salmond, "Re-building Ships at Sea," 221.

How might these displaced witnesses impact how climate-justice preaching is practiced in pulpits presently insulated from disaster? I offer the following exhortations as places to begin.

Navigate Scripture with Communal Courage

Given the history of Scripture's co-optation by colonial exploitation and by those who diminish creation's value, it is no surprise that many Western preachers committed to climate justice approach the text with a certain trepidation. Questions of contemporary application also loom. What do ancient narratives have to do with sea-level rise and threshold temperatures? In gathering sermons for a course on preaching in the climate crisis, I found that the vast majority of environmental sermons were based on one of two texts: Genesis 1:1 or Romans 8:22. Both are fine scriptures. But it seemed as if ecologically minded preachers were not sure where else to turn for biblical wisdom. The witnesses of the displaced communities in this project point preachers to large swaths of the Scriptures that remain underexplored for the purposes of environmental justice, particularly those texts dealing with covenantal fidelity and exilic ruin. Indeed, the ache of place's destruction and the fury of imperial force throb throughout the biblical narratives. The more the colonial roots of the environmental crisis come into view, the more relevant the biblical text becomes. In a crisis fueled by broken relationships with others, creation, and God, Scripture becomes an especially apt conversation partner in reframing the stories we tell about ourselves.

I am not innocent, however, to the text's dangers.

Hulisani Ramantswama describes how the Bible has been both "a problem and a solution" in relation to the colonial legacies that plague the African continent, and this ambivalent history is not solved by excising troubling passages.[96] There is no safe biblical text, unable to be co-opted by greed and power—just as there is no safe ocean passage. Moreover, difficult texts may be the ones most crucial for those who benefit from colonial legacies to hear. In her sermon on Isaiah 65, Ellen Davis notes that verses 10–16 have been removed from the Revised Common Lectionary, presumably because of their fierce depiction of divine judgment. "Surely," she ironizes, "these words

[96] Hulisani Ramantswama, "Negotiation and/or Conquest of the Land: Reading the Land of Promise Motif in the Hexateuch through Decolonial Lenses," in Yeo and Green, *Theologies of Land*, 98–131.

were not meant for us."[97] The cosmic promise of divine restoration—the "new heavens and a new earth" described in verses 17–25—sound so much easier on our ears. But how does that promise sound "to those of us who are not in exile . . . settled as we are in the American empire, situated at the helm of the transnational, industrial empire"? Becoming granular in her mapping of the place she is preaching, Davis notes that "Duke University was founded with the profits of industrialism . . . [and] those same profits contribute to my salary and retirement account." And then she asks a version of the question that echoes throughout this book: "From our position of empire, can we share with [the displaced] the prophetic word of consolation and hope?" Where, she wonders, is her seminary congregation standing in this passage? Davis does not preach in simple reductions. God's hope is extended to all—but the hope of the "new heavens and a new earth" might look precisely like judgment for those who have reaped benefits from the world that is. To navigate Scripture is to courageously locate one's place in the story, which has everything to do with locating one's place in the contemporary world.[98] The same night-sky constellations signal different trajectories to different voyagers, calling for journeys of different distance. The meaning of a constellation depends on the place one's voyage begins—and where it leads.

Such navigation is not done alone. Navigation requires communities of accountability and the "vulnerability that relational commitments entail."[99] These relational commitments impact who joins us on our journeys and whose navigational wisdom we honor—factors that may fundamentally alter our trajectory to an "elsewhere" unseen.[100] Not every constellation is visible from the hemisphere a community inhabits, suggesting

97 Ellen Davis, untitled sermon on Isaiah 65, Service of Word and Table in collaboration with the Anglican Episcopal House of Study, Goodson Chapel, Duke University, Durham, N.C., November 9, 2022.

98 Daniel Smith-Christopher acknowledges the difficulty of such work, noting the dangers of simplistic comparisons between distinct histories. He also forecloses simplistic abdications of that comparative work. One cannot both say that there is "too little evidence for comparison and that [contexts] are too different to compare": "Reading Exile Then," 144.

99 Amiria Salmond, "Transforming Translations (Part 2): Addressing Ontological Alterity," *HAU: Journal of Ethnographic Theory* 4, no. 1 (2014): 155–87 [180].

100 Tuck and Yang, "Decolonization Is Not a Metaphor," 36. The authors insist that decolonization may require "elsewhere" understandings of justice, rather than extensions of Western conceptions of "rights."

the necessity of a global conversation.[101] But such relational commitments impact more than the breadth of one's interlocutors. They also impact one's approach to the biblical text itself.[102] They advise a relinquishing of control, even as genuine sources of conflict are negotiated. Such honest engagements of difference and disagreement, with the biblical text and with those who recognize God's Word through that text, allow for the slow growth of trust and responsible action. Upolu Lumā Vaai describes the necessary vulnerability that such approaches to Scripture involve, holding together calm and storm. Such interpretive risk gifts a formational "becoming," braiding together communal commitment, attention, and expectancy.[103] Each are critical qualities in a season of rapid ecological change. And each is necessary for preachers who navigate the biblical text for the purpose of climate justice.

The time is urgent, so I will speak plainly to my preaching colleagues. Do not be afraid of these scriptures, which have so often been the place of God's revelatory call for your community. They testify to where you stand and what will be required of your congregation in this season, particularly when interpreted in conversation with those who know place's loss. Your covenantal response to Scripture is covenantally tethered to the world God loves—a world in present danger.

Tack toward Scripture's Big Questions

At the start of a course on climate-justice preaching, it is common for participants to rattle off the resistances in their congregations to climate-justice sermons. Climate change may be politically polarizing in their communities. Certain members might distrust climate science. Perhaps a congregation is engaging other pressing challenges through its local witness,

[101] Such a conversation might be especially needed for deeply contextual forms of rhetoric like sermonic speech. The "push toward better-informed perspectives of rhetoric and writing as global enterprises" is championed by feminist and Afro-feminist rhetoricians Gesa Kirsch and Jacqueline Jones Royster in *Feminist Rhetorical Practices: New Horizons for Rhetoric, Composition, and Literary Studies* (Carbondale: Southern Illinois University Press, 2012), 25.

[102] Lisa Thompson notes that ethical praxis is the "first consideration" in sermon preparation, which includes self-reflexivity, discernment of "who benefits" from one's reading, and an analysis of which "canons" shape one's assumptions: "A Different Renewal of Sorts: Black Women's Lives, Praxis, and Disciplinary Futures," keynote address at the Academy of Homiletics Virtual Meeting, December 2021.

[103] Upolu Lumā Vaai, "*Motu ma le Taula*: Towards an Island 'Let Be' Hermeneutics," *Pacific Journal of Theology* 2, no. 53 (2015): 25–40 [32].

like patterns of racialized violence. Even for congregations who regularly hear sermons on climate catastrophe, resistances are many. There can be a "been there, done that" familiarity with the topic, or a growing sense of despair. It takes preachers several weeks to own their deeper fears. Given the breadth and depth of creation's plight, they have no idea what to say. *How does one preach in the face of the unthinkable?* The losses on the horizon are vast, and false hope seems the only hope to be had.

It is precisely here where the Scriptures testify most clearly. This is not the first time that God's people have asked the fundamental question, *Can the creation and covenant of a good God come undone?* The prophets' answer to that question is a discomfiting "yes." Jeremiah witnesses an undoing of creation resulting from his community's stubborn hearts (Jer 4:23–26), and Isaiah names his people's exchange of the divine "everlasting covenant" for a covenant with death (Isa 24:5; 28:15).[104] Ezekiel rejects false hope again and again, insisting that it is a dangerous game to treat God's promise like a pocketed possession (e.g., Ezek 13:10–15; 33:23–25). YHWH is no object to be manipulated, and covenantal relationship is no bill of sale. Surprisingly, this difficult word often grounds preachers in their task. I see their backbones strengthen and their spirits settle. They are still in a fight for life, and no simple answers have emerged. But they are not alone. God's people have asked these hard questions before, and the book in their hands does not ask them to pretend. It articulates the material and theological depth of our present need.

To preach climate sermons that are biblically faithful is to tack into these difficult waters. I encourage environmental preachers to use robust theological language. "Use the big words," I tell them, by which I mean the most resonant, familiar words in our faith. To name the crisis' theological roots integrates creation's well-being into the soil of a congregation's conviction. Creation care becomes more than a passing fad or one more good idea crowded out by many others on the congregational calendar.

This will look different in distinct communities, just as the theological framing of creation takes on different emphases in various portions of the biblical text. This project has emphasized the biblical and theologi-

[104] In Jer 4:23–26, the prophet describes creation's unraveling through rough reversal of the Gen 1 account, and Isa 24:5 accuses humanity of breaking God's "everlasting covenant," referencing the Noahic covenant of Gen 9. Instead, God's people have made a covenant with death (Isa 28:15). My thanks to Ellen Davis for these insights: "Divine Sovereignty and Covenant," course lecture for Hope for Creation? An Exilic Perspective, Duke University, February 7, 2023.

cal concept of covenant, reframing it in light of Indigenous and biblical understandings of place, but there are other scriptural streams relevant to climate engagement. How does the Incarnation impact a covenantally connected world?[105] How does the climate crisis intersect with God's atoning work?[106] What is the significance of Christ's resurrection life and new creation?[107] Scripture's questions are often so much braver than the questions that preachers ask, though displaced communities provide a regular exception to this norm.[108] Rev. Iliesa Nasedra's community and its multi-year prayer vigil did not ask God how to stop the rising tides.[109] They asked harder and more biblical questions: "Who is God?" and "What is the purpose of our lives?"[110]

How might that reframing shift the climate conversation in North America? It might change a congregation's sense of its identity and purpose, certainly. But it might also create surprising points of resonance with others whose relation to faith communities is ambivalent. When Lisa Wells describes "the problem of other people, [and] of how to live together"[111] as one of the central quandaries of the climate crisis, she does not explicitly speak in the language of faith. But she does ask *who we are* and *who we are not*—and if there is any hope of standing alongside one another in solidarity. She asks questions fundamental to the biblical text and the preacher's task.

[105] Elizabeth Johnson's discussion of "deep incarnation" in *Creation and the Cross: The Mercy of God for a Planet in Peril* (Maryknoll, N.Y.: Orbis Books, 2018) is an accessible description of divine incarnation that includes all flesh.

[106] Sylvia Keesmat and Brian Walsh reframe Romans' understanding of justification in light of the book's imperial context in *Romans Disarmed: Resisting Empire, Demanding Justice* (Grand Rapids: Brazos, 2019).

[107] Barbara Rossing's "Reimagining Eschatology: Toward Healing and Hope for a World at the Eschatos" images creation as the "little daughter" in Mark 5 at the edge of death: in *Planetary Solidarity: Global Women's Voices on Christian Doctrine and Climate Justice*, ed. Grace Ji-Sun Kim and Hilda Koster (Minneapolis: Fortress, 2017), 325–47.

[108] Speaking to white, mainline churches in North America, Walter Brueggemann notes, "It is likely that our theological problem . . . is that our gospel is a story believed, shaped and transmitted by the dispossessed; and we are now a church of possessions. . . . We are left to see if it is possible for us again to embrace solidarity with the dispossessed": *The Land: Place as Gift, Promise, and Challenge in Biblical Faith*, 2nd ed. (Minneapolis: Fortress, 2002), 206.

[109] As described in chapter 3.

[110] Iliesa Nasedra, personal interview, January 18, 2022.

[111] Wells, *Believers*, 296.

Weave Creation Justice with Human Justice

The interconnection between human and nonhuman thriving has been a recurring theme in this volume, as has the connection between the exploitation of land and the exploitation of people. The introduction of Rob Nixon's *Slow Violence and the Environmentalism of the Poor* begins with a confidential memo written by Lawrence Summers while he was working at the World Bank. "Shouldn't the World Bank be encouraging more migration of the dirty industries to the Least Developed Countries?" Summers asks. "The economic logic behind dumping a load of toxic waste in the lowest-wage country is impeccable."[112] Such a sentiment may be shocking to see in print, but it is no real surprise. Summers names a reality we already know: those on the front lines of climate devastation are people of color, the forcibly displaced, and those nations stripped of resources during the European colonial project. Womanist ecotheologian Melanie Harris delineates the ways in which "parallel oppressions are shared by women of African descent and the earth,"[113] an assertion that Heather McTeer Toney fleshes out. "Who exactly do you think is on the front lines every single day," Toney asks, "fighting to keep our communities safe from industries, polluters, and those seeking to harm our kids? Who do you think is fighting that filthy incinerator at the end of the neighborhood block, that coal-ash pile on the way to work, or that cement plant next door to the church? . . . Black women are everyday environmentalists; we are climate leaders. We just don't get the headlines too often."[114]

Climate-justice work is anti-racism work, gender-equity work, and decolonial work—and vice versa. The goals of each are woven together. But congregations need preachers to make this connection. Political rhetoric has often described human and nonhuman interests as being at odds, creating a false choice between human justice and environmental advocacy. This has resulted in congregations able to talk a good environmental game that never engages disenfranchised communities in equitable relationships of solidarity. On the other hand, it has resulted in congregations ready to march against racialized violence, but resigned to the slow

[112] Rob Nixon, *Slow Violence and the Environmentalism of the Poor* (Cambridge, Mass.: Harvard University Press, 2011), 1. Summers' confidential memo is dated December 12, 1991.

[113] Melanie Harris, *Ecowomanism: African American Women and Earth-Honoring Faiths* (Maryknoll, N.Y.: Orbis Books, 2017), 38.

[114] Toney, "Collards Are Just as Good as Kale," 80–81.

violence of water and air pollution's disproportionate impact on neighborhoods of color.[115]

Preachers are especially equipped to explain why neither of these responses is sufficient. The biblical text is full of examples of how the dehumanization of persons and environmental devastation go hand in hand. Davis notes that Egypt is "the biblical archetype of the industrial society . . . ceaseless in its demand for slave labor (the cheapest fuel of the ancient industrial machine)."[116] The book of Revelation describes the judgment of the merchant ships of "Babylon"—a euphemism for Rome—for carrying luxury cargo like gold and ivory alongside "human lives" (Rev 18:13).[117] In contrast, Ezekiel 48:18–19 describes the restoration of the land alongside allocations of material provision for the workers of the city. Creation's thriving brings hope and sustenance to ordinary people. Rev. Dr. Howard-John Wesley's 2019 sermon "I Can't Breathe" makes the connections between chattel slavery, contemporary expressions of racialized violence, and environmental injustice explicit through a reframing of the creation account in Genesis 2.[118] The connections are there, and they run throughout Scripture—but preachers have to notice and name them.

Noticing and naming are, of course, preliminary steps in weaving creation justice with human justice. Justice also necessitates refiguring "how the seats are arranged" in the decision-making of local and inter-

[115] For an account of environmental injustice linked not only to socioeconomic class but to race, see Steven Bouma-Prediger's account of Rev. Benjamin Chavis' report, "Toxic Wastes and Race in the United States: A National Report on the Racial and Socio-Economic Characteristics of Communities with Hazardous Waste Sites," Commission for Racial Justice (New York: United Church of Christ, 1987), and his larger discussion of environmental racism in *Creation Care Discipleship: Why Earthkeeping Is an Essential Christian Practice* (Grand Rapids: Baker Academic, 2023), 138–41.

[116] Ellen Davis, *Scripture, Culture, and Agriculture: An Agrarian Reading of the Bible* (Cambridge: Cambridge University Press, 2009), 69.

[117] Barbara Rossing discusses the connections between Rome's luxury economy, deforestation, and imperial force in "Lost Land: Visualizing Deforestation and Eschatology in the Apocalypse of John and the Column of Trajan in Rome," in *People and Land: Decolonizing Theologies*, ed. Jione Havea (Minneapolis: Fortress Academic, 2019), 159–74. See also Presian Burroughs, *Creation's Slavery and Liberation: Paul's Letter to Rome in the Face of Imperial and Industrial Agriculture* (Eugene, Ore.: Cascade, 2022).

[118] Rev. Dr. Howard-John Wesley, "I Can't Breathe," sermon preached at Alfred Street Baptist Church, Alexandria, Va., August 11, 2019, https://www.youtube.com/watch?v=fc0KaXUhwK8.

national communities. It involves, in other words, redistributions of political influence, wealth, and place for the sake of justice.[119] This may mean that wealthier, polluting nations pay "reparations and compensation"[120] for the irresponsible consumption of others' resources. But it may also mean that local congregations will need to change the ways they participate in ecological advocacy, foregrounding collaboration with others and systemic engagement.[121] In her work as a U.S. Environmental Protection Agency (EPA) regional administrator, Heather McTeer Toney visited a community green space that the West Atlanta Watershed Alliance (WAWA) had developed in coordination with the EPA, local community organizations, and churches. A large suburban church had wanted to come and teach this urban neighborhood about ecologically friendly planting practices, but seemed less interested in interacting with the community itself and working together on the community's concerns. The WAWA members "wanted access to the same opportunit[ies as others] to create the green pastures in their own neighborhoods with their own hands," Toney remembers. They wanted "access to burned-out homes and overgrown lots: real transfer of land opportunities that could be used to elevate the viability of the neighborhood while teaching the community about land preparation, growing, sustainability and resiliency."[122] But changes in access and redistributions of land would require more than the suburban congregation's planting of trees on a Saturday

[119] Timothy Eberhart, "Sustaining the Planetary Household: Methodist Contributions to an Ecological Economy," in *Methodist Revolutions: Evangelical Engagements of Church and World*, ed. Joerg Rieger and Upolu Lumā Vaai (Nashville: Wesley's Foundery Books, 2021), 117–48 [138–39].

[120] Sheryl Johnson sees such "climate debt" reparations as a necessary part of reconciliations work in "Truth, Reconciliation, and Climate Justice: Applications of Truth and Reconciliation Processes to Climate Debt and the Climate Crisis," in Mendoza and Zachariah, *Decolonizing Ecotheology*, 131–43 [135]. Manann Donoghoe and Andre M. Perry describe the increasingly prominent conversations around climate reparations at COP27: "The Case for Climate Reparations in the United States," Brookings Institution, March 2023, https://www.brookings.edu/articles/the-case-for-climate-reparations-in-the-united-states/. Michael Northcott discusses the inequities of the current crisis in *A Moral Climate: The Ethics of Global Warming* (London: Christian Aid Publishing, 2007), 55–56.

[121] Bale's description of "listening to one another" is not only important wisdom for researchers and preachers. It is also a significant skill set for churches working in collaboration.

[122] Heather McTeer Toney, *Before the Streetlights Come On: Black America's Urgent Call for Climate Solutions* (Minneapolis: Broadleaf Books, 2023), 151–52.

morning. It would require legislative change and local economic investment. Environmental justice would require systemic thinking.

Ezekiel's temple vision is just one of many scriptures helpful to preachers when teaching congregations how natural, social, and spiritual ecosystems of communities intersect. This is another of the biblical text's notable gifts in relation to the climate-justice conversation. Given the exilic context of the canon's construction, Scripture regularly points churches interested in environmental work toward questions of power and necessary conversations about how the seats in their communities are arranged.

Strengthen Relational Ties

There is no meaningful "rearranging of seats" without genuine relational investment. When Bishop Clarence Laney first came to the Hayti community as a young pastor in the 1990s, he made it a point to visit the homes of residents in the neighborhood, whether they were part of his congregation or not. It was from these third- and fourth-generation residents that Bishop Laney learned the history of Hayti and began to envision the distinct challenges and privileges of preaching in this place.[123] Over decades, Bishop Laney deepened these relationships, working to "center the voices" of a particularly invested group of neighborhood women in the decision-making processes of Durham's local government.[124] Working together with community-organizing networks like Durham CAN (Congregations, Associations and Networks), Bishop Laney pressured city officials to walk the neighborhood with these local residents and attend to their concerns. In 2017, Durham CAN received public commitments from the Durham City Council to reacquire a vacant nineteen-acre Hayti lot from a private, out-of-state developer—insisting that the community play a significant role in deciding how best to use the land. "We have to stand together," said one of the local leaders from Hayti's Grant Street neighborhood. "There are only a few of us," said another, but now that "we have someone behind us, we will fight."[125]

[123] Bishop Laney describes the revitalizing hope he continues to witness in the community in "Hayti History," interview by Student U Durham, October 20, 2018, https://www.youtube.com/watch?v=xBFlHi_FcKU.

[124] I am grateful for Bishop Clarence Laney and Cullen McKenney, cochairs of the Durham CAN Strategy Team, for sharing their wisdom with students in my Preaching Place: The Challenge and Promise of a Global Gospel course, Duke Divinity School, Monument of Faith Church, Durham, N.C., November 7, 2023.

[125] Lauren Greenspan, "Hayti: Past and Present," October 26, 2017, https://www.youtube.com/watch?v=Dr_v5bPgIZ8.

This work of equipping, connecting, and amplifying local concerns is critical in a climate-changed world. In a presentation to local clergy, Dr. Elizabeth Albright, an environmental ethicist at Duke's Nicholas School, laid out sobering statistics from scholars across her field that indicate widening wealth gaps between white communities and communities of color after each natural disaster—regardless of education and relief funding.[126] Climate change is reifying the ugliest legacies of U.S. history. Then she shared a mitigating factor that has made a statistically significant difference. If communities build meaningful relationships across difference *before the storm*, recovery looks different.[127] Albright did not offer quick, reductive fixes to the entrenched problems she had named. Ecumenical luncheons and community food banks do not solve systemic racism. But the data is consequential. Building relationships is a concrete way to move the needle toward justice—and it is something that local churches have the capacity and expertise to do.

For Fijian pastors, relational networks are framed as covenantal commitments critical to strengthening a community's climate resilience. I asked Vani Catanasiga how the village of Cogea knew to contact the Fiji Council of Social Services (FCOSS) for help when the community realized its relocation was imminent. How did this rural, isolated village know who to trust after a storm? Catanasiga told me that FCOSS had been active in the area since 2007, when the organization protested an internationally owned bauxite mine impacting local food and water security. Since that time, FCOSS had been working with local leaders after each disaster to keep records of recovery efforts. "We used the traditional structure, not the governmental structure," Catanasiga told me—referencing village and ecclesial institutions. "This decentralization was a strength." When Cyclones Yasa and Ana hit and the village decided that it needed to move, they were not dependent on the decisions of a single outside developer or a single governmental office. They had a relational network they could activate.[128]

126 Elizabeth Albright, "Floods and Fire," presentation at the Pastoral Care for Climate: Weaving Science and Theology for Justice conference, Duke Marine Lab, Beaufort, N.C., June 14, 2023. See, for example, Junia Howell and James Elliot, "As Disaster Costs Rise, So Does Inequality," *Socius: Sociological Research for a Dynamic World* 4 (2018): 1–3.

127 For an example of a study that links social capital with climate resilience, see Daniel P. Aldrich and Michelle A. Meyer, "Social Capital and Community Resilience," *American Behavioral Scientist* 59, no. 2 (2015): 254–69.

128 Vani Catanasiga, personal interview at FCOSS Suva offices, July 1, 2023.

Sermonic education on climate-related legislative issues or scientific data is valuable. But preachers who limit climate-justice preaching to these conversations define the category too narrowly. One of the most significant roles preachers can play in the climate struggle is engaging communities across difference and encouraging congregations to strengthen those relational ties. Climate-justice preaching reminds communities that they are woven together with the created world, which includes their neighbors across the street.

Honor the Material Impact of Spiritual Practices

Faith leaders reading this book are likely convinced that responsible relation to creation requires more than spoken platitudes. Environmental commitments are enfleshed in the material realities of budgets, food choices, and energy use.[129] Such leaders may be less convinced by Fijian preachers' recurring insistence that spiritual practices—practices like prayer and fasting—vitally impact these material realities. But the witnesses of the displaced communities in this volume suggest that spiritual practices provide more than comfort. They give meaning and musculature to material acts of creation care, breaking down simplistic bifurcations between the holy and mundane.

On November 26, 2023, Kia Island commemorated the first Christian sermon preached on Fijian soil. In 1809, a European trading ship enroute to Australia ran aground on Kia's reef. A missionary on board conducted a worship service when the crew was safely ashore. The text of the sermon was Hebrews 13:5: "Keep your lives free from the love of money . . . for [God] has said, 'I will never leave you or forsake you.'" It was preached to a group of shipwrecked sandalwood loggers, rather than to the local community.[130] Regardless, Rev. Esala Nacuva, minister to the congregations of the Kia Circuit, believes the Scripture has contemporary significance

[129] For faith leaders eager to have concrete conversations with their congregations about material changes that impact energy consumption, carbon footprints, food justice, and environmental pollutants, several accessible places to begin conversations include Mike Berners-Lee, *There Is No Planet B: A Handbook for the Make or Break Years* (Cambridge: Cambridge University Press, 2021); Paul Hawken, *Drawdown: The Most Comprehensive Plan Ever Composed to Reduce Global Warming* (New York: Penguin, 2017); and Katharine Hayhoe, *Saving Us: A Climate Scientist's Case for Hope and Healing in a Divided World* (New York: One Signal, 2021).

[130] Jone Salusalu, "Kia Island Rewrites History 214 Years On," *Fiji Sun*, November 30, 2023, https://www.pressreader.com/fiji/fiji-sun/20231130/281616720126556.

for Fijian communities today, particularly in light of the region's globalizing economy and rapidly changing ecosystem. Kia's three villages spent a year preparing for the sermon's commemoration, and during that time the villages decided to create a *tabu* (prohibition) against catching large fish caught on a nearby reef—a staple of local diets and incomes. Part of this decision was practical. They did not want to overburden the ecosystem when providing fish for the visitors that would attend the celebration. But the *tabu* was also a spiritual practice akin to fasting, honoring the biblical injunction against insatiable monetary desire in ways that shaped the community's common life. In recounting their decision, Nacuva underscores the importance of the community's sacrificial, sustainable care for its environment as an act of spiritual discipline.[131] Presumably, European sandalwood traders did not show such care. Most of Fiji's original sandalwood stores were harvested by 1820.[132]

Rev. Tevita Kete notes that spiritual practices like denominational offerings are another way that communities support material environmental change. Like Nacuva, Rev. Kete expresses respect for the sustainable fishing practices of the Methodist fishermen in his community. For these men and their families, allowing the reef's fish population to replenish through intermittent fishing restrictions is a costly Christian witness. But Kete believes these fisherman should receive more than ecclesial respect. He argues that their commitments invite the larger church to examine its relational responsibilities to those who sacrifice for the environmental good of the whole. *How might the church stand in financial solidarity with these families?* he wonders. "Christ's yoke is something we carry together."[133] North American preachers don't often think about financial stewardship or denominational giving[134] as climate-justice work, but Kete insists that spiritual practices like tithing lead to interconnected ways of understanding the world, deepening the strength of our commitments to human and nonhuman creation alike.

Spiritual practices also sustain communities when material solutions to intractable problems are not readily seen. Peniana Tinaimana, a Sunday school teacher and prayer leader in Cogea, speaks of the changes that

131 Esala Nacuva, personal interview, Daku, Fiji, July 4, 2023.

132 Fijian Ministry of Forestry, "Sandalwood: A Fragrant Tree with Great Potential," *Forestry Focus*, November 2021, 1–2, https://www.forestry.gov.fj/docs/newsl/ForestFocus_NovIssue_2021.pdf.

133 Kete, personal online interview, January 27, 2022.

134 Apportionments, for example, in the United Methodist Church.

the communities' communal prayer commitments have made as the village prepares for relocation. "I used to have sleepless nights," she testifies. "Now I have peace. Peace that the fear will go."[135] Jale Samuwai, the village's *dauvunau* (local preacher), notes that the community is designing spiritual "rituals to cherish the memories" of the land the community is leaving.[136] For Bishop Laney, ministry in Durham's Hayti community has also required careful attention to how spiritual practices sustain material resistance over time. The land that the city repurchased from out-of-state developers in 2017 has yet to become a hub of community revitalization. One community member recently noted that the fenced-in lot "looks worse than when the [outsiders] owned it."[137] The lot's twenty-year vacancy would seem an absurdity given its downtown location if it weren't for the grim logic of Durham's long racial wound. In 2020, Bishop Laney and Durham CAN organized a silent candlelight vigil around the site, bearing witness to the land's continuing history of broken promises, chronic disinvestment, and neglect—and testifying to God's presence in haunted, holy places.

One of the most significant ways spiritual practices impact the material realities of the climate crisis is through their formation of preachers themselves. The work of climate justice is not easy. It is humble work, often taking years to bear fruit. For all of their importance, the strengths of relational networks are rarely seen over the short term. The benefits of walking neighborhoods, serving meals, and worshipping together with those who look different or speak a different language emerge through time and consistency. And that consistency can be difficult to sustain given societal forces that isolate, numb, and divide. Climate restoration plays out over decades, and environmental advocacy literature is increasingly attending to the spiritual fortitude and perseverance required.[138]

135 Peniana Tinaimana, personal interview, Cogea, Fiji, July 5, 2023.

136 Jale Samuwai, personal interview, Cogea, Fiji, July 5, 2023.

137 Durham CAN, "A Call for Awareness, Agitation, and Action: Durham CAN to Hold Silent Vigil Walk around Fayette Place Site," October 8, 2020, https://www.durhamcan.org/a_call_for_awareness_agitation_and_action_durham_can_to_hold_silent_vigil_walk_around_fayette_place_site.

138 See, for example, Schade and Bullitt-Jonas, *Rooted and Rising*, which includes spiritual practices, and invitations to reflection at the end of each of its sections. See also Rebecca Randall, "Spirituality, Global Warming, and Grief: How Clergy Can Tackle Climate Anxiety," Grist, June 9, 2023, https://grist.org/culture/spirituality-global-warming-and-grief-how-clergy-can-help-tackle-climate-anxiety/. For an Oceanic call for liturgical resources to meet the communal grief of climate displacement, see Seforosa Carroll, "Displacing Liturgy: A Pacific Exploration," *Liturgy* 36, no. 2 (2021): 26–35.

Bishop Laney is a powerful preacher. At the 2017 community meeting with city officials where the Durham Housing Authority agreed to include the Hayti community in decisions about the land's future, he gave a powerful sermonic reflection on Ezekiel 37, calling those present to bone-rattling change.[139] Such preaching is necessary in moments of decision. But Bishop Laney's larger work in Hayti bears witness to the necessity of spiritual commitment over time. Revitalizing places for "ordinary use" (Ezek 48:15) often requires sermons that look less like Ezekiel 37 and more like Ezekiel 48.[140] It invites God's people into consistent, everyday practices of community building and shared governance—practices that require spiritual resources to sustain.[141]

To those who minister to a fraying world, I offer the testimony of the Fijian preachers in this volume, and I add my own. Your prayer matters. Your confession matters. Your lament matters—as does your joy. Your breaking bread over kitchen tables and at the Lord's table matters. And your preaching matters. These practices are places of divine encounter, teaching the church who we are and who we are not. They are places where God continues to mark God's people with covenantal accountability and train them in hope.

Renounce False Hope and Reject Despair

Hope, of course, is a tricky word. Environmental preaching has been plagued by a naïve optimism that is little more than a denial of the problem's depth or an overestimation of human capacity to undo the damage it has caused. Such preaching assures congregations that goodwill, small tweaks, and a little extra science funding will save the day. For Miguel De La Torre, these sermons are as dangerous as those that absolve

139 Recounting the 2017 Durham CAN meeting with city officials, Cullen McKenney describes the power of Bishop Laney's sermonic reflection on Ezek 37 in an oral presentation to students in a course by Neal, Preaching Place.

140 In Ezek 48's attention to "ordinary" life, Michael Lyons emphasizes its spiritual vision. It is "attempting to convince the exiles of a reality deeper than that they can currently see": "Envisioning Restoration: Innovations in Ezekiel 40–48," in *"I Lifted My Eyes and Saw": Reading Dream and Vision Reports in the Hebrew Bible*, ed. Elizabeth Hayes and Lena-Sofia Tiemeyer (New York: Bloomsbury T&T Clark, 2014), 71–83 [79].

141 John McClure describes multiple models of ethical preaching in *Ethical Approaches to Preaching: Choosing the Best Way to Preach about Difficult Issues* (Eugene, Ore.: Cascade, 2021). While some models arc toward a paradigm shift in perspective, others work toward expansions of shared understanding or strengthened networks of social capital across difference.

congregations of any responsibility for the crisis. Optimistic sermons provide "quick and easy fixes that temporarily soothe the conscience of the privileged but do not offer a more just social structure based on the empowerment of the world's used and abused."[142] If that is "hope," De La Torre doesn't want it. He prefers the hopelessness of desperation to hope that "short-circuits the struggle" for justice with complacency.[143] Desperation, at least, can lead to action.

But complacency comes in many guises. Hopeless desperation is one thing; hopeless despair is another. Cynthia Moe-Lobeda notes that economically privileged, progressive communities in the United States, communities that claim to want more equitable, ecologically sustainable lives, "are at the same time profoundly inactive or ineffective in seeking it and are effectively resigned to the 'the way things are.'"[144] The present organization of the world and its resources seems inevitable. Even those who put their faith in the salvific power of human ingenuity acknowledge the devastating limits of the nation's anemic political will—and their own.

For Charles Campbell, the problem of hope in the climate crisis presents a particular challenge for preachers who have wedded themselves to the "comfortable homiletical pattern of disruption-resolution" in norming the shape of the Christian gospel.[145] The climate crisis is complex, sprawling, and rooted in centuries of hubris. Preachers attempting to settle the issue with a twenty-minute quick fix can reinforce the same escapism, anthropocentrism, and mechanisms of control that created the mess. Campbell turns to the biblical prophet Joel for a different model of hope—a hope that leaves room for the open question of God's action. Act righteously, Joel tells his community, and "who knows?" (Joel 2:14). God may relent. It is a hope that Campbell describes as "grotesque" in its exchange of idealized closure for the open-ended grit of communal repentance.[146]

142 Miguel De La Torre, *Embracing Hopelessness* (Minneapolis: Fortress, 2017), 139.

143 De La Torre, *Embracing Hopelessness*, 141.

144 Cynthia D. Moe-Lobeda, "Decolonizing the Privileged: Resisting and Rebuilding the New Economy," in Mendoza and Zachariah, *Decolonizing Ecotheology*, 174–98 [182].

145 Charles L. Campbell, *The Scandal of the Gospel: Preaching and the Grotesque* (Louisville: Westminster John Knox, 2021), 67.

146 For Campbell's description of a grotesque aesthetic, see *Scandal of the Gospel*, 46–48.

I would describe such hope as exilic in character. Exilic hope makes space for lament—and even anger—at the forces that tear at creation's covenantal threads. It refuses to lie about what has been lost and complicates compromises that save the lives of some at the expense of others. The God of exilic hope is no puppet on a string. And yet, this God might still remove our hearts of stone and give us hearts of flesh (Ezek 36:26).

Such hope would result in a far humbler vision of the church, and perhaps a far humbler vision of preaching—but a more stalwart one. Debra Rienstra notes that "God loves to work in small . . . hidden places."[147] An ecological refugium is a crevice or hidden crack where seeds of new life take root after devastation. Rienstra wonders what it would mean for the church to be such a "microcounterculture" where life endures and "where we prepare for new ways of living."[148] It is a calling that would require trust, commitment, and patience—quiet virtues that are surprisingly potent when leveraged in communal efforts to sustain life. A Rocha, a faith-based environmental advocacy organization, has been working with Ghanaian communities living in and near the Atewa Forest for over ten years to protect water resources and species diversity from the pressures of large-scale commercial mining. Daryl Bosu, the deputy national director, describes how grassroots work with local communities committed to protecting the forest is gaining traction with the government, how the voices of local leaders are being amplified through international partnerships, and how sustainable, economic alternatives to mining are being found.[149] It is not that patient commitment to local communities makes no difference. But it requires preachers to proclaim a hope rooted in God—by which I mean a gospel that looks less like a silver bullet (or a list of shiny bullet points) and more like a community rooted in the water that flows from God's sanctuary (Ezek 47:12). Such a gospel would be evidenced by a community oriented in sacrificial, covenantal solidarity with the world God loves.

[147] Debra Rienstra, *Refugia Faith: Seeking Hidden Shelters, Ordinary Wonders, and the Healing of the Earth* (Minneapolis: Fortress, 2022), 5.

[148] Rienstra, *Refugia Faith*, uses the example of Mount Saint Helens' unexpectedly rapid reforestation to describe the significance of natural refugia in seeding and protecting life after devastation (3–4).

[149] Daryl Bosu's interview at COP26 and a description of A Rocha's work in Ghana can be found at "Protecting Atewa Forest: Conservation of Ghana's Upland Treasure," A Rocha Ghana, accessed December 18, 2023, https://old-ghana.arocha.org/projects/protecting-atewa-forest/.

Testify to Place's Apocalypse

The Fijian villages currently slated for relocation are "living in the cultural graveyard of their forebears," Tevita Nawadra Banivanua tells me. "It is not an easy thing moving out of our villages to somewhere else because the tide keeps coming up and . . . digging this grave. But that is our reality."[150] Attending to, naming, and grappling with this loss is a critical aspect of a congregation's covenantal solidarity and faithful response. I pass to preachers the exhortation that Banivanua, qase levu vakacegu of the Methodist Church in Fiji, gave to me: "*Tell your people climate change is real.*"

But telling a congregation the factual information about species extinction and sea-level rise is only part of a preacher's task. Preachers must testify to God's abiding relationship with all creation and its sacred value. In the words of poet Lia Purpura, "To understand ruin, know first what is being ruined."[151] Purpura asks her readers to listen to place with a poet's ear. "Listen archaeologically . . . navigate constellationally," she says, ". . . so a singular moment overrides its fixed place . . . [and] nothing is still or unto itself."[152] Listen to place, in other words, as the particular, relational, moving being that it is—a being connected to other places and marked by other times. As Annie Dillard notes in her classic essay "Teaching a Stone to Talk," such listening requires extended attention "when the world . . . has gone silent" in the hearing of so many Western preachers.[153] Many have forgotten how to hear the preaching of place, which makes the witness of those who do remember—in the world and in the biblical text—especially significant conversation partners. Listening requires the humility to be taught.

Listening also requires courage. In their book *Healing Haunted Histories*, Elaine Enns and Ched Myers describe discipleship practices for communities formed by settler-colonial narratives that move "past abstraction and exoneration" in their engaging of colonial history.[154] The layers and losses of the colonial project can be framed as "too complicated for us to deal with" or a history "from which we are conveniently disconnected

[150] Banivanua, personal interview, January 2, 2023.

[151] Lia Purpura, *All the Fierce Tethers* (Louisville: Sarabande Books, 2019), 59.

[152] Purpura, *All the Fierce Tethers*, 59.

[153] Annie Dillard, "Teaching a Stone to Talk," in *Teaching a Stone to Talk: Expeditions and Encounters* (New York: Harper & Row, 1982), 85–94 [88–89].

[154] Elaine Enns and Ched Myers, *Healing Haunted Histories: A Settler Discipleship of Decolonization* (Eugene, Ore.: Cascade, 2021), 43.

and to which we are therefore unaccountable."[155] But Enns and Myers insist that responsibility, confession, reparation, and connection are what the healing of place requires. Without this holistic work, environmental preaching becomes a mechanism for making places comfortable for some and unlivable for others. Richard Voelz begins his class Proclaiming Justice in the Church and Public Square with an analysis of his seminary's "liminal" placement in Richmond, Virginia, "a literal dividing line between racial and class differences in Richmond's Northside."[156] These geographic, demographic markers speak to deeper specters: the history of Richmond's slave trade, "in addition to the infamous monuments to Confederate figures on Monument Avenue."[157] What does it mean, Voelz presses his students, to preach in such a place? And what does it mean to preach in ours?

For the Rocky Mountain Synod of the Evangelical Lutheran Church in America, it meant returning a Denver church and its land to the Four Winds American Indian Council, an organization that had been assisting its community and ministering in the space since 1986. The return of land happened after extended reflection and the clear articulation of the Native community that it would not be displaced. "We're not going anywhere," George "Tink" Tinker, an elder in the Four Winds community from the Osage nation and professor emeritus at Iliff School of Theology, remembers the community telling the synod. But clear articulations by Native communities have rarely resulted in just responses. Tinker, who is also an ordained Lutheran pastor, reflects on the outcome of this dialogue with wonder. "To give the land back to the Four Winds American Indian

[155] Enns and Myers' process of helping congregations remember the "landlines, bloodlines, and songlines" that have formed their places and persons is worth reading in full. For a summary of their method, see *Healing Haunted Histories*, 22–24.

[156] Richard Voelz, "Wrestling with Whiteness in Homiletic Pedagogy: A Reflection on Teaching 'Proclaiming Justice in the Church and Public Square,'" in *Unmasking White Preaching: Racial Hegemony, Resistance, and Possibilities in Homiletics*, ed. Lis Valle-Ruiz and Andrew Wymer (Lanham, Md.: Lexington Books, 2022), 141–54 [143–44]. Voelz notes how the seminary's neighborhood has been impacted by white flight in the 1950s, when predominantly Black neighborhoods displaced by the construction of Interstate 95 relocated to the area.

[157] Voelz, "Wrestling with Whiteness," 143. Shelly Rambo speaks of such monuments as part of the "master story," raising the question, "Whose America?" in "Exiling in America: The American Myth and the Spectral Christ," in *Interpreting Exile: Displacement and Deportation in Biblical and Modern Contexts*, ed. Brad E. Kelle, Frank Ames, and Jacob Wright (Atlanta: SBL, 2011), 409–22 [416, 418].

Council . . . was a stunning thing . . . a role model to the rest of the euro-colonial world, the colonizer world. What churches in the United States are not on Indian land? What universities are not on Indian land?" The return of land was "remarkable."[158]

In testifying to place's apocalypse—in discerning the devastations that mark place's past and present—something profound is revealed. Rossing notes that "apocalypses not only diagnose imperial pathology . . . they also give a counter-imperial vision of hope and call us to come out of Empire."[159] Binaries that have become the rigid scaffolding of congregational paralysis stretch and bend when communities attend to the holy ground beneath them. Hope and lament can be held in each hand. Dependence and agency are claimed together. Place's fragility and its sacred strength connect to our own. The climate crisis is unveiled as the unraveling of the world—and a revelation of God.

Where Is God Standing?

There was a massive tree that grew outside the chapel windows of Fiji's Davuilevu seminary. It was a home to birds and bats, insects and frogs, a whole ecosystem. When Cyclone Winston hit in 2016, it fell. It was just before Lent. And all through February and March, I sat, listening to our morning sermons while gazing at the gash created by that tree's destruction. The upturned root system, rotting in the sun. The torn ground, and the crushed life beneath its weight.

To preach in the climate crisis is not only to ask *where we are standing* in the face of increasing devastation. It is to ask where God is standing. Through that long Lenten season of penitence, the Fijian church grappled with what it had done to deserve such a fate. Had Christians not prayed enough? Or worshipped enough? Had they not fasted enough? I suggested at one point that perhaps it was the Western countries pouring pollutants into the sky that needed to do more fasting—a line of thought that seemed deeply unfair to my Fijian colleagues. Why would God punish their people for Western apathy and greed? I had no answer for that.

But as our worshipping community neared the week of the Passion, something different came into view. Or perhaps, something different

158 Terra Brockman, "Decolonized Sacred Land: How a Church Became the Home of an American Indian Organization," *Christian Century* 137, no. 6 (2020): 22–25 [24].

159 Barbara R. Rossing, "Waters Cry Out: Water Protectors, Watershed Justice, and the Voice of the Waters," in Mendoza and Zachariah, *Decolonizing Ecotheology*, 39–57 [54].

was unveiled. The tree became less of a scar indicating divine disapproval, and more a sign of God's costly, cruciform solidarity. The God of Scripture is not a God who stands outside earth's storms, unscathed. That is not Jesus' God. And neither is it Ezekiel's. To center covenant in one's theology is to embrace "a larger creational schema through which God becomes involved with and affected by the world."[160] A covenanting God is an "affectable" God. The God of Scripture is displaced alongside creation, bearing the blows, fire, and floods of broken covenant, even when God's roots are ripped from the ground. A tree that had preached to me of God's absence became a Holy Week testimony to God's presence.

Stephen Procopio's artistic rendering of the Ascension is a study of Christ's nail-pierced feet. They are recognizably particular, and decidedly strange. Jesus' flesh is made of vines and soil and sprouting leaves. His feet are the stuff of earth, even as they leave the ground. Procopio's artistic rendering does not suggest that leaves and grass, algae and grubs, are in themselves divine. Instead, he depicts a nail-pierced God so marked by covenantal solidarity that the whole world is held in the singularity of God's flesh.[161] Place is gathered and held in a holy embrace. Place's suffering is Christ's suffering, as place's redemption is the prayer of God's Spirit (Rom 8:26).[162] To preach in the climate crisis is to preach about the place of Christ's body—displaced and nail-scarred—and our place within it. Emplaced in that flesh, we are woven with creation, tissue and bone. It is a costly connection, as any covenant of consequence is. To be held in the place of Christ's body is to be one with the world God loves. It is to make one's life about more than oneself.

160 William Briggs, "Idols and Land Grabs, Ancient and Modern: Creation and Ecotheology in Ezekiel 6, 35:1–36:15," *Horizons in Biblical Theology* 40, no. 1 (2018): 41–64 [63].

161 Stephen Procopio, "Ascension," Procopio Draws, accessed December 27, 2023, https://procopiodraws.com/d7hh7jj64p6os2l3v459ibq6zjnmp9. A published version of "Ascension" can also be found in Procopio's illustrated retelling of John's gospel, *John: Come See a Man* (Richmond, Va.: Fish Coin Press, 2023).

162 Elizabeth Johnson argues that, in becoming flesh, Jesus reaffirms God's covenant with creation (Gen 9:15–17) and takes into himself the entire created order. Scientists have noted that human flesh contains matter from flora, fauna, and even the stars. She writes, "Seen through the lens of deep incarnation, Calvary graphically shows that the God of suffering love abides in solidarity with all creatures, bearing the cost of new life . . . from the extinction of whole species to, yes, every sparrow that falls to the ground" (*Creation and the Cross*, 189).

Not all trees lead to place's healing. But some do. On November 14, 2021, volunteers from the Hayti community and the broader Durham area planted trees together in Hayti. The community took the lead. Forty-seven species of saplings were chosen and planted in conversation with a neighborhood stakeholder board, comprised of individuals from local businesses, churches, schools, and community organizations.[163] Changing Durham's tree canopy is the work of decades, but the work has begun. In Fiji, the work continues. On Easter morning 2023, youths from the interior of Vanua Levu reforested three hectares of local land under the direction of Rev. Ilitomasi Nute as part of their congregation's investment in creation care.[164] The Fijian Methodist Church has asked every one of its members to plant a tree each quarter—four a year.[165] Many of these trees are mangrove seedlings, planted at Fiji's shoreline to protect it from further coastal erosion.

Leaves of healing grow in these salt-saturated waters, testifying to place's resistance to the broken relationships of centuries. They testify to a restoration not yet seen, even as they testify to the presence of a covenanting God.

They ask preachers to join them.

[163] Key, "Medical Student-Led Initiative."

[164] Fijian Ministry of Forestry, "Sorokoba and Nayarabale Youths Reforest 3 Hectares of Land, Minister Applauds Act," April 15, 2023, https://www.forestry.gov.fj/pressdetail.php?id=171.

[165] Jeremaia Waqainabete, *Christian Environmentalism: An Ecotheological Approach to Earthkeeping* (Suva: WWF Pacific, 2018), 22, https://wwfasia.awsassets.panda.org/downloads/theological_mindset_english_2018_sep.pdf.

Appendix

Sermons for Conversation

Sermons are places of encounter, movement, and change. They set their trajectories in shifting waves and watch for stars beyond the storm. They do not reinforce a community's listing masts or reweave its tattered sails from the safety of an academic harbor. They "rebuild their ships at sea."[1] I offer these sermons as my own provisional attempts to navigate the climate crisis from the place of three particular Western pulpits. These sermons are not models, but honest acknowledgments of the challenge this book describes.

For preachers scrambling across the deck of weekly sermon preparation, my hope is that these offerings might provide accessible ways to start conversations about this book's themes with congregations and clergy covenant groups. The sermons echo many of the stories, quotes, and ideas found in these pages. For homileticians, the sermons allow classrooms and scholars to compare what I have said in these chapters to what I have done in the sermon event. Practice always complicates theory; this is preaching's dreadful joy. One tries, fails, and then discovers transcendence waiting in failure's wake. I would be grateful for colleagues to chasten this work, to point out its contradictions and hedges, and to press it toward holiness. It is a season where all hands are needed on deck.

Mostly, I offer these sermons in response to Marika Bale's insistence that listening carries responsibility. Our research is meant to change us. It is not to be held at a remove from the muddy soil of the church; it is meant to impact what we do and how we do it. The Fijian Methodist preachers in this volume offered their sermons to this project because they believed

1 Amiria Salmond, "Re-building Ships at Sea: Ontological Innovation in Action," in *Environmental Transformations and Cultural Responses: Ontologies, Discourses, and Practices in Oceania*, ed. Eveline Dürr and Arno Pascht (New York: Palgrave Macmillan, 2017), 215–26 [221].

the climate crisis worth the risk of speech. They chose the vulnerability of engagement over the comfort of silence. I want them to know that their speech changed me. These sermons are tokens of that change.

There is another project for another day that engages climate-crisis preaching in local, Western congregations where resistance to the conversation is high and shared knowledge of the subject matter is shallow. The sermons offered here were preached on different ground. They were preached in conferences on climate justice and in those high-steeple pulpits that bridge church and academy. The questions around climate in these sanctuaries are less about the "facts on the ground" and more about priorities in the struggle, the nearness of despair, and the seeming irrelevance of spiritual practices in the fight. But underneath the particularity of these concerns, the sermons address broader, *exilic* questions.

Can the creation and covenant of a good God come undone? What does one preach in the face of the unthinkable? What does covenantal holiness require?

These are questions God's people have been asking for a very long time. To navigate them in our present crisis is holy ground.

Sermon One

"Wading in Water"
Ezekiel 40:1–4; 47:1–12

Just Creation Conference
Opening Worship
Columbia Theological Seminary
March 17, 2023

(The opening worship service began with a moment of silence for Manuel "Tortuguita" Páez Terán. Manuel had been protesting the construction of the Atlanta Public Safety Training Center—nicknamed "Cop City"—when they were killed by police in the Weelaunee Forest. When they were shot, Manuel's hands were raised.)

Ezekiel is a hard book to read. Even the hopeful chapters are haunted. The resurrections rattle with bones. It is a book about knowing the value of land and sanctuary and home. And losing it. About places stolen, and then destroyed. Babylon had environmental tactics for keeping the territories it defeated in line. Trees cut down. Villages burned. Soil salted. Laying waste to land is particularly effective urban warfare—and often the cruelest way to break a people's spirit.

Ezekiel's people had witnessed that cruelty, but to the frustration of many readers, the prophet speaks relatively little about Babylon. Nothing like the verbal takedown of empire we hope for. We want him to hold the oppressors accountable. And instead, he speaks again and again of the sin of God's people and, to my own profound discomfort, the sovereignty of God.

It's a hard book to read.

Ezekiel can sound like some televangelist parody, blaming hurricanes on Mardi Gras—pinning the fault for climate catastrophe and forced dislocation on those least responsible. And we could just write him off. Except that his prophecy doesn't come from some mauve-upholstered couch in a climate-controlled television studio. It comes from the banks of an irrigation ditch called the Chebar Canal—an actual ditch dug by exiles to grow food those exiles would never eat. A ditch dug to fund imperial temples that stood largely empty all year until Babylon's annual military festival, when the emperor reclaimed his role as conqueror of all creation. Exiles were regularly marched through the capitol during these massive military parades—their flesh providing stripped, bound counter-testimony to the empire's faith in its own civilized, civilizing benevolence.

So, don't let Ezekiel 47 fool you. This lovely vision of a temple built with mandala-like symmetry, this river—clear and miraculous—growing deeper as it flows from God's altar, restoring all that has been taken away . . . the poisoned soil, the famine, the sores of the body, and the scars of the heart. This is not the only river in Ezekiel. There is also the river of Ezekiel's opening chapters: an engineered canal hundreds of miles from Judah's hills, commissioned by an empire adept at using agricultural technology to centralize its power. It was dug by exiles who had lost family and home, exiles whose questions about the character and commitments of God roared in their ears. It's on the banks of the Chebar that Ezekiel receives his vision of the tree-lined river described in our Scripture today. We see him wading in waters of healing while the muck of Babylon sucks at his heels.

So. Here is my question: What good is Ezekiel's vision of wholeness and holiness when the world is being laid waste? More to the point: What good does it do to enter this comfortable sanctuary filled with light and order and harmony at a conference about climate justice—and hear of a God who can restore all that has been taken away?

It's a deadly serious question for people of faith who care about creation. We know how religion has been co-opted for complacency. We know how spiritual platitudes of some future glory have lulled entire communities—even entire nations—to sleep.

How is this passage anything more than a naïve fantasy, filling bellies with false hope? God will restore, Ezekiel tells us, his vision echoing the stories of Israelite faith: trees of life, rivers of promise, temples like tabernacles, where God dwells with creation rather than conquering it. It's beautiful. I don't deny it.

But what good is any of it when the world is coming undone?

For three years, my family and I lived and taught at Davuilevu Theological College, the oldest pastoral training school in the Fiji islands. When we left, the president of the Fijian Methodist Church gave us this charge: *When you return to your people, tell them climate change is real.* He was deadly serious—and we knew why. We had seen the bleached coral and the eroding shorelines. We had prayed with ministers who poured out communion wine on the sand in front of their sanctuaries to keep the tides from rising. We had heard the howls of Cyclone Winston, the strongest cyclone ever to make landfall in the Southern Hemisphere.

I had been teaching systematic theology the spring that Winston hit, and the class had just heard of a massacre of faithful churchgoers that had taken place the year before in a far-off city called Charleston, inside a different Methodist church. We'd looked at the photographs of those murdered, and we'd spoken of the crisis that evil raises for anyone who believes in a God who is both good and powerful. Theologians have a word for that. At some point, I had written the word *theodicy* on the board.

The week after the cyclone hit, there was a knock at our door. A student who rarely spoke in class stood outside, his eyes averted. He hadn't heard anything yet from his village or his family. He spoke in a low voice, with something between grief and fury. "I have a question," he said to me, ". . . about theodicy."

I panicked. I dodged. I assigned blame to globalizing greed and contemporary empires. I spoke of melting ice caps and warming oceans, as if science could soothe his ache. I pointed to proposals from COP20, COP21, COP22 (that's as far as the numbers went at the time)—anything, anywhere that sounded like hope.

But the student sat unimpressed. He had seen Western decision-makers try to fix things before. He wasn't placing his faith in the civilized, civilizing benevolence of the global north. Nor was he placing his fate in their duplicity. He was wading in bigger questions. *Who was this God we worshipped—and who were we when the places that shape us wash away?*

These are the questions that someone who has loved land and lost it knows to ask. And, for those with ears to hear, these are the questions that Ezekiel asks in our passage today. The text is no opiate. It's resistance literature—refusing the empire's credo with three very basic pieces of information.

First and foremost, Ezekiel 47 tells us *who we are not*. We are not immortal or invincible. And we are certainly not God only wise. The text is unabashedly theocentric. There are no builders of that visionary temple; there are no diggers of that river. God's altar is holy all by itself, and the water that flows from that altar needs no engineer to deepen or widen its impact. God is God, and we are not. That is Ezekiel's message. Which, if you've felt the careless cruelty of empire, is more than a Sunday school platitude—or a theology of resignation. It's a theology of defiance.

Nebuchadnezzar II had a royal title that asserted his authority. He was—here is the exact quote—"to wield a just scepter over all populated regions and make humanity thrive."[2] And to that colonizing claim, Ezekiel 47 protests: no mortal has such power, and there is no thriving under the scepter of anyone who thinks otherwise.

Amitav Ghosh's *The Nutmeg's Curse* includes a scathing indictment of colonial legacies in Western conservation movements, and he includes this disturbing quote from the director of Survival International, an advocacy group for the rights of tribal people. The director states that over decades of confrontations with governments and corporations about their abuse of tribal people's rights, "none have been as duplicitous as big conservation NGOs."[3] Not because of their work for conservation. Because of their hubris—a kind of mechanistic universalizing approach to climate justice that ignores the communities most impacted. For Samoan theologian Faafetai Aiava, such an approach to creation's devastation tries to "solve a problem with the same level of consciousness that created it."[4]

What does the work of climate advocacy look like for those who know they are not gods? Most certainly, such work would require commitments to prayer and lament—and to urgent, materially consequential repentance. There are treasures of creation that have been destroyed and are being destroyed that no human technology, no global accord, no conference workshop will be able to undo. And we have a God to whom we

2 VAB 4 112 (Nbk 12=Zyl III, 5) I 13–17, quoted in Leo Perdue and Warren Carter, *Israel and Empire: A Postcolonial History of Israel and Early Judaism* (London: Bloomsbury, 2015).

3 Amitav Ghosh, *The Nutmeg's Curse: Parables for a Planet in Crisis* (Chicago: University of Chicago Press, 2021), 226.

4 Faafetai Aiava, "Talking God in a Divided House: Rebuilding Spirituality in God's Pacific Household," paper presentation at Iliff School of Theology, Denver, Colo., October 22, 2021.

answer. This is not a mess that we will be able tweak or tap-dance our way out of. We are limited, mortal beings. Not gods.

A year before the COVID epidemic, Jonathan Franzen wrote an article entitled, "What If We Stopped Pretending?" in which he argued that to adequately prepare for the climate catastrophes that are coming, we need to stop believing that we can avoid them.[5] Franzen is not a person of faith, but he sounds positively Ezekiel-like in his analysis. His point is not that we should be passive in the face of a traumatic future. His point is that we should be working harder than ever with different priorities, growing trust and accountability in our politics, amplifying voices previously excluded from decision-making processes, strengthening support networks to receive climate-refugee communities. For Franzen, the limits of our capacity are not cause for resignation; they are the reason we build relationships.

Which brings us to the second basic piece of information that Ezekiel 47 has for us. It does not only tell us *who we are not*. It tells us *who we are*. We are, in a word, *creatures*—which means that we have been created to be dependent on and connected to the world God made.

Ezekiel 40–48 can seem like a dissociative, golden-hazed dream, but there is a detailed practicality to it. It envisions the healing of salinized soil and new freshwater sources. In these waters, fish of all kinds swim. Ezekiel sees the spreading of nets and the careful preservation of salt marsh. He sees trees that sprout with healing leaves, producing fruit as regular sources of food. The river that flows from God's glory swarms with abundance, a reference to the "swarming waters" of Genesis 1. It connects land, tree, field, and sea with the altar of God's worship—and it connects all of that to Ezekiel.

What does it mean to live woven into creation—leaning on it, listening to it, learning from it? For Ezekiel it means wading into relationship with a river he cannot cross. This priest-in-training turned ditch-digging prophet has spent decades of his life in service of an empire that tames, rations, controls, and leverages the flow of water. *And now he is the one being carried.* Up to his chin in an ancient creation story. Remembering what it means to live wet in the world, remembering who he is. Dripping counter-testimony to the empire's univocal echo chamber.

5 Jonathan Franzen, "What If We Stopped Pretending?" *New Yorker*, September 8, 2019, https://www.newyorker.com/culture/cultural-comment/what-if-we-stopped-pretending.

Such remembering has material consequence—economic, political, ecological consequence. When artist Cannupa Hanska Luger learned that the Standing Rock Sioux Reservation, where he had grown up, was under threat from the Dakota Access Pipeline, he collaborated with Rory Wakemup to create a performance to remind the Water Protectors—and the militarized law enforcement officers that opposed them—*who they were*. Filmed from above, the Water Protectors lifted rectangles of wood covered with reflective Mylar over their heads and moved in a spiral, creating a dancing water serpent that looked for all the world like the Missouri River they were safeguarding.[6]

The image shone—though one might wonder what practical good it did. Today, despite court-ordered environmental-impact reports that were never completed, that pipeline runs.

For Luger, part of the answer to that question lies in the reflective quality of that Mylar. Those silver, wooden rectangles were not only intended as visual props to help the community enact its fully submerged intimacy with the earth. They were also mirrored shields that could be held up to those who displaced the community with rubber bullets and riot gear—so they could see themselves. The mirrors could show them who they had become . . . and *where they were standing* in this fight.

The book of Ezekiel doesn't end with trees of healing. It ends with the reallocation of land, giving extra honor to smaller, less influential tribes. It ends with limits being placed on the land grabs of ruling officials. It ends with economic resources and legal standing granted to foreigners, immigrants, aliens. I don't actually think Western, white, middle-class readers, readers like me, snub Ezekiel 47 because his pretty language obscures real-world problems. I think we avoid Ezekiel 47 because it *implicates* us in real-world problems we would rather not face. It is a scriptural mirror that shows us *where we are standing* in this story.

What is the insulation that protects us?

In what riot gear have we put our trust?

On whose land are we standing?

And whose histories have we silenced in the echo chambers of our worship?

A conference on climate justice has no business dismissing Ezekiel as naïve—or uncomfortably changing the subject when communities

[6] Cannupa Hanska Luger, "The Mirror Shield Project" (2016), accessed October 15, 2023, https://www.cannupahanska.com/social-engagement/mirror-shield-project.

stripped of their land want to have a serious conversation about "theodicy." This prophet knows *who he is not*, *who he is*, and *where he stands*—and now, he wants to know where God is standing.

There is a moment in the book of Ezekiel that breaks my heart. It's the book's very last line. Throughout the entire final nine chapters of the book, Ezekiel has been fully present in this vision. He walks the halls of the temple, he sees the city and its borders, he wades in the water of the river. And then the moment comes to reveal the name of that city. Ezekiel 48:35: "And the name of the city is . . ." And we are expecting it to be, The Lord Is Here. But it doesn't say that. It says, "The name of the city is The Lord Is There." And just like that, we remember that Ezekiel is still standing on the banks of that ugly canal. He has been speaking a far-off promise—about a world that is not yet, and a world that may never be. He has no delusions about where he is. He has always known. He is here. And the Lord is . . . there. The name aches with protest.

Tracy K. Smith's poem "Wade in the Water" holds a similar ache. She describes that familiar African American Spiritual as a "blood-deep song," haunted by hope, but also terror. The words speak of a Red Sea crossing, but were used by enslaved fugitives as code to "wade in water" when slave-catcher dogs were on their trails.

The song "dragged us to those banks," she writes, "and cast us in. . . . O Woods—O Dogs—O Tree—O Gun—*O Girl, run*—O Miraculous Many gone."[7]

The unvoiced question swirls like river mud: *Where were you, God? Where was the troubling of the waters you promised?*

What good are the waters of ancient story or the songs of future liberation, if God is not *here*? If God's holy trouble never touches down in the Missouri, or the cyclone, or the endangered forest of the Weelaunee?

O Miraculous Many, . . . so many . . . too many . . . gone.

And then, a whisper comes—like the beating of wings—or the turn of a wheel within a wheel.

7 Tracy K. Smith, "Wade in the Water," in *Wade in the Water* (Minneapolis: Graywolf Press, 2018), 15–16 (emphasis in original).

What if the name of the city Ezekiel saw is not a promise of what will be, but a reflective mirror, speaking what is? What if the city is a sermon and its name a blessing spoken by the city itself, rebounding over a broken world? "The name of God's city is . . . The Lord Is *There*." Not here, bathed in golden haze. But *there*, Ezekiel, with you—standing on the banks of the Chebar. Muck sucking at divine heels. *There*. Creeping through Georgia's marshes with fugitives, held breath, scarred back, and frantic heart. *There*. In the cyclone's flood. Ankle-deep in rising tides as ancestral graves wash away. *There*. Dancing in circles, Mylar rectangle pressed to the sky. *There*. Weeping in a Weelaunee shelter, knees pressed into bloody ground, torn between grief and fury.

Here. Wading in water. Even now.

Smith ends her poem with an unresolved question, addressed to God:

"Is this love the trouble You promised?"

That's a question only God can answer.

But Ezekiel's unresolved question is addressed to us:

What does such Love require?

Sermon Two

"A Question of Consequence"
Ezekiel 18:1–4, 25–32; Matthew 21:23–32

World Communion Sunday
Founders' Day
Duke University Chapel
October 1, 2023

(This sermon, preached on World Communion Sunday, was also preached on Duke University's Founders' Day. Founders' Day has often provided Duke Chapel preachers the opportunity to reflect on the university's vocational calling. The day also marked the one-year anniversary of Duke University's Climate Commitment.)

I don't actually think it's a bad proverb: "The parents eat sour grapes, and the children's teeth are set on edge." I know that the point of our Ezekiel passage today is that the proverb is wrong. God gets very worked up about it. One's actions, not one's ancestors, determine one's fate. We aren't punished for the sins of others. "The one who sins is the one who dies," the text concludes, with grim, almost elegant, simplicity.

I'm not so sure about that.

This summer, as part of a research project on South Pacific villages being displaced by climate change, I sat with a group of Fijian women whose homes had been washed away two years before by back-to-back cyclones that made landfall in the span of three weeks. I had come to hear stories about rebuilding and the community's decision-making processes around relocation. Leaving one's land is a complicated business for Indigenous Fijian communities—it is a grief-stricken business. But the village had come to consensus. They could not stay. The storms were coming stronger and more frequently. The last storm had washed away decades of effort in this remote rainforest inlet. "Everything we've worked for, for years, is gone," a woman said to me. "It's gone."

There used to be a story that scholars told about this morning's Ezekiel passage. The story was that this text was about individual accountability. That it was a turning point in theological ethics—from the murky world where the consequences of sin were passed down from generation to generation, to a bright, new future where each person had only to account for their personal actions and their personal failings. And for the most part, that story was told as if this atomized way of understanding sin and consequence was progress and enlightenment. Everyone gets a clean slate—a fresh start in a new race. Win or lose, it's in your hands.

But the climate crisis presses into our palms a more difficult reality. It witnesses to a connected world, over time and across space. We are woven together, and when the sins of ancestors tear that weave, generations unravel. When nations on the east side of the Pacific ignore their impact on the temperature of the ocean, a village on the west side of the Pacific weeps. It's not just that climate disasters are indiscriminate. They call for *communal* accountability. They press the sour grapes of environmental catastrophe into a common cup, and they ask us to taste the grief of others. Climate change tells us that others' cups of sorrow—their sour fruit—*should* set our teeth on edge. Because it is rarely the case that those beset by environmental loss are the ones to blame. And innocents die all the time.

What those scholars of the past miss about Ezekiel 18—and what we could miss today if we're not careful—is that this passage also has *communal* accountability at its heart. In the passage's last seven verses, God addresses God's people collectively four times. "House of Israel." "House of Israel." "House of Israel." "House of Israel." "Get yourselves a new heart and new spirit!" Notice: not "new hearts" and "new spirits"—as if we all had our own. But get yourselves (plural) a new collective heart (singular). As if all of God's children shared a body. Shared a common life. Individual action matters in this passage, yes. But not because we are insulated and atomized from each other. It matters precisely because we are not. We are bound together—for good and for ill. This isn't a passage that's about assigning blame on the basis of tragedy. Anyone who's read Job knows that suffering doesn't follow an algorithm. *This is a passage about accepting responsibility*—for the consequences of one's actions, yes. But more fundamentally, for the consequence, the weight, the significance of life.

Life is precious, not in spite of its ties that bind, but because of them.

And life is worth fighting for.

More than that. Life *can* be fought for.

This summer was the hottest on record in many parts of the country—and when news outlets interviewed those sweltering in their kitchens, a weary fatalism was whispering. A fatalism not unlike the weary shrug of Ezekiel's proverb. "Nothing can be done." "The die has already been cast." "Others are to blame." One woman spoke of "going all doomsday bunker" in her thinking—resistance giving way to resignation. Her aim had become very insular: *how do I save me and mine?*

Ezekiel was also composed during a time when destructive forces outside of his people's control threatened to overwhelm their collective hope. They had been conquered by empire. Their leaders exiled. Their temple razed. The loss was great enough that they had started to believe in a chillingly deterministic world. Nothing really mattered. Certainly not the insignificant choices of their lives. It is *that* chill that the God of Ezekiel refuses to abide.

I hear an echo of that holy rebuttal of fatalism in the voice of one of the Indigenous Fijian pastors whom I interviewed in this project on climate resilience. His village was being impacted by the government's careless logging practices, which, when combined with changing weather conditions, created extended seasons of drought, punctuated by dangerous landslides when rain finally arrived. "Our village used to be known for having the sweetest mangoes in Fiji," he said. "But when I arrived as pastor in the village, the mango harvest was small—and they were hard and dry. . . . I decided to start preaching about our relationships with each other," he continued. "There were long-standing grudges in the community. Sisters not speaking to each other. Resentments between families that went back years. We needed to improve our connection to God and our connections within the village if we wanted to heal our relationship with the land."

I will be very honest with you. This would not have been my first line of attack in trying to preserve the mangos. What difference do individual relationships make in a world of global economies and international developers and weather patterns formed by the rising temperatures of oceans? It sounds rather fuzzy about how cause and effect relate. Preaching is a roundabout method for changing something as concrete as crops. But I'll tell you what this pastor's plan was not. It was not the fatalism of that despairing, heat-sick woman who was ready to give up on local relationships altogether. *This was not resignation.* He didn't throw up his hands in despair . . . or place the fate of his people at the feet of a different empire on a different continent half a world away. He was fighting for the life of his village every way he knew how. And this pastor, more than most, knew what he was up against. He had been an influential government official before he accepted a call to ministry.

Which is why I asked him a follow-up question:

"Pastor, you know that climate change isn't just about a broken relationships between sisters. It's about the broken relationships between nations—about the broken relationships of history and colonialism—it's about broken legislation in my country and broken accountability structures in yours. . . . How do the relationships of a small village make any difference to problems that big?"

He took a breath—a long breath. The question wasn't new for him. "Jerusha," he said, "Change happens when we start small. We must start somewhere. I am a person of faith. And that is what I believe. I must start somewhere."

It's Founders' Day at Duke University, a full year after the school announced its Climate Commitment. Our formidable resources, our reach of global research partnerships, and the depth of our roots in the North Carolina landscape make us seem anything but small. We are gathered under towering buttresses, approaching our centennial celebration. And I am grateful that this institution has rejected the undertow of fatalism in a fight for life that will define the next generation—and perhaps many more. To those institutional leaders present today—thank you.

But even with all our resources, one only has to scratch the surface of this crisis to feel the smallness of our hands . . . how little we control. The concerted, coordinated efforts of this R1 research institution are dwarfed by the gathering storm.

I tell you today what this Fijian pastor told me—because he thought someone should know. *The mangos in his village got sweeter*. Sour fruit became sweet through restored relationships. That's his testimony.

It would take a whole other research project to figure out what that means. It might have been the rainfall was different that year . . . but there are other possibilities. Maybe better village relationships led to more cohesion in advocacy efforts with the local government; maybe the village men that had seasonal jobs in the forestry service started organizing for change; maybe community women spent more time together caring for the village's mango groves . . . or maybe there is a far more complicated connection between cause and effect than is dreamt of in our philosophies. Whatever the reason, the idea that *small networks of local relationships make a difference in the climate fight*—that idea is not new. And it's supported by data.

Dr. Elizabeth Albright is an associate professor of environmental ethics at Duke's Nicholas School of the Environment. She studies how natural disasters impact the long-term well-being of local communities. I first heard about her research at an interdisciplinary teaching and learning event that brought together natural and social scientists, public policy specialists, and theologians to engage local clergy about how faith communities could further the work of climate justice. The Duke Marine Lab was our host. And this year, the gathering was funded by the Nicholas School's Climate Research Innovation Seed Program (CRISP). Not all in that conversation identified as people of faith—but every last one of us was convinced that faith communities were necessary to the work.

Dr. Albright laid out sobering statistics about how wealth inequalities worsen every time a natural disaster hits. She had chart after chart from scholars across her field showing how the wealth gap widens between white communities and communities of color after each storm—regardless of education, regardless of relief funding.[8] Climate change was reifying the ugliest legacies of this nation's history. Then she said this. "There *is* a mitigating factor that has made a statistically significant difference. If communities build relationships *before the storm*, recovery looks different."[9] You would have thought the clergy in the room grew two inches taller all at once. Here was a bona fide scientist telling them that the very work they knew how to do—keeping track of shut-ins, attending ecumenical council meetings, organizing a community food bank with a church that looks different and worships different, serving sandwiches with the women's group in the fellowship hall, getting coffee with the Spanish-speaking pastor down the street (or the English-speaking pastor if you're the one who speaks Spanish)—all of those things . . . those small things . . . have consequence.

That's not so different from the Fijian pastor's testimony, or from what God told God's people through the prophet Ezekiel. Your choices. Your relationships. Your commitment to life and its ties that bind matter. I

[8] See, for example, Junia Howell and James Elliot, "As Disaster Costs Rise, So Does Inequality," *Socius: Sociological Research for a Dynamic World* 4 (2018): 1–3.

[9] For an example of a study that links social capital with climate resilience, see Daniel P. Aldrich and Michelle A. Meyer, "Social Capital and Community Resilience," *American Behavioral Scientist* 59, no. 2 (2015): 254–69.

wonder if at the end of the day the most significant ways that Duke will move the needle in its climate commitment are through the investments it makes in its relationships. Its relationships with local communities. The interdisciplinary relationships between its scholars. And its relationships with an increasingly connected world. That's slow, humble work. Relationship building *accepts responsibility* for the damage caused by the institution, whether in this generation or in generations prior, and it works to undo that damage. It works to rebuild trust from the ground up, which is not easy. But it may finally be the work that is most consequential.

Today is not only Founders' Day—it is World Communion Sunday. A day when we remember a world held in God's body by the work of God's Spirit. The God of Ezekiel not only wanted God's people to know that their lives mattered. God wanted them to know that life matters to God. This table, to extend Anathea Portier-Young's beautiful reflection on Ezekiel 18, is not just a place for "looking back" but for "looking around."[10] It is a place where the responsibilities of relationship are pressed into our hands . . . but also a place where we are pressed into God's hand . . . a God who presses past all of our excuses to fight for the life of the world God loves. And a God who asks us to do the same.

Interestingly, our Gospel reading today also references grapes—or at least a vineyard. There are two brothers, one who pays lip service to the work required, who says all the right things and never shows up, and another brother who, for whatever reason, tells his father that sour fruit is not his responsibility. I see that straightforward, "tell-it-like-it-is" brother on the road, convinced in his decision. And then I see him pause . . . and turn. It's no small thing, that turning. It trusts that a different path is possible. That change is possible.

The communion table requires just such a turn. So come, heart pressed to your chest. Breath warm in your body. It's not too late in the day—even now, when so many have died.

You are alive, connected to a living world and Living God.

And life is a matter of consequence.

[10] Anathea Portier-Young, "Commentary on Ezekiel 18:1–4, 25–32," Working Preacher, September 28, 2008, https://www.workingpreacher.org/commentaries/revised-common-lectionary/ordinary-26/commentary-on-ezekiel-181-4-25-32-3.

Sermon Three

"Glory"

Matthew 3:13–17; Psalm 29

Remembrance of Baptism Service
Goodson Memorial Chapel
Duke Divinity School
January 9, 2020

(This first worship service of the divinity school term and calendar year was a remembrance of baptism service. It also took place during the cataclysmic bushfires of January 2020. Denominational divisions and a difficult election were on the horizon, but no one in the room anticipated how much courage 2020 would require. Just over two months from this service, the COVID-19 pandemic would close the campus.)

Sometimes, it's not about us. This book that we read, this Scripture that we pore over. Sometimes, we are really beside the point. Psalm 29 spends ten of its eleven verses describing the voice of the Lord as untranslatable storm. God's voice does not speak in Hebrew in this passage or in any human language at all. It roars. It rips. It whirls. Its argument is its effect, its ability to shake the earth, demanding that lesser gods ascribe to the God of Israel power, sovereignty, and glory. And they do. God sits enthroned by psalm's end . . . and we have nothing to do with it.

And in that light, verse 11 seems small—and brave. Spoken under the thunder's growl. Smothered by an ocean of sky. "May the Lord give strength," the psalmist says. "May the Lord bless God's people with peace."

Where does one get that kind of courage? When one's face is pelted by rain and lit by lightning, when one's hands are empty before a cosmic throne. Evacuation is the order of the day—not encounter. "Find shelter," we holler back. "This storm is bigger than you." Yet there the psalmist stands.

This week, as millions of acres of Australia burned, the moderator of the United Church of Christ wrote a similarly small and empty-handed prayer: "In this world of changed and changing climate, God, we ask that you would help us to help those who are in immediate danger; . . . call us to repent of all the ways we continue to make this world a place of death. . . . Help us, God . . . because we don't seem to be able to do it on our own."[11]

It's an extrapolation of verse 11. "Give strength. Grant peace."

And it is brave. It is a prayer from one who knows that we do not really deserve an answer. We who have set up idols in God's place. We who have refused to hear the voice of God thundering in creation's tremor. We who have done all we can to convince ourselves of God's impotence when it served our purposes . . . and then turned around and used arguments about God's power to relieve us of our responsibility. We are too clever by half. And now the world burns. Nations rage. The laws of the land swirl like leaves in a gale. For we find that we are not, in fact, the center of the world, and there are storms we cannot weather. We can repent—we can equivocate, but it is not finally about us. The world will have its way. More to the point, God will have God's way. We do not hold God hostage by our sin, for all we are held accountable for it. And we do not force God's hand. Human repentance is not enough, for all of its necessity.

John the Baptist understood this with wilderness-born clarity. In the verses before our Gospel passage today, he makes it plain. This is not about him—nor is it about his listeners. "I baptize you with water for repentance, but One who is more powerful than I is coming after me . . . He will baptize you with the Holy Spirit and with fire. His winnowing fork is in his hand, and he will clear his threshing floor. The wheat he will gather; the chaff he will burn." The fiery imagery of Psalm 29 undergirds John the Baptist's Christology. A God enthroned, whose voice "flashes forth flames," and the baptism he describes is terrifying. A storm we cannot weather—a "Glory" we would rather not see. For who among us doesn't have chaff hoarded like treasure, and dry as tinder?

It is hard to read these fiery Scriptures after the week we've had. When missile strikes and environmental catastrophes fill our news feeds. When cedar-like denominations crack. To talk about God's power on weeks like this raises the ugliest of theological questions. If God is powerful, why the fires? Why the lives lost? The half a billion animals, choked and dying? Why does God stand silent? But more terrifying than even the mystery of God is the certainty of sin. For the drumbeat of war is calling, and we find

[11] The prayer was written by the Rt. Rev. Richard Bott, moderator of the United Church of Canada, for the people and animals of Australia.

the winnowing fork of violent retribution is not just in the hands of a Savior. It is in our own. Not every baptism is godly. Not every fire is holy. And the violence of John's words and Psalm 29, the violence of nature itself, can spark our own attempts to thresh the world—our own desire to play God.

I have complete sympathy with John when he stands shocked, staring into the face of Jesus. "I need to be baptized by you," John says. "Do you come to me? I am the one who needs threshing. I am the one whose repentance is not enough. I am the one who sees vipers among your people and hopes for their destruction. I am the one who has nothing but a cracked voice and dirty river water to offer the living God. And you come to me?"

"Let it be so now," Jesus replies. Not forever. "Now. For us to fulfill all righteousness." *And this is the mystery of Psalm 29 and Matthew 3*. It is not a theodicy puzzle. It is a reckoning and a revelation. It is about a human preacher who, with fear and trembling, presses the God-Man beneath the Jordan's surface, foreshadowing the baptism of death that Jesus would undergo by human hands. A human preacher whose own baptism of repentance—the best thing he has to offer God—testifies to the hardness of the human heart with whispers of Golgotha. But John agrees to do it. Because he knows that the first step in "fulfilling all righteousness" is owning one's sin. Seeing it . . . standing in its swirling deep. He may baptize Jesus with water, but make no mistake, in that very river, John is *being* baptized with fire. For the mystery is double-edged. And the mystery isn't really about him or humankind or you and me. The mystery is about a God enthroned above the waters, *who chooses to go underwater instead*.

There was a massive tree that grew outside the chapel windows of Fiji's Davuilevu seminary. It was a home to birds and bats and frogs and insects, a whole ecosystem, and it was beautiful. When Cyclone Winston hit in 2016, the fiercest cyclone in recorded history to hit the Southern Hemisphere, it fell. It was just before Lent. And all through February and March, I sat, listening to our morning sermons while gazing at the gash created by that tree's destruction. The upturned root system, rotting in the sun. The torn ground, the crushed life beneath its weight.

That whole Lent, Fiji grappled with what it had done to deserve such a fate. Had Christians not prayed enough? Or worshipped enough? Had they not fasted enough? I suggested at one point that perhaps it was the Western countries who poured pollutants into the sky who needed to do more fasting. But that line of thought seemed deeply unfair to my Fijian colleagues. Why would God punish them for Western greed? I had no answer for that.

But as Lent progressed to the week of the Passion, the tree became less of a scar indicating divine disapproval, and more a sign of God's costly cruciform solidarity. The God of Scripture is not a God who stands outside the earth's storms, unscathed. Not even the God of the Old Testament is that God. How else does the writer of Psalm 29 stand and pray for strength and peace except through an abiding faith in a God that swings low over creation? A God who breaks cedars, yes. But also a God whose servant, in the words of Isaiah 42, "will not break . . . a bruised reed." Isaiah 42 is where Jesus' title of "Beloved" comes from.

No, the God of Scripture enters the storm, leaves the throne, dives into the abyss. Revealed in the person of Jesus, bearing the baptism of our sin—even when the blows, the bombs, the fire and flood rip his roots from the ground. The God of Glory is the God of Golgotha. And so that broken tree became for me a testimony of God's presence.

And that is good news. But the good news of the gospel is greater still.

God does more than bear our baptism.

God becomes our baptism.

To be baptized Christian is not to be baptized in a storm. It is to be baptized in a Body. We are incorporated into the flesh of the One who bore the baptism of creation, and so we are connected to creation through him, woven together tissue and bone. This is how "righteousness is fulfilled." This is how our God bends low. Not by protecting us from harm, but by making us one with the world that Christ embraces. *And in His body, we learn again that this redemption—this groaning—this story is not really about us, even as it includes us.* Even as it calls forth our deepest, fiercest commitment. God's redemption is about the world. The choking oceans, the silent nests, the migrant children, and the border patrol guard. The living soil and the bleached coral—the mourning widow and the orphaned son. We see more than our sin in the swirling Jordan. We see our connection. Which means that this world of ours matters immensely. And we matter immensely. Our action matters. Our choices matter. Our speech matters—as does our silence.

I've been praying about the word to leave with you on this Communion Day. This Remembrance of Baptism Day. This first day of chapel in a new decade.

And I've struggled with this text, because the blessing that Jesus receives, the blessing of a "Beloved" in whom God is well pleased, feels very far from where we are. I need a more concrete word, a word that offers more than imputed righteousness. I want righteousness imparted.

Because the challenges of this next year have sharp edges. An election year in an angry season, a burning planet and the specter of war. A denomination discerning a divided future. And that doesn't name the personal storms in every human heart in this room. There will be all manner of threshings in the year to come.

And a story came back to me from my youth.

In *The Voyage of the Dawn Treader*, C. S. Lewis paints a frightening picture of a ship that becomes lost in a nightmare of darkness.[12] It loses its way, and a panicked madness begins to spread among the sailors. Then a pinprick of light appears in the sky. It gets closer, and begins to look like a cross—and then it becomes clear that the cross is a bird . . . an albatross. It circles the mast, bringing the crew to their senses. But only Lucy, a young girl standing on deck, hears the words the albatross whispers as it brushes by. "Courage, dear heart," it says. And that is my word to you. Resonating from the minor chord of Psalm 29 and Matthew 3. Courage, dear hearts.

The God who has borne your baptism has become your baptism. And that same God blesses your baptism. For the Christ that was buried beneath the waves came up from them, covered with water that flashed like flame. Carrying the world in his flesh . . . and carrying *you*, if you would be one with the world—if you would make your life about something other than yourself.

Courage. For the darkness will come. And you will be asked for more than you thought you could give. Courage. Do not be afraid. Not because the world is safe. I will not lie to you—and these texts do not lie. But, courage. Because the same God who carries you under the water is the God enthroned above. Give that God everything you have. And then stand in good company, empty-handed in the storm. Stand with that psalmist who prays for *strength* and prays for *peace*, under a weight of glory. Courage.

Because this story is bigger than you.

[12] C. S. Lewis, *The Voyage of the Dawn Treader*, book 5 of The Chronicles of Narnia (New York: Harper Trophy, 1994), 187.

Bibliography

Ahiamadu, Amadi. "A Postcolonial Critical Assessment of the *Imago Dei* in Gen. 1:26–28 in Nigerian Perspective." *Scriptura* 103 (2010): 97–106.

Aiava, Faafetai. "Eleele Interrupts the Eden Wedding: From Mother Earth to Mistress." In *Decolonizing Ecotheology: Indigenous and Subaltern Challenges*, edited by S. Lily Mendoza and George Zachariah, 93–107. Eugene, Ore.: Wipf and Stock, 2022.

———. "Talking God in a Divided House: Rebuilding Spirituality in God's Pacific Household." Paper presented at Iliff School of Theology, Denver, Colo., October 22, 2021.

Albertz, Ranier. "From Aliens to Proselytes: Non-priestly and Priestly Legislation Concerning Strangers." In *The Foreigner and the Law: Perspectives from the Hebrew Bible and the Ancient Near East*, edited by Reinhard Achenbach, Ranier Albertz, and Jakob Wöhrle, 53–69. Wiesbaden: Harrassowitz, 2011.

Albright, Elizabeth. "Floods and Fire." Presentation at the Pastoral Care for Climate: Weaving Science and Theology for Justice Conference, Duke Marine Lab, Beaufort, N.C., June 14, 2023.

Aldrich, Daniel P., and Michelle A. Meyer. "Social Capital and Community Resilience." *American Behavioral Scientist* 59, no. 2 (2015): 254–69.

Anderson, Bernard. *Rediscovering the Bible.* New York: Haddam House Books, 1951.

Bacon, J. M. "Settler Colonialism as Eco-social Structure and the Production of Colonial Ecological Violence." *Environmental Sociology* 5, no. 1 (2019): 59–69.

Baleidelabaci "Bale" Tamanikaigavo, Marika. Personal email correspondence, September 5, 2023.

———. Personal email correspondence, November 5, 2023.

———. Personal online interview, January 25, 2022.

———. "Racial Integration in Transition: A *Magimagi* Historical Assessment of the Methodist Church in Fiji's Response to Identity and Race Issues in Fiji." Unpublished master of theology thesis, Pacific Theological College, September 2021.

——. "Silences and Representations: A Postcolonial-Indigenous Historiographical Exploration of the Histories of Christianity in Oceania Using the *Mata ni Katuba* Concept." PhD diss., Pacific Theological College, Suva, forthcoming 2025.

Baleloa, Eric. "What Must I Do to Ensure Blessing upon My *Vanua*: Luke 15:8–10." Sermon preached to the Buca Circuit, Fiji, February 2, 2020.

Balenaikorodawa, Taniela. "God the Creator, Recreating." Sermon preached in Cogea village, Fiji, January 29, 2021.

——. Personal online interview, January 3, 2023.

——. "Sosolevaki: A Collaborative Learning Model for the Methodist Church in Fiji, to Enable Transformational and Sustainable Christian Education among the Laity." PhD diss., Pacific Theological College, Suva, forthcoming 2025.

Balenaikorodawa, Taniela, and Jerusha Matsen Neal. "Anti-colonial Homiletic Strategies for Combating Climate Change in the Fijian Methodist Church." Paper presentation at the World Christianity Conference, Princeton Theological Seminary, March 6, 2021.

Ballantyne, Tony. "Genesis 1:28 and the Languages of Colonial Improvement in Victorian New Zealand." *Victorian Review* 37, no. 2 (2011): 9–13.

Banivanua, Tevita Nawadra. "Kairos, God's Plan and Wilderness Formation in the New Exodus." Sermon preached at Davuilevu Theological College, Nausori, Fiji, November 2016.

——. "Mapping the *Lako Yani Vou* [New Exodus]." Keynote address, Divisional Superintendents and Divisional Stewards Conference, Centenary Church, Suva, March 10, 2015.

——. Personal interview, Suva, Fiji, January 2, 2023.

——. Personal online interview, October 31, 2021.

——. "Signposts of the New Exodus." Keynote address, 2016 MCIF Annual Conference, Centenary Church, Suva, August 15, 2016.

——. "The Status of the Methodist Church in Fiji." Keynote address, 2015 MCIF Annual Conference, Centenary Church, Suva, August 14, 2015.

——. "Staying the Course in the Winds of Change." Keynote address to Davuilevu Theological College graduates, Baker Memorial Church, Nausori, Fiji, November 24, 2017.

Barstad, Hans. *The Myth of the Empty Land: A Study in the History and Archaeology of Judah during the "Exilic" Period.* Oslo: Scandinavian University Press, 1996.

Bartholomew, Craig. *Where Mortals Dwell: A Christian View of Place for Today.* Grand Rapids: Baker Academic, 2011.

Basso, Keith. *Wisdom Sits in Places: Landscape and Language among the Western Apache.* Santa Fe: University of New Mexico Press, 1996.

——. "Wisdom Sits in Places: Notes on a Western Apache Landscape." In *Senses of Place*, edited by Steven Feld and Keith Basso, 53–90. Santa Fe, N.Mex.: School of American Research Press, 1996.

Bau, Netani. "Respect and Sustain the Work of God's Hands: Luke 18:35–43." Sermon preached in Drana village, Fiji, May 19, 2019.

Becking, Bob. "Global Warming and the Babylonian Exile." In *By the Irrigation Canals of Babylon: Approaches to the Study of the Exile*, edited by John Ahn and Jill Middlemas, 49–62. London: T&T Clark, 2012.

Bennet, Zöe, Elaine Graham, Stephen Pattison, and Heather Walton. *An Introduction to Research in Practical Theology*. New York: Routledge, 2018.

Bergmann, Sigurd. "Theology in Its Spatial Turn: Space, Place and Built Environments Challenging and Changing the Images of God." *Religion Compass* 1, no. 3 (2007): 353–79.

Berkes, Fikret. *Sacred Ecology*. 3rd ed. New York: Routledge, 2021.

Berners-Lee, Mike. *There Is No Planet B: A Handbook for the Make or Break Years*. Cambridge: Cambridge University Press, 2021.

Bernstein, Ellen. "Rereading Genesis: Human Stewardship of the Earth." In *Righteous Indignation: A Jewish Call for Justice*, edited by Or N. Rose, Jo Ellen Green Kaiser, and Margie Klein, 55–59. Woodstock, Vt.: Jewish Lights, 2008.

Berry, Wendell. "Discipline and Hope." In *A Continuous Harmony: Essays Cultural and Agricultural*, 83–164. Berkeley, Calif.: Counterpoint, 2021.

Bhagwan, James. "Ezekiel 37 Bible Study." 15th UCA Triennial Assembly, Victoria, Australia, July 2018.

———. "Gratitude." Girmitiya welcome service sermon, Nabudrau village, Rewa, Fiji, October 9, 2016.

———. Interviewed on *Speak Your Mind*, FBCTV, episode 163, April 8, 2015. https://www.youtube.com/watch?v=hrRMrC4TNvQ.

———. "Methodist Church in Fiji President Sails on Traditional Voyaging Canoe—Reflects on Climate Change." *First Friday Newsletter: World Methodist Council*, September 28, 2017. https://firstfridayletter.worldmethodistcouncil.org/2017/09/methodist-church-in-fiji-president-sails-on-traditional-voyaging-canoe-reflects-on-climate-change/.

———. Personal email correspondence, June 20, 2023.

———. Personal interview, Suva, January 7, 2023.

———. Personal online interview, September 28, 2021.

———. "Sailing to the Island of Hope: A Wesleyan Ethical Framework for 21st Century Fiji." Unpublished master of theology thesis, Methodist Theological University, Seoul, 2013.

Bird, Cliff. 2016. "God and Natural Disasters." Facebook, Methodist Church in Fiji and Rotuma. February 29, 2016. https://www.facebook.com/notes/685264578781425/.

Blenkinsopp, Joseph. *Ezekiel*. Louisville: Westminster John Knox, 1990.

Block, Daniel I. *Beyond the River Chebar: Studies in Kingship and Eschatology in the Book of Ezekiel*. Eugene, Ore.: Cascade, 2013.

———. *The Book of Ezekiel: Chapters 1–24*. Grand Rapids: Eerdmans, 1997.

———. *The Book of Ezekiel: Chapters 25–48*. Grand Rapids: Eerdmans, 1998.

——. "Divine Abandonment: Ezekiel's Adaptation of an Ancient Near Eastern Motif." In *The Book of Ezekiel: Theological and Anthropological Perspectives*, edited by Margaret S. Odell and John T. Strong, 15–42. Atlanta: SBL, 2000.

——. "In Search of Theological Meaning: Ezekiel Scholarship at the Turn of the Millennium." In *Ezekiel's Hierarchical World: Wrestling with a Tiered Reality*, edited by Stephen L. Cook and Corrine L. Patton, 227–40. Atlanta: SBL, 2004.

Bosman, Hendrik. "Myth, Metaphor, or Memory? Allusions to Creation and Exodus in Isaiah 51:9–11 as a Theological Response to Suffering during the Exile." In *Exile and Suffering: A Selection of Papers Read at the 50th Anniversary Meeting of the Old Testament Society of South Africa, Pretoria, August 2007*, edited by Bob Becking and Dirk Human, 71–82. Leiden: Brill, 2009.

Bosu, Daryl. "Protecting Atewa Forest: Conservation of Ghana's Upland Treasure." A Rocha Ghana. Accessed December 18, 2023. https://old-ghana.arocha.org/projects/protecting-atewa-forest/.

Bouma-Prediger, Steven. *Creation Care Discipleship: Why Earthkeeping Is an Essential Christian Practice*. Grand Rapids: Baker Academic, 2023.

Bourdieu, Pierre. "Social Space and Symbolic Space." *Poetics Today* 12, no. 4 (1991): 627–38.

Bowler, Kate. *Blessed: A History of the American Prosperity Gospel*. Oxford: Oxford University Press, 2013.

Bradfield, Abraham. "Decolonizing the Intercultural: A Call for Decolonizing Consciousness in Settler-Colonial Australia." *Religions* 10, no. 469 (2019): 1–21.

Brett, Mark. *Political Trauma and Healing: Biblical Ethics for a Postcolonial World*. Grand Rapids: Eerdmans, 2016.

Briggs, William. "Idols and Land Grabs, Ancient and Modern: Creation and Ecotheology in Ezekiel 6, 35:1–36:15." *Horizons in Biblical Theology* 40, no. 1 (2018): 41–64.

Brinkman, J. A. "Babylonian Royal Land Grants, Memorials of Financial Interest, and Invocation of the Divine." *Journal of the Economic and Social History of the Orient* 49, no. 1 (2006): 1–47.

Brockman, Terra. "Decolonized Sacred Land: How a Church Became the Home of an American Indian Organization." *Christian Century* 137, no. 6 (2020): 22–25.

Brown, Joel. "Durham's Hayti Community: A Black Business Mecca Crushed by Empty Promises Reckons with Its Past." ABC Eyewitness News, February 22, 2021. https://abc11.com/durham-hayti-community-freeway-history-black/10362504/.

Brown, Sally A. *Sunday's Sermon for Monday's World: Preaching to Shape Daring Witness*. Grand Rapids: Eerdmans, 2020.

Bruce, Susan. "Utopian Justifications: More's 'Utopia,' Settler Colonialism, and Contemporary Ecocritical Concerns." *College Literature* 42, no. 1 (2015): 23–43.

Brueggemann, Walter. *Interpretation and Obedience: From Faithful Reading to Faithful Living*. Minneapolis: Fortress, 1991.

———. *The Land: Place as Gift, Promise, and Challenge in Biblical Faith*. 2nd ed. Minneapolis: Fortress, 2002.

———. "The Loss and Recovery of Creation in Old Testament Theology." *Theology Today* 53, no. 2 (1996): 177–90.

———. "A Shape for Old Testament Theology, I: Structure Legitimation." *Catholic Biblical Quarterly* 47 (1985): 28–46.

Burroughs, Presian. *Creation's Slavery and Liberation: Paul's Letter to Rome in the Face of Imperial and Industrial Agriculture*. Eugene, Ore.: Cascade, 2022.

Cagle, Nicki, Ryan Emanuel, Malinda Lowery, and members of the Duke Native American Student Alliance. "Land Acknowledgements: An Exploration." Diversity, equity, and inclusion at Duke's Nicholas School of the Environment. Accessed December 2, 2023. https://sites.nicholas.duke.edu/diversity/expand-your-knowledge/teaching/land-acknowledgements-an-exploration/.

Calvert, James. *Fiji and the Fijians: Mission History*. London: Alexander Heylin, 1858.

Cameron, John. "Some Implications of Malpas's *Place and Experience* for Place Ethics and Education." *Newsletter of the Society for Phenomenology and Architecture* 15, no. 2 (2004): 5–9.

Campbell, Charles L. "Preaching Gospel: Four Theological Tones." In *Preaching Gospel: Essays in Honor of Richard Lischer*, edited by Charles L. Campbell, Clayton Schmit, Mary Hinkle Shore, and Jennifer E. Copeland, 1–15. Eugene, Ore.: Cascade, 2016.

———. *Preaching Jesus: The New Directions for Homiletics in Hans Frei's Postliberal Theology*. Eugene, Ore.: Wipf and Stock, 1997.

———. *The Scandal of the Gospel: Preaching and the Grotesque*. Louisville: Westminster John Knox, 2021.

———. *The Word Before the Powers: An Ethic of Preaching*. Louisville: Westminster John Knox, 2002.

Campbell, John. "Climate Induced Community Relocation in the Pacific: The Meaning and Importance of Land." In *Climate Change and Displacement: Multidisciplinary Perspectives*, edited by Jane McAdam, 57–79. New York: Bloomsbury, 2010.

Carley, Keith. "From Harshness to Hope: The Implications for Earth of Hierarchy in Ezekiel." In *Ezekiel's Hierarchical World: Wrestling with a Tiered Reality*, edited by Stephen L. Cook and Corrine L. Patton, 109–26. Atlanta: SBL, 2004.

Carr, David. *Holy Resilience: The Bible's Traumatic Origins*. New Haven: Yale University Press, 2014.

Carr, Wylie, Michael Patterson, Laurie Yung, and Daniel Spencer. "The Faithful Skeptics: Evangelical Religious Beliefs and Perceptions of Climate Change." *Journal for the Study of Religion, Nature and Culture* 6, no. 3 (2012): 276–99.

Carroll, Seforosa. "Displacing Liturgy: A Pacific Exploration." *Liturgy* 36, no. 2 (2021): 26–35.

Carvalho, Corrine. "Finding a Treasure Map: Sacred Space in the Old Testament." In *Touching the Altar: The Old Testament for Christian Worship*, edited by Carol M. Bechtel, 123–51. Grand Rapids: Eerdmans, 2008.

Casey, Edward. *The Fate of Place: A Philosophical History*. Berkeley: University of California Press, 1998.

———. *Getting Back into Place*. Bloomington: Indiana University Press, 1993.

Casimira, Aisake. *Who Do You Say That I Am? Interaction between Bible and Culture*. Suva: ECREA Press, 2008.

Catanasiga, Vani. Personal correspondence, January 16, 2022.

———. Personal correspondence, November 22, 2023.

———. Personal interview, Suva, January 11, 2023.

———. Personal interview, FCOSS offices, Suva, July 1, 2023.

Césaire, Aimé. *Discourse on Colonialism*. Translated by Joan Pinkham. New York: Monthly Review Press, 2000.

Chao, Sophie, and Dion Enari. "Decolonising Climate Change: A Call for Beyond-Human Imaginaries and Knowledge Generation." Special issue, "Tropical Imaginaries and the Climate Crisis," *eTropic* 20, no. 2 (2021): 32–54. http://dx.doi.org/10.25120/etropic.20.2.2021.3796.

Charleston, Steven. *We Survived the End of the World: Lessons from Native America on Apocalypse and Hope*. Minneapolis: Broadleaf Books, 2023.

Chavis, Benjamin. "Toxic Wastes and Race in the United States: A National Report on the Racial and Socio-Economic Characteristics of Communities with Hazardous Waste Sites." Commission for Racial Justice. New York: United Church of Christ, 1987.

Chen, Ruoxue. "Master's Project Investigates Health and Environmental Disparities in Durham, N.C." Duke Nicholas School of the Environment, April 8, 2022. https://nicholas.duke.edu/news/masters-project-investigates-health-and-environmental-disparities-durham-nc.

Childs, Brevard. *Biblical Theology in Crisis*. Philadelphia: Westminster, 1970.

Chiu, Crystal. "Reimagining the Sermon: Stories of Immigration." Directed by Miriam Cho, July 2, 2021. https://www.youtube.com/watch?v=isnn2ke_HJY&t=44s.

Christie, Douglas. "Place-Making as Contemplative Practice." *Anglican Theological Review* 91, no. 3 (2009): 347–72.

Clark, Natalie. "Shock and Awe: Trauma as the New Colonial Frontier." *Humanities* 5, no. 14 (2016): 1–16.

Clarke, David. "Heidegger, Hermeneutics and History: Undermining Jeff Malpas's Philosophy of Place." *Philosophia* 42, no. 3 (2014): 571–91.

Clements, Ronald E. *Ezekiel*. Louisville: Westminster John Knox, 1996.

Cloete, Karla. "To Protect and Protest: Artist Cannupa Hanska Luger and the Mirror Shield Project." Arts Help, August 28, 2022. https://www.artshelp.com/mirror-shield-project/.

Collier, Jane. "A Chief Does Not Rule Land: He Rules People (Luganda Proverb)." In *Law and Empire in the Pacific: Fiji and Hawai'i*, edited by Sally Engle Merry and Donald Brenneis, 35–60. Santa Fe, N.Mex.: School of American Research Press, 2003.

Collins, John J. *The Apocalyptic Imagination: An Introduction to Jewish Apocalyptic Literature*. 2nd ed. Grand Rapids: Eerdmans, 1998.

———, ed. *The Origins of Apocalypticism in Judaism and Christianity*. Vol. 1 of *Encyclopedia of Apocalypticism*, edited by Bernard McGinn, John J. Collins, and Stephen J. Stein. New York: Continuum, 1998.

Cook, Stephen L. "Burgeoning Holiness: Fecundity Let Loose in Ezekiel 34–36." In *Ezekiel: Current Debates and Future Directions*, edited by William A. Tooman and Penelope Barter, 345–59. Tübingen: Mohr Siebeck, 2017.

———. *Ezekiel 38–48*. New Haven: Yale University Press, 2018.

———. "Ezekiel's God Incarnate!" In *The God Ezekiel Creates*, edited by Paul Joyce and Dalit Rom-Shiloni, 107–31. London: T&T Clark, 2015.

———. "Ezekiel's Recovery of Premonarchic, Tribal Israel." In *Ezekiel: Current Debates and Future Directions*, edited by William A. Tooman and Penelope Barter, 360–76. Tübingen: Mohr Siebeck, 2017.

———. *Prophecy and Apocalypticism: The Postexilic Social Setting*. Minneapolis: Fortress, 1995.

Cook, Stephen L., and Corrine L. Patton. "Hierarchical Thinking and Theology in Ezekiel's Book." Introduction to *Ezekiel's Hierarchical World: Wrestling with a Tiered Reality*, edited by Stephen L. Cook and Corrine L. Patton, 1–26. Atlanta: SBL, 2004.

Cooley, Jeffrey L. "Psalm 19: A Sabbath Song." *Vetus Testamentum* 64, no. 2 (2014): 177–95.

Copeland, Rebecca. "'Their Leaves Shall Be for Healing': Ecological Trauma and Recovery in Ezekiel 47:1–12." *Biblical Theology Bulletin* 49, no. 4 (2019): 214–22.

Cotterill, Aden. "The Dialectics of Place and Pastoral Practices." *Ecclesiology* 19, no. 2 (2023): 186–204.

Cox, John, Glen Finau, Romitesh Kant, Jope Tarai, and Jason Titifanue. "Disaster, Divine Judgement, and Original Sin: Christian Interpretations of Tropical Cyclone Winston and Climate Change in Fiji." *Contemporary Pacific* 30, no. 2 (2018): 380–411.

Craddock, Fred B. *As One without Authority*. Nashville: Abingdon, 1979.

Craft, Jennifer. *Placemaking and the Arts: Cultivating the Christian Life*. Downers Grove, Ill.: IVP Academic, 2018.

Crampton, David. "Fiji's Government Again Cancels Methodist Conference." *Christian Century*, August 25, 2011. https://www.christiancentury.org/article/2011-08/fiji-government-again-cancels-methodist-conference.

Cunningham, Chris. "Indigenous by Definition, Experience, or Worldview." *British Medical Journal* 327 (2003): 403–4.

Curuki, Semisi Nabiri. Personal interview, Cogea, Fiji, July 6, 2023.

Danford, Iva. "Discussion to Be Held on Deaconesses Being Given the Same Pay as Ordained Ministers." Fijivillage, August 21, 2017. https://fijivillage.com/news/Discussion-to-be-held-on-deaconesses-being-given-the-same-pay-as-ordained-ministers--sr2k95/.

Davidson, Steed Vernyl. "Lost Paradises: Tracing the Imperial Contours of Modern Tourism on Land and Peoples." In *People and Land: Decolonizing Theologies*, edited by Jione Havea, 15–34. Minneapolis: Fortress Academic, 2019.

Davis, Eddie. "West Durham Baptist Church—Fifty Years at 1901 Athens Avenue." West Durham Baptist Church. Accessed December 3, 2023. https://westdurhambaptist.org/jubilee-athens-avenue/.

Davis, Ellen. *Biblical Prophecy: Perspectives for Christian Theology, Discipleship, and Ministry*. Louisville: Westminster John Knox, 2014.

———. "Divine Sovereignty and Covenant." Course lecture for Hope for Creation? An Exilic Perspective, Duke University, February 7, 2023.

———. *Opening Israel's Scriptures*. Oxford: Oxford University Press, 2019.

———. *Scripture, Culture, and Agriculture: An Agrarian Reading of the Bible*. Cambridge: Cambridge University Press, 2009.

———. "Seeing God in Exile." Public lecture at New York Avenue Presbyterian Church, Washington, D.C., October 21, 2018. https://www.youtube.com/watch?v=n6V_BJWLjSU.

———. *Swallowing the Scroll: Textuality and the Dynamics of Discourse in Ezekiel's Prophecy*. Sheffield: Almond Press, 1989.

———. Untitled sermon on Isaiah 65. Sermon preached at a Service of Word and Table in collaboration with the Anglican Episcopal House of Study, Goodson Chapel, Duke University, Durham, N.C., November 9, 2022.

Davis, Richard A. "Decolonial Settler Theology: A *Pakeha* Theological Proposition." Public lecture at Wesley House Cambridge, January 26, 2023. https://www.youtube.com/watch?v=kzWNYjHANpY.

De Boer, Martinus C. "Paul, Theologian of God's Apocalypse." *Interpretation* 56, no. 1 (2002): 21–33.

Deifelt, Wanda. "And G*d Saw That It Was Good—*Imago Dei* and Its Challenge to Climate Justice." In *Planetary Solidarity: Global Women's Voices on Christian Doctrine and Climate Justice*, edited by Grace Ji-Sun Kim and Hilda P. Koster, 119–34. Minneapolis: Fortress, 2017.

Delai, Samuela. Personal email correspondence, June 29, 2023.

———. Personal online interview, January 23, 2022.

———. "Spiritual Views on Climate Change: John 3:16." Sermon preached for the Saqele Circuit, Fiji, in honor of Fiji Day, October 10, 2021.

De La Torre, Miguel. *Embracing Hopelessness*. Minneapolis: Fortress, 2017.

Deleuze, Gilles, and Félix Guattari. *A Thousand Plateaus: Capitalism and Schizophrenia*. Translated by Brian Massumi. London: Continuum, 2004.

Desai, Bharat, and Moumita Mandal. "Role of Climate Change in Exacerbating Sexual and Gender-Based Violence against Women: A New Challenge for International Law." *Environmental Policy and Law* 51 (2021): 137–57.

Desmond, Matthew. *Poverty by America*. New York: Penguin Random House, 2023.

Dillard, Annie. "Teaching a Stone to Talk." In *Teaching a Stone to Talk: Expeditions and Encounters*, 85–94. New York: Harper & Row, 1982.

Donner, Simon. "Domain of the Gods: An Editorial Essay." *Climate Change* 85 (2007): 231–33.

Donoghoe, Manann, and Andre M. Perry. "The Case for Climate Reparations in the United States." Brookings Institution, March 2023. https://www.brookings.edu/articles/the-case-for-climate-reparations-in-the-united-states/.

Donu, Viliame. "Profound Spirituality Illuminates Our Perception of the Natural World: Job 1:6-22." Sermon preached for the Nakorotubu Circuit, Fiji, February 2021.

Duba, Mary Emily Briehl. "God Is Here: A Theology of Place and Displacement." PhD diss., University of Chicago Divinity School, 2018.

DuBois, W. E. B. "The Upbuilding of Black Durham: The Success of the Negroes and Their Value to a Tolerant and Helpful Southern City." *The World's Work* 23 (January 1912): 334–38.

Duke Today staff. "What Durham's Tree Canopy Tells Us about Its Racial and Class Divides." *Duke Today*, July 2, 2021. https://today.duke.edu/2021/07/what-durhams-tree-canopy-tells-us-about-its-racial-and-class-divides.

Dunlap, Susan. *Shelter Theology: The Religious Lives of People without Homes*. Minneapolis: Fortress, 2021.

Durham CAN. "A Call for Awareness, Agitation, and Action: Durham CAN to Hold Silent Vigil Walk around Fayette Place Site." October 8, 2020. https://www.durhamcan.org/a_call_for_awareness_agitation_and_action_durham_can_to_hold_silent_vigil_walk_around_fayette_place_site.

Eastman, Susan. "Apocalypse and Incarnation: The Participatory Logic of Paul's Gospel." In *Oneself in Another: Participation and Personhood in Pauline Theology*, 3–20. Eugene, Ore.: Cascade, 2023.

Eberhart, Timothy. "Sustaining the Planetary Household: Methodist Contributions to an Ecological Economy." In *Methodist Revolutions: Evangelical Engagements of Church and World*, edited by Joerg Rieger and Upolu Lumā Vaai, 117–48. Nashville: Wesley's Foundery Books, 2021.

Egan, Andrea. "Heart of the Spirit: GEF-SGP UNDP 25th Anniversary Stories." *Uto ni Yalo* Trust. Accessed June 10, 2023. https://undp.shorthandstories.com/gef-sgp-heart-of-the-spirit/.

Eliasson, Olafur. "Ice Watch, 2014." Images from the Tate Museum Installation, London, 2018. https://olafureliasson.net/artwork/ice-watch-2014/.

Enns, Elaine, and Ched Meyers. *Healing Haunted Histories: A Settler Discipleship of Decolonization*. Eugene, Ore.: Cascade, 2021.

Ernst, Manfred. *Globalization and the Reshaping of Christianity in the Pacific Islands*. Suva: Pacific Theological College, 2006.

———. *Winds of Change: Rapidly Growing Religious Groups in the Pacific Islands*. Suva: Pacific Council of Churches, 1994.

Evans, John. *You Shall Know that I Am Yahweh: An Inner-Biblical Interpretation of Ezekiel's Recognition Formula*. University Park, Pa.: Eisenbrauns, 2019.

Fache, Elodie, and Hannah Fair. "Turning Away from Wicked Ways: Christian Climate Change Politics in the Pacific Island Region." *Anthropological Forum* 30, no. 3 (2020): 233–53.

Fair, Hannah. "Three Stories of Noah: Navigating Religious Climate Change Narratives in the Pacific Island Region." *Geo: Geography and Environment* 5, no. 2 (2018): 1–15.

Farin, Ingo. "Place and World: On Jeff Malpas's Second Edition of *Place and Experience*." *International Journal of Philosophical Studies* 28, no. 2 (2020): 254–66.

Feld, Steven, and Keith H. Basso, eds. *Senses of Place*. Santa Fe, N.Mex.: School of American Research Press, 1996.

Ferreira da Silva, Denise. *Toward a Global Idea of Race*. Minneapolis: University of Minnesota Press, 2007.

Fijian Ministry of Forestry. "Sandalwood: A Fragrant Tree with Great Potential." *Forestry Focus*, November 2021, 1–2. https://www.forestry.gov.fj/docs/newsl/ForestFocus_NovIssue_2021.pdf.

———. "Sorokoba and Nayarabale Youths Reforest 3 Hectares of Land, Minister Applauds Act." April 15, 2023. https://www.forestry.gov.fj/pressdetail.php?id=171.

Fohrer, G. *Ezechiel*. Tübingen: Mohr, 1955.

Franzen, Jonathan. "What If We Stopped Pretending?" *New Yorker*, September 8, 2019. https://www.newyorker.com/culture/cultural-comment/what-if-we-stopped-pretending.

Fretheim, Terence. *Exodus*. Louisville: Westminster John Knox, 1991.

———. *God and World in the Old Testament: A Relational Theology of Creation*. Nashville: Abingdon, 2005.

———. "The Plagues as Ecological Signs of Historical Disaster." *Journal of Biblical Literature* 110 (1991): 385–96.

Frost, Natasha. "In Fiji's Election, One Former Coup Leader Replaces Another." *New York Times*, December 20, 2022. https://www.nytimes.com/2022/12/20/world/asia/fiji-election.html.

Gadamer, Hans-Georg. *Truth and Method*. Edited by Garrett Barden and John Cumming. New York: Seabury Press, 1975.

Galambush, Julie. "God's Land and Mine: Creation as Property in the Book of Ezekiel." In *Ezekiel's Hierarchical World: Wrestling with a Tiered Reality*, edited by Stephen L. Cook and Corrine L. Patton, 91–108. Atlanta: SBL, 2004.

Ganzel, Tova. *Ezekiel's Visionary Temple in Babylonian Context*. Berlin: De Gruyter, 2021.

García-Johnson, Oscar. "Faith Seeking for Land: A Theology of the Landless." In *Theologies of Land: Contested Land, Spatial Justice, and Identity*, edited by K. K. Yeo and Gene L. Green, 38–68. Eugene, Ore.: Cascade, 2021.

Ghosh, Amitav. *The Great Derangement: Climate Change and the Unthinkable*. Chicago: University of Chicago Press, 2016.

———. *The Nutmeg's Curse: Parables for a Planet in Crisis*. Chicago: University of Chicago Press, 2021.

Giddens, Heather. "Neolithic Meshworks: A Multi-scalar Approach to Understanding Social Relations within the LBK." PhD diss., School of History, Archaeology and Religion, Cardiff University, 2015.

Gilbert, Kenyatta R. *Exodus Preaching: Crafting Sermons about Justice and Hope*. Nashville: Abingdon, 2018.

Godette, Gerald. Interviewed in "Faithful Resilience: Stories from North Carolina." Creation Justice Ministries, August 26, 2021. https://www.youtube.com/watch?v=7OKCe6lT9Yc.

Godette, Lynne. "Day 4: Disaster Relief Climate Action." Interview by Avery Davis Lamb, co-executive director of Creation Justice Ministries. EcoAmerica, April 8, 2022. https://www.youtube.com/watch?v=yU1f9tU2noA.

———. Presentation at the Pastoral Care for Climate: Weaving Science and Theology for Justice Conference. Duke Marine Lab, Beaufort, N.C., June 15, 2021.

Goeman, Mishuana. "Disrupting a Settler-Colonial Grammar of Place: The Visual Memoir of Hulleah Tsinhnahjinnie." In *Theorizing Native Studies*, edited by Audra Simpson and Andrea Smith, 235–65. Durham, N.C.: Duke University Press, 2014.

Gordon, Benjamin. *Land and Temple: Sacred Real Estate and the Second Temple Priesthood*. Munich: De Gruyter, 2020.

Goto, Courtney. *Taking on Practical Theology*. Leiden: Brill, 2018.

Graham, Elaine. "The State of the Art: Practical Theology Yesterday, Today, and Tomorrow." *Theology* 120, no. 3 (2017): 172–80.

Grantland, Elizabeth. "The Sunday After." Unpublished reflection as part of "Preaching and Protest: Sermons from Duke Chapel during the Civil Rights Era." Sponsored by the John Hope Franklin Humanities Institute at Duke University, 2018. https://www.youtube.com/watch?v=kQYA5Oxflus.

Green, Michael. "Contested Territory." *Nature: Climate Change* 6 (2016): 817–20.

Greenberg, Moshe. "The Design and Themes of Ezekiel's Program of Restoration." *Interpretation* 38, no. 2 (1984): 181–208.

——. *Ezekiel 1–20*. New York: Doubleday, 1983.

——. *Ezekiel 21–37*. New York: Doubleday, 1997.

Greene, Christina. *Our Separate Ways: Women and the Black Freedom Movement in Durham, NC*. Chapel Hill: University of North Carolina Press, 2004.

Greenspan, Lauren. "Hayti: Past and Present." October 26, 2017. https://www.youtube.com/watch?v=Dr_v5bPgIZ8.

Griffiths, Sean. "In the Name of Placemaking, Architects Are Often Complicit in Social Cleansing." *dezeen*, January 25, 2018. https://www.dezeen.com/2018/01/25/sean-griffiths-opinion-place-making-social-cleansing/.

Habel, Norman. "Guiding Ecojustice Principles." *Spiritan Horizons* 11, no. 11 (2016): 92–109.

——. *The Land Is Mine: Six Biblical Land Ideologies*. Minneapolis: Fortress, 1995.

——. "The Silence of the Lands: The Ecojustice Implications of Ezekiel's Judgement Oracles." In *Ezekiel's Hierarchical World: Wrestling with a Tiered Reality*, edited by Stephen L. Cook and Corrine L. Patton, 127–40. Atlanta: SBL, 2004.

Hadian, Amir Sasan. "Toward a Placemaking Approach to Building Architecture and Structures." *International Journal of Architectonic, Spatial, and Environmental Design* 13, no. 4 (2019): 41–56.

Halapua, Winston. *Waves of God's Embrace: Sacred Perspectives on the Ocean*. Norwich: Canterbury Press, 2008.

Halperin, David J. *Seeking Ezekiel: Text and Psychology*. University Park: Pennsylvania State University Press, 1993.

Harris, Melanie. *Ecowomanism: African American Women and Earth-Honoring Faiths*. Maryknoll, N.Y.: Orbis Books, 2017.

Harrison, Peter. "'Fill the Earth and Subdue It': Biblical Warrants for Colonization in Seventeenth Century England." *Journal of Religious History* 29, no. 1 (2005): 3–24.

Harsha, Dan. "Returning Land to Native Nations Is about Righting Historical Wrongs—and Also Averting Future Environmental Disaster." Harvard Kennedy School, Ash Center for Democratic Governance and Innovation, October 27, 2022. https://ash.harvard.edu/articles/returning-land-to-native-nations-is-about-righting-historical-wrongs-and-also-stemming-future-environmental-disaster/.

Hasel, Michael G. *Military Practice and Polemic: Israel's Laws of Warfare in Near Eastern Perspective*. Berrien Springs, Mich.: Andrews University Press, 2005.

Hauerwas, Stanley. *War and the American Difference: Theological Reflections on Violence and National Identity*. Grand Rapids: Baker, 2011.

Havea, Jione. "Drifting Homes." In *The Future of a Biblical Past: Envisioning Biblical Studies in a Global Key*, edited by Fernando Segovia and Roland Boer, 195–206. Atlanta: SBL, 2012.

———. "Holy Baptism, Troubled Waters." *Liturgy* 34, no. 2 (2019): 3–11.

———. "The Land Has Colors." In *People and the Land: Decolonizing Theologies*, edited by Jione Havea, 1–14. London: Lexington Books, 2020.

———. "The Politics of Climate Change: A *Talanoa* from Oceania." *International Journal of Public Theology* 4 (2010): 345–55.

Hawken, Paul. *Drawdown: The Most Comprehensive Plan Ever Composed to Reduce Global Warming*. New York: Penguin, 2017.

Hayhoe, Katharine. "The Most Important Thing You Can Do to Fight Climate Change: Talk about It." TEDWomen 2018. https://www.ted.com/talks/katharine_hayhoe_the_most_important_thing_you_can_do_to_fight_climate_change_talk_about_it?language=en.

———. *Saving Us: A Climate Scientist's Case for Hope and Healing in a Divided World*. New York: One Signal, 2021.

Heaney, Robert. "Prospects and Problems for Evangelical Postcolonialisms." In *Evangelical Postcolonial Conversations: Global Awakenings in Theology and Practice*, edited by Kay Smith, Jayachitra Lalitha, and L. Daniel Hawk, 29–44. Downers Grove, Ill.: IVP Academic, 2014.

Heaney, Seamus. "The Sense of Place." In *Preoccupations: Selected Prose, 1968–1978*. London: Faber & Faber, 1984.

Heidegger, Martin. "Letter on Humanism." In *Basic Writings*, edited by David Krell, 189–242. New York: Harper & Row, 1977.

Held, Shai. "Bere'shit #2: Created in God's Image." In *The Heart of Torah*, vol. 1, *Essays on the Weekly Torah Portion: Genesis and Exodus*, 7–11. Omaha: University of Nebraska Press, 2017.

Hereniko, Vilsoni. "The Deep, Beautiful but Dangerous Sea." Program for the film premiere of *Moana: The Rising of the Sea*, University of the South Pacific, Suva, 2014. https://catalog.library.vanderbilt.edu/discovery/fulldisplay/alma991043676880703276/01VAN_INST:vanui.

Hermann, Elfriede, and Wolfgang Kempf. "Introduction to Relations in Multicultural Fiji: The Dynamic of Articulations, Transformations, and Positionings." *Oceania* 75, no. 4 (2005): 319–20.

Hermes, Karin Louise. "Why I Quit Being a Climate Activist." *Vice*, February 6, 2020. https://www.vice.com/en_us/article/g5x5ny/why-i-quit-being-a-climate-activist.

Hill, Harriet, Margaret Hill, Dick Baggé, and Pat Miersma. *Na I Wali Ni Mavoa Se Tadoka Ni Yalo*. Translated by Apenisa Lewantoro. Suva: Magi Magi, 2013.

Hill, Wesley. "Apocalyptic Theology." In *The New Cambridge Companion to Christian Doctrine*, edited by Michael Allen, 315–29. Cambridge: Cambridge University Press, 2023.

Hīroa, Te Rangi. *The Coming of the Maori*. Wellington: Maori Purposes Fund Board, 1949.

Holton, M. Jan. "Imagining Hope and Redemption: A Salvation Narrative among the Displaced in Sudan." In *Interpreting Exile: Displacement and Deportation in Biblical and Modern Contexts*, edited by Brad E. Kelle, Frank Ames, and Jacob Wright, 217–34. Atlanta: SBL, 2011.

Howell, Junia, and James Elliot. "As Disaster Costs Rise, So Does Inequality." *Socius: Sociological Research for a Dynamic World* 4 (2018): 1–3.

Hulkenberg, Jara. "Fijian Kinship: Exchange and Migration." In *Living Kinship in the Pacific*, edited by Christina Toren and Simonne Pauwels, 60–85. New York: Berghahn Books, 2015.

Iati, Marisa. "The US Once Forced This Native American Tribe to Move. Now They Are Getting Their Land Back." *Washington Post*, September 20, 2019. https://www.washingtonpost.com/religion/2019/09/19/us-once-forced-this-native-american-tribe-move-now-theyre-getting-their-land-back/.

Idestrom, Rebecca. "Echoes of the Book of Exodus in Ezekiel." *Journal for the Study of the Old Testament* 33, no. 4 (2009): 489–510.

"Indigenous." In the *Concise Oxford Dictionary of Current English*, edited by R. E. Allen. London: Clarendon, 1990.

Inge, John. *A Christian Theology of Place*. New York: Routledge, 2003.

Ingold, Tim. *Being Alive: Essays on Movement, Knowledge, and Description*. New York: Routledge, 2011.

———. "Globes and Spheres: The Topology of Environmentalism." In *The Perception of the Environment: Essays on Livelihood, Dwelling, and Skill*. London: Routledge, 2000.

———. "On Human Correspondence." *Journal of the Royal Anthropological Institute* 23, no. 1 (2017): 9–27.

Ingold, Tim, Ana Letícia de Fiori, José Agnello Alves Dias de Andrade, Adriana Queiróz Testa, and Yuri Bassichetto Tambucci. "Wayfaring Thoughts: Life, Movement and Anthropology." *Ponto Urbe* 11 (2012): 1–15. https://doi.org/10.4000/pontourbe.341.

Intergovernmental Panel on Climate Change. "Climate Change, Widespread, Rapid, and Intensifying." August 9, 2021. https://www.ipcc.ch/2021/08/09/ar6-wg1-20210809-pr/.

iTaukei Land Trust Board. "Land Ownership in Fiji." Accessed July 30, 2022. https://www.tltb.com.fj/getattachment/Media/Brochures/Land-Ownership-in-Fiji-Booklet-(1).pdf.aspx?lang=en-US.

Iutzi, Fred. "We Were Born for This." The Land Institute. *Land Report* 125 (Fall 2019): 4–11.

Jackson, Michael. "A Walk on the Wild Side: The Idea of Nature Revisited." In *Diversity and Dominion: Dialogues in Ecology, Ethics, and Theology*, edited by Kyle Van Houtan and Michael Northcott, 46–63. Eugene, Ore.: Cascade, 2010.

Jackson, Wes. *Becoming Native to This Place*. New York: Counterpoint, 1996.

Jacobsen, Thorkild, and Robert M. Adams. "Salt and Silt in Ancient Mesopotamian Agriculture." *Science* 128, no. 3334 (1958): 1251–58.

Jenkins, Willis, Evan Berry, and Luke Beck Kreider. "Religion and Climate Change." *Annual Review of Environment and Resources* 43 (2018): 85–108.

Jennings, Willie James. *The Christian Imagination: Theology and the Origins of Race*. New Haven: Yale University Press, 2011.

Jobe, Sarah. "A Hermeneutics of Trauma: Trauma-Informed Care as Method for Biblical Interpretation." In *A Hermeneutics of Trauma*. Sheffield: Sheffield Phoenix Press, forthcoming 2025.

Joerstad, Mari. *The Hebrew Bible and Environmental Ethics: Humans, Nonhumans, and the Living Landscape*. Cambridge: Cambridge University Press, 2019.

Johnson, Elizabeth. *Creation and the Cross: The Mercy of God for a Planet in Peril*. Maryknoll, N.Y.: Orbis Books, 2018.

Johnson, Sheryl. "Truth, Reconciliation, and Climate Justice: Applications of Truth and Reconciliation Processes to Climate Debt and the Climate Crisis." In *Decolonizing Ecotheology: Indigenous and Subaltern Challenges*, edited by S. Lily Mendoza and George Zachariah, 131–43. Eugene, Ore.: Wipf and Stock, 2022.

Jonker, Gerdien. *The Topography of Remembrance*. Leiden: Brill, 1995.

Joyce, Paul. *Divine Initiative and Human Response in Ezekiel*. Sheffield: JSOT Press, 1989.

———. "Temple and Worship in Ezekiel 40–48." In *Temple and Worship in Biblical Israel*, edited by John Day, 145–63. New York: T&T Clark, 2005.

Kamikamica, Savenaca. "God Continues to Restore and Renew His Creations: Genesis 6; Mark 5:35–41." Sermon preached in Vitawa village, Fiji, March 3, 2021.

Kang, Namsoon. "Envisioning Postcolonial Theological Education: Dilemmas and Possibilities." In *Handbook of Theological Education in World Christianity: Theological Perspectives, Ecumenical Trends, and Regional Surveys*, edited by Dietrich Werner, David Esterline, Namsoon Kang, and Joshva Raja, 30–41. Eugene, Ore.: Wipf and Stock, 2010.

Kanyogo, Mumbi. "Native American Students." In *Activating History for Justice at Duke: A Report by Constructing Memory at Duke*, final report of the Constructing Memory at Duke Bass Connections Project, Durham, N.C., 2017. https://www.activatinghistoryatduke.com/uploads/8/9/2/3/89234082/print_activatinghistoryatduke_final_1up_highres.pdf.

Kaplan, Martha. "Promised Lands: From Colonial Lawgiving to Postcolonial Takeovers in Fiji." In *Law and Empire in the Pacific: Fiji and Hawai'i*, edited by Sally Engle Merry and Donald Brenneis, 153–86. Santa Fe, N.Mex.: School of American Research Press, 2003.

Katongole, Emmanuel. *Born from Lament: The Theology and Politics of Hope in Africa*. Grand Rapids: Eerdmans, 2017.

Katz, Eric. "Judaism and the Ecological Crisis." In *Worldviews and Ecology*, edited by Mary E. Tucker and John A. Grin, 55–70. Maryknoll, N.Y.: Orbis Books, 1994.

Kay, James F. "The Word of the Cross at the Turn of the Ages." *Interpretation* 53, no. 1 (1999): 44–56.

Keesmat, Sylvia, and Brian Walsh. *Romans Disarmed: Resisting Empire, Demanding Justice*. Grand Rapids: Brazos, 2019.

Kelle, Brad E. *Ancient Israel at War 853–586 BC*. Essential Histories 67. Oxford: Osprey, 2007.

———. "Dealing with the Trauma of Defeat: The Rhetoric of the Devastation and Rejuvenation of Nature in Ezekiel." *Journal of Biblical Literature* 128, no. 3 (2009): 469–90.

———. "An Interdisciplinary Approach to Exile." In *Interpreting Exile: Displacement and Deportation in Biblical and Modern Contexts*, edited by Brad E. Kelle, Frank Ames, and Jacob Wright, 5–40. Atlanta: SBL, 2011.

Kelly, John D. "Gordon Was No Amateur: Imperial Legal Strategies in the Colonization of Fiji." In *Law and Empire in the Pacific: Fiji and Hawai'i*, edited by Sally Engle Merry and Donald Brenneis, 61–100. Santa Fe: University of New Mexico Press, 2003.

Kelly, John D., and Martha Kaplan. *Represented Communities: Fiji and World Decolonization*. Chicago: University of Chicago Press, 2001.

Kempf, Wolfgang. "Climate Change, Christian Religion and Songs: Revisiting the Noah Story in the Central Pacific." In *Environmental Transformations and Cultural Responses: Ontologies, Discourses, and Practices in Oceania*, edited by Eveline Dürr and Arno Pascht, 19–48. New York: Palgrave Macmillan, 2017.

———. "Climate Change, Migration, and Christianity in Oceania." In *Climate Change and Human Mobility*, edited by Kirsten Hastrup and Karen Fog Olwig, 243–47. Cambridge: Cambridge University Press, 2012.

Kempf, Wolfgang, and Elfriede Hermann. "Reconfigurations of Place and Ethnicity: Positionings, Performances, and Politics of Relocated Banabans in Fiji." *Oceania* 75, no. 4 (2005): 368–86.

———. "Uncertain Futures of Belonging: Consequences of Climate Change and Sea-Level Rise in Oceania." In *Belonging in Oceania: Movement, Place-Making and Multiple Identifications*, edited by Elfriede Hermann, Wolfgang Kempf, and Toon van Meijl, 189–213. New York: Berghahn Books, 2014.

Kempf, Wolfgang, Toon van Meijl, and Elfriede Hermann. "Movement, Place-Making and Cultural Identification: Multiplicities of Belonging." In *Belonging in Oceania: Movement, Place-Making and Multiple Identifications*, edited by Elfriede Hermann, Wolfgang Kempf, and Toon van Meijl, 1–24. New York: Berghahn Books, 2014.

Kenan Institute for Ethics. "Facing the Anthropocene: A Conversation with Willie Jennings." March 11, 2021. https://www.youtube.com/watch?v=SGMXeE6p-tQ.

Kennedy, George. *On Rhetoric: A Theory of Civic Discourse*. Oxford: Oxford University Press, 1991.

Kete, Tevita. Personal online interview, January 27, 2022.

———. "Your *Yalava*: Keep It or You Will Destroy It: Genesis 1:26–28; Genesis 2:15–17." Sermon preached for the Korotubu Circuit, Fiji, June 7, 2015.

Key, Lindsey. "Medical Student-Led Initiative Plants Change for Environmental Justice in Durham." *Magnify*, Duke School of Medicine, January 12, 2022. https://medschool.duke.edu/stories/medical-student-led-initiative-plants-change-environmental-justice-durham.

Kim-Cragg, HyeRan. *Interdependence: A Postcolonial Feminist Practical Theology*. Eugene, Ore.: Pickwick, 2018.

———. *Postcolonial Preaching: Creating a Ripple Effect*. Lanham, Md.: Lexington Books, 2021.

Kimmerer, Robin Wall. *Braiding Sweetgrass: Indigenous Wisdom, Scientific Knowledge, and the Teachings of Plants*. Minneapolis: Milkweed Editions, 2015.

Kiran, Sashi. "For Peace Which Is True." *Fiji Times*, September 29, 2016. http://friendfiji.com/wp-content/uploads/2016/12/160929-For-peace-which-is-true-Fiji-Times-Online.pdf.

———. "Home, Where They Belong." *Fiji Times*, July 9, 2016. https://www.fijitimes.com.fj/home-where-they-belong/.

Kirsch, Gesa, and Jacqueline Jones Royster. *Feminist Rhetorical Practices: New Horizons for Rhetoric, Composition, and Literary Studies*. Carbondale: Southern Illinois University Press, 2012.

Kirsch, Stuart. "Changing Views of Place and Time along the Ok Tedi." In *Mining and Indigenous Lifeworlds in Australia and Papua New Guinea*, edited by Alan Rumsey and James Weiner, 182–207. Wantage, UK: Sean Kingston, 2004.

Kohn, Risa Levitt. *A New Heart and a New Soul: Ezekiel, Exile, and Torah*. Sheffield: Sheffield Academic Press, 2002.

Koya-Vaka'uta, Cresantia Frances. "The Line and the Cord: A Response to Blood in the Kava Bowl." In *Relational Hermeneutics: Decolonising the Mindset and the Pacific Itulagi*, edited by Upolu Lumā Vaai and Aisake Casimira, 115–17. Suva: University of the South Pacific Press, 2017.

———. "Rethinking Research as Relational Space in the Pacific." In *Relational Hermeneutics: Decolonising the Mindset and the Pacific Itulagi*, ed. Upolu Lumā Vaai and Aisake Casimira, 65–84. Suva: University of the South Pacific Press, 2017.

Krishant, Naveel. "Methodist Church of Fiji Apologises to the Descendants of the Girmityas for the Hurt of the Past." Fijivillage, May 14, 2023. https://www.fijivillage.com/news/Methodist-Church-of-Fiji-apologises-to-the-descendants-of-the-girmityas-for-the-hurt-of-the-past-5rf4x8/.

Kuruppu, Natasha, and Reenate Willie. "Barriers to Reducing Climate Enhanced Disaster Risks in Least Developed Country-Small Islands through Anticipatory Adaptation." *Weather and Climate Extremes* 7 (2015): 72–83.

Kühner, Angela. Kollektiv Traumata—Annahmen, Argumente, Konzepte: Eine Bestandsawfnahme nach dem 11. September. Berghof Report 9. Berlin: Berghof Forschungszentrum für Konstruktive Konflicktbearbeitung, 2002.

Lakhani, Nina. "'Shameful' UN Silencing Indigenous Voices, Say Banned COP27 Activists." *Guardian*, November 18, 2022. https://www.theguardian.com/environment/2022/nov/18/shameful-un-silencing-indigenous-voices-say-banned-cop27-activists.

Lalagavesi, Epeli. "COP26: Calling for Greater Ambition and Climate Justice." Peace Boat, November 16, 2021. https://peaceboat.org/english/news/cop26-greater-ambition-and-climate-justice.

Laney, Clarence. "Hayti History." Interview by Student U Durham, October 20, 2018. https://www.youtube.com/watch?v=xBFlHi_FcKU.

Lange, Raeburn. *Indigenous Ministers: Indigenous Leadership in Nineteenth Century Pacific Islands Christianity*. Christchurch: Pandanus Books, 2005.

Lapsley, Jacqueline. *Can These Bones Live? The Problem of the Moral Self in the Book of Ezekiel*. New York: De Gruyter, 2000.

Larsen, Soren C., and Jay T. Johnson. "The Agency of Place: Toward a More-Than-Human Geographical Self." *GeoHumanities* 2 (2016): 149–66.

LaRue, Cleophus. *I Believe I'll Testify*. Louisville: Westminster John Knox, 2011.

Lasawa Sedra, Jolame. Personal interview, Nausori, Fiji, January 5, 2023.

Lassiter, Luke Eric. *The Chicago Guide to Collaborative Ethnography*. Chicago: University of Chicago Press, 2005.

Latour, Bruno. *We Have Never Been Modern*. Boston: Harvard University Press, 1993.

Lawless, Elaine. "'I Was Afraid Someone Like You . . . an Outsider . . . Would Misunderstand': Negotiating Interpretive Differences between Ethnographers and Subjects." *Journal of American Folklore* 105, no. 417 (1992): 302–14.

Lee, Peace Pyunghwa, and David Stark. "The Towering Sermon: Duke Chapel as Monument to White Supremacy." In *Unmasking White Preaching: Racial Hegemony, Resistance, and Possibilities in Homiletics*, edited by Lis Valle-Ruiz and Andrew Wymer, 45–54. London: Lexington Books, 2022.

Lee, Peace Pyunghwa, and Miriam Cho, project managers. "When I Was a Stranger: Immigration, Preaching, and Religious Imagination (2020–2021)." Bass Connections program at Duke University, Summer 2020 through Spring 2021. https://bassconnections.duke.edu/project-teams/when-i-was-stranger-immigration-preaching-and-religious-imagination-2020-2021.

Leithart, Peter. *Revelation 1–11*. London: T&T Clark, 2018.

Lenski, Gerhard. *Power and Privilege: A Theory of Social Stratification*. New York: McGraw-Hill, 1966.

Leveen, Adriane. *Biblical Narratives of Israelites and Their Neighbors: Strangers at the Gate*. New York: Routledge, 2017.

———. "Returning the Body to Its Place: Ezekiel's Tour of the Temple." *Harvard Theological Review* 105, no. 4 (2012): 385–401.

Levenson, Jon Douglas. *Theology of the Program of Restoration of Ezekiel 40–48*. Missoula, Mont.: Scholars Press, 1976.

Lewis, C. S. *The Voyage of the Dawn Treader*. Book 5 of The Chronicles of Narnia. New York: Harper Trophy, 1994.

Lied, Liv. *The Other Lands of Israel: Imaginations of the Land in 2 Baruch*. Leiden: Brill, 2008.

Ligavatu, Eparama. "Keeping the Navolau Circuit *Drua* in Good Shape to Withstand Rough Weather: It Depends on Our Diligence in Keeping the Covenant: Genesis 6:5–18; 9:8–11." Sermon preached for the Navolau Circuit, Fiji, April 7, 2019.

———. Personal interview, January 20, 2022.

Lilburne, Geoffrey. *A Sense of Place: A Christian Theology of the Land*. Nashville: Abingdon, 1989.

Lipschits, Oded. *The Fall and Rise of Jerusalem: Judah under Babylonian Rule*. Winona Lake, Ind.: Eisenbrauns, 2005.

Lischer, Richard. *A Theology of Preaching: The Dynamics of the Gospel*. Nashville: Abingdon, 1981.

Lolohea, Petero. Personal interview, July 7, 2023.

———. "Respect and Care for Our Environment." Sermon given in Nasinu village, Ovalau, Fiji, June 4, 2000.

Long, Maebh. "Vanua in the Anthropocene: Relationality and Sea Level Rise in Fiji." *symploke* 26, nos. 1–2 (2018): 51–70.

Lose, David. "Whither Hence, New Homiletic?" Archived paper of the Academy of Homiletics' 35th Meeting, Perkins School of Theology, Southern Methodist University, Dallas, December 1, 2000.

Low, Setha M., and Denise Lawrence-Zuniga. "Locating Culture." In *The Anthropology of Space and Place: Locating Culture*, edited by Setha M. Low and Denise Lawrence-Zuniga, 1–47. Malden, Mass.: Blackwell, 2003.

Luger, Cannupa Hanska. "The Mirror Shield Project (2016)." Accessed October 15, 2023. https://www.cannupahanska.com/social-engagement/mirror-shield-project.

Luger, Cannupa Hanska, and Rory Erler Wakemup. "Water Serpent." Video edited by Sheila Regan, Oceti Sakowin Camp, December 21, 2016. https://www.youtube.com/watch?v=78epQ8GspaQ. Accessed October 15, 2023.

Lust, John. "Exodus 6,2–8 and Ezekiel." In *Studies in the Book of Exodus: Redaction, Reception, Interpretation*, edited by Marc Vervenne, 209–24. Leuven: Leuven University Press, 1996.

Lustgarten, Abrahm. "The Barbados Rebellion: An Island Nation's Fight for Climate Justice." *New York Times Magazine*, July 27, 2022. https://www.nytimes.com/interactive/2022/07/27/magazine/barbados-climate-debt-mia-mottley.html.

Lyons, Kate. "How to Move a Country: Fiji's Radical Plan to Escape Rising Sea Levels." *Guardian*, November 8, 2022. https://www.theguardian.com/environment/2022/nov/08/how-to-move-a-country-fiji-radical-plan-escape-rising-seas-climate-crisis.

Lyons, Michael. "Envisioning Restoration: Innovations in Ezekiel 40–48." In *"I Lifted My Eyes and Saw": Reading Dream and Vision Reports in the Hebrew Bible*, edited by Elizabeth Hayes and Lena-Sofia Tiemeyer, 71–83. New York: Bloomsbury T&T Clark, 2014.

McClure, John. *Ethical Approaches to Preaching: Choosing the Best Way to Preach about Difficult Issues*. Eugene, Ore.: Cascade, 2021.

———. *Otherwise Preaching: A Postmodern Ethic for Homiletics*. St. Louis: Chalice Press, 2001.

McCray, Donyelle. "Playing in Church: Insights from the Boundaries of the Sermon Genre." *Liturgy* 36, no. 2 (2021): 11–17.

McFague, Sally. *The Body of God: An Ecological Theology*. Minneapolis: Augsburg Fortress, 1993.

———. *Models of God: Theology for an Ecological, Nuclear Age*. Minneapolis: Fortress, 1987.

McKeating, H. "Ezekiel the 'Prophet Like Moses'?" *Journal for the Study of the Old Testament* 61 (1994): 97–109.

Maas, Steve. "Searching for the Origins of Redlining Black Neighborhoods." *National Bureau of Economic Research Digest* 2 (February 2021). https://www.nber.org/digest/202102/searching-origins-redlining-black-neighborhoods.

MacDonald, Nathan. *Priestly Rule: Polemic and Biblical Interpretation in Ezekiel 44*. Berlin: De Gruyter, 2015.

Machinist, Peter. "On Self-Consciousness in Mesopotamia." In *The Origins and Diversity of Axial Age Civilizations*, edited by Shmuel N. Eisenstadt, 183–202. Albany, N.Y.: SUNY Press, 1986.

Malpas, Jeff. *The Ethics of Place*. London: Routledge, forthcoming 2025.

———. "Is There an Ethics of Place?" *Localities* 2 (2012): 7–31.

———. *Place and Experience: A Philosophical Topography*. 2nd ed. New York: Routledge, 2018.

———. "Place and Human Being." *Environmental and Architectural Phenomenology Newsletter* 20, no. 3 (2009): 19–23.

———. "The Remembrance of Place." In *Exploring the Work of Edward S. Casey: Giving Voice to Place, Memory, and Imagination*, edited by Azucena Cruz-Pierre and Donald Landes, 63–72. New York: Bloomsbury Academic, 2013.

Mar, T. Banivanua. "Settler-Colonial Landscapes and Narratives of Possession." *Arena Journal* 37/38 (2012): 176–98.

Marlow, Hilary. "Creation, Humanity, and Hubris in the Hebrew Bible." *Practical Theology* 15, no. 5 (2022): 491–503.

Martin, Piérick C. M., Patrick D. Nunn, Javier Leon, and Neil Tindale. "Responding to Multiple Climate-Linked Stressors in a Remote Island Context: The Example of Yadua Island, Fiji." *Climate Risk Management* 21 (2018): 7–15.

Martyn, J. Louis. "Epistemology at the Turn of the Ages." In *Theological Issues in the Letters of Paul*, 89–110. Nashville: Abingdon, 1997.

———. *Galatians*. New York: Doubleday, 1997.

Matthews, Jeanette, and Matt Tomlinson. "Conversations between Theology, Anthropology, and History." Introduction to "Anthropology, Theology, and History in Conversation," special issue, *St. Mark's Review* 244 (June 2018): 1–8.

Mays, James Luther. *The Lord Reigns: A Theological Handbook to the Psalms*. Louisville: Westminster John Knox, 1994.

Mein, Andrew. *Ezekiel and the Ethics of Exile*. Oxford: Oxford University Press, 2001.

Mendoza, S. Lily. "Transdiasporic Indigeneity and Decolonizing Faith: Recovering Earth Spirituality in a Settler Colonial Context." In *Decolonizing Ecology: Indigenous and Subaltern Challenges*, edited by S. Lily Mendoza and George Zachariah, 259–80. Eugene, Ore.: Wipf and Stock, 2022.

Mendoza, S. Lily, and George Zachariah. Introduction to *Decolonizing Ecotheology: Indigenous and Subaltern Challenges*, edited by S. Lily Mendoza and George Zachariah, 1–16. Eugene, Ore.: Wipf and Stock, 2022.

Methodist Church in Fiji. "Loloma ni Jisu sa dau Voliti Au, #65." In *Ai vola ni sere ni Lotu Wesele e Viti*. Suva: Methodist Church in Fiji, 2003.

———. "Methodist Church in Fiji Connexional Plan." Accessed June 10, 2023. https://www.academia.edu/6620913/Methodist_Church_Connexional_Plan_First_Draft_.

———. "Methodist President Banivanua Sails on the *Uto ni Yalo*." *Fiji Sun*, September 13, 2017. https://fijisun.com.fj/2017/09/13/methodist-president-banivanua-sails-on-uto-ni-yalo/.

Methodist Church in Fiji and Rotuma. "Newly Inducted Talatala Qase of the Matuku Division in the Lau Group . . . Reflects on the Significance of His Induction." Interview with Talatala Qase Jope Navuki, divisional superintendent of Matuku. September 19, 2017. https://www.facebook.com/watch/?v=1533311446758788.

Middleton, J. Richard. "The Image of God in Ecological Perspective." In *The Oxford Handbook of the Bible and Ecology*, edited by Hilary Marlow and Mark Harris, 284–98. Oxford: Oxford University Press, 2022.

———. "Is Creation Theology Inherently Conservative? A Dialogue with Walter Brueggemann." *Harvard Theological Review* 87, no. 3 (1994): 257–77.

———. *The Liberating Image: The* Imago Dei *in Genesis 1*. Grand Rapids: Brazos, 2005.

Milbank, Alison, John Hughes, and Arabella Milbank, eds. *Preaching Radical and Orthodox*. London: SCM Press, 2017.

Miller, Terri. "LWF Council Meeting in Medan Indonesia Opens with Traditional Worship." Lutheran World Federation, June 12, 2014. https://lutheranworld.org/blog/lwf-council-meeting-medan-indonesia-opened-traditional-worship.

Mills, Mary. *Urban Imagination in Biblical Prophecy*. London: T&T Clark, 2012.

Moe-Lobeda, Cynthia D. "Decolonizing the Privileged: Resisting and Rebuilding the New Economy." In *Decolonizing Ecotheology: Indigenous and Subaltern Challenges*, edited by S. Lily Mendoza and George Zachariah, 174–98. Eugene, Ore.: Wipf and Stock, 2022.

Monture-Angus, Patricia. *Journeying Forward*. Halifax: Fernwood, 1999.

Moyo, Fulata Lusungu. "'Ukugqiba inkaba'—Burying the Umbilical Cord: An African Indigenous Ecofeminist Perspective on Incarnation." In *Planetary Solidarity: Global Women's Voices on Christian Doctrine and Climate Justice*, edited by Grace Ji-Sun Kim and Hilda P. Koster, 179–92. Minneapolis: Fortress, 2017.

Mylonas, Natalie. *Jerusalem as Contested Space in Ezekiel: Exilic Encounters with Emotion, Space, and Identity Politics*. New York: T&T Clark, 2023.

Nabobo-Baba, Unaisi. *Knowing and Learning: An Indigenous Fijian Approach*. Suva: University of the South Pacific Press, 2006.

———. "The Mutual Implication of Kinship and Chiefship in Fiji." In *Living Kinship in the Pacific*, edited by Christina Toren and Simonne Pauwels, 15–35. New York: Berghahn Books, 2015.

Nacuva, Esala. "The Holy Spirit Is Our Source of Strength." Sermon preached in Nadaravakawalu Circuit, Saumakia village, Fiji, June 2019.

———. Personal interview, Daku, Fiji, July 4, 2023.

———. Personal online interview, January 17, 2022.

Najman, Hindy. *Losing the Temple and Recovering the Future*. New York: Cambridge University Press, 2013.

Nakamura, Kate. "COP26: Five Indigenous Activists and Movements You Should Know About." Global Citizen, November 9, 2021. https://www.globalcitizen.org/en/content/cop26-indigenous-activists-organizations-movements/.

Narsey, Wadan. "Why the Fiji Regime Rejected the Draft Constitution." The Interpreter (blog). Lowy Institute for National Policy, Sydney, January 15, 2013. https://www.lowyinstitute.org/the-interpreter/why-fiji-regime-rejected-draft-constitution.

Nasedra, Iliesa. "Jesus Christ Is Our Hope: Mark 4:35–41." Sermon preached in Malake Methodist Church, October 2021.

———. Personal interview, January 18, 2022.

Nash, R. "Rounding Out the American Revolution: Ethical Extension and the New Environmentalism." In *Deep Ecology*, edited by Michael Tobias, 170–81. New York: Pfeiffer, 1985.

Navuki, Jope. "Facing Climate Change in Matuku." Methodist Church in Fiji Communications, September 25, 2017. https://www.facebook.com/watch/?v=1538505886239344.

Neal, Jerusha Matsen. "Decolonizing 'Dominion': Oceanic Methodist Women Read Genesis 1." In *Decoloniality and Methodism: A Commitment to Ecological Justice*, edited by Upolu Lumā Vaai. Oxford: Oxford Institute of Methodist Studies Publications, forthcoming 2024.

———. "The Edge of Water: Preaching Sovereignty in Rising Tides." *Interpretation: A Journal of Bible and Theology* 75, no. 2 (2021): 112–22.

———. "Exodus of Exile: Hermeneutic Shifts in a Shifting Fijian Methodist Church." *International Journal of Homiletics* 1, no. 2 (2019): 16–25.

———. "Oceanic (Fijian)." In *The Bible in Folklore Worldwide*, vol. 2, *A Handbook of Biblical Reception in Folklores of Africa, Asia, Oceania, and the Americas*, edited by Eric Ziolkowski, 183–97. Berlin: De Gruyter, 2023.

———. *The Overshadowed Preacher: Mary, the Spirit, and the Labor of Proclamation*. Grand Rapids: Eerdmans, 2020.

Nelson, R. David. "The Word of the Cross and Christian Theology: Paul's Theological Temperament for Today." *Theology Today* 75, no. 1 (2018): 64–76.

Newland, Lynda. "From the Land to the Sea: Christianity, Community and State in Fiji—and the 2014 Elections." In *The People Have Spoken: The 2014 Elections in Fiji*, edited by Steven Ratuva and Stephanie Lawson, 109–34. Canberra: ANU Press, 2016.

———. "Imagining Nationhood: Narratives of Belonging and the Question of a Christian State in Fiji." *Global Change, Peace & Security* 25 (2013): 227–42.

———. "The Lost Tribes of Israel—and the Genesis of Christianity in Fiji: Missionary Notions of Fijian Origin from 1835 to Cession and Beyond." *Oceania* 85, no. 3 (2015): 256–70.

———. "Religion and Politics: The Christian Churches and the 2006 Coup in Fiji." In *The 2006 Military Takeover in Fiji: A Coup to End All Coups?* edited by Jon Fraenkel, Stewart Firth, and Brij V. Lal, 187–207. Canberra: ANU Publications, 2007.

Niditch, Susan. "Ezekiel 40–48 in a Visionary Context." *Catholic Biblical Quarterly* 48 (1986): 208–24.

Nixon, Rob. *Slow Violence and the Environmentalism of the Poor*. Cambridge, Mass.: Harvard University Press, 2011.

North Carolina Central University. "The Hayti Restorative Justice Report, Spring 2023." Social Justice and Racial Equity Insitute, April 15, 2023.

Northcott, Michael. *A Moral Climate: The Ethics of Global Warming*. London: Christian Aid Publishing, 2007.

Nunn, Patrick D., William Aalbersberg, Shalini Lata, and Marion Gwilliam. "Beyond the Core: Community Governance for Climate-Change Adaptation in Peripheral Parts of Pacific Island Countries." *Regional Environmental Change* 14, no. 1 (2013): 221–35.

Odell, Margaret. *Ezekiel*. Macon, Ga.: Smyth and Helwys, 2005.

O'Donovan, Oliver, and Joan Lockwood O'Donovan. *Bonds of Imperfection: Christian Politics, Past and Present*. Grand Rapids: Eerdmans, 2004.

Office of International Religious Freedom. "2020 Report on International Religious Freedom: Fiji." U.S. State Department, May 12, 2021. https://www.state.gov/reports/2020-report-on-international-religious-freedom/fiji/.

Ottoni-Wilhelm, Dawn. "New Hermeneutic, New Homiletic and New Directions: A U.S.-North American Perspective." *Homiletic* 35, no. 1 (2010): 17–31.

Park, Aubrey. *Degei's Descendants: Spirits, Place and People in Pre-Cessation Fiji*. Edited by Matthew Spriggs and Deryck Scarr. Canberra: ANU Press, 2014.

Patton, Corrine L. "Priest, Prophet, and Exile: Ezekiel as a Literary Construct." In *Ezekiel's Hierarchical World: Wrestling with a Tiered Reality*, edited by Stephen L. Cook and Corrine L. Patton, 73–90. Atlanta: SBL, 2004.

Perdue, Leo, and Warren Carter. *Israel and Empire: A Postcolonial History of Israel and Early Judaism*. London: Bloomsbury, 2015.

Perkins, Anna Kasafi. "Resisting Definitive Interpretation: Seeing the Story of the Exodus through Caribbean(ite) Eyes." *Caribbean Quarterly* 51, no. 2 (2005): 53–66.

Petersen, David, L. "Creation and Hierarchy in Ezekiel: Methodological Perspectives and Theological Prospects." In *Ezekiel's Hierarchical World: Wrestling with a Tiered Reality*, edited by Stephen L. Cook and Corrine L. Patton, 169–78. Atlanta: SBL, 2004.

Phan, Peter. "God, the Beginning and End of Migration: A Theology of God from the Perspective of Migrants." In *Christian Theology in an Age of Migration: Implications for World Christianity*, edited by Peter Phan, 101–28. Lanham, Md.: Lexington Books, 2020.

Phillips, Elizabeth. "Narrating Catastrophe, Cultivating Hope: Apocalyptic Practices and Theological Virtue." *Studies in Christian Ethics* 31, no. 1 (2018): 17–33.

———. *Political Theology: A Guide for the Perplexed*. London: T&T Clark, 2012.

Pikor, Wojciech. *The Land of Israel in the Book of Ezekiel*. New York: T&T Clark, 2018.

———. "The Subjecthood of the Land of Israel in the Book of Ezekiel." *Biblical Annals* 3, no. 1 (2013): 29–46.

Portier-Young, Anathea. *Apocalypse against Empire: Theologies of Resistance in Early Judaism*. Grand Rapids: Eerdmans, 2011.

———. "Commentary on Ezekiel 18:1–4, 25–32." Working Preacher, September 28, 2008. https://www.workingpreacher.org/commentaries/revised-common-lectionary/ordinary-26/commentary-on-ezekiel-181-4-25-32-3.

———. "Drinking the Cup of Horror and Gnawing on Its Shards: Biblical Theology through Biblical Violence, Not around It." In *Beyond Biblical Theologies*, edited by Heinrich Assel, Stefan Beyerle, and Christfried Böttrich, 387–408. Tübingen: Mohr Siebeck, 2012.

Poser, Ruth. "No Words: The Book of Ezekiel as Trauma Literature and a Response to Exile." In *Bible through the Lens of Trauma*, edited by Elizabeth Boase and Christopher Frechette, 27–48. Atlanta: SBL, 2016.

Powery, Luke. *Dem Dry Bones: Preaching, Death, and Hope*. Minneapolis: Fortress, 2012.

Procopio, Stephen. "Ascension." In *John: Come See a Man*. Richmond, Va.: Fish Coin Press, 2023.

———. "Ascension." Procopio Draws. Accessed December 27, 2023. https://procopiodraws.com/d7hh7jj64p6os2l3v459ibq6zjnmp9.

Proust, Katrina. "Salinity in Colonial Irrigation British India and South-Eastern Australia." *Australian Geographer* 39, no. 2 (2008): 131–47.

Purifoy, Christie. *Placemaker: Cultivating Places of Comfort, Beauty, and Peace*. Grand Rapids: Zondervan, 2019.

Purpura, Lia. *All the Fierce Tethers*. Louisville: Sarabande Books, 2019.

Qarase, Laisenia. *Prisoner 302*. Self-published, Amazon Digital Services, 2022. Kindle.

Raboteau, Albert J. "African Americans, Exodus, and the American Israel." In *Religion and American Culture: A Reader*, edited by David Hackett, 73–88. 2nd ed. New York: Routledge, 2003.

Ramantswama, Hulisani. "Negotiation and/or Conquest of the Land: Reading the Land of Promise Motif in the Hexateuch through Decolonial Lenses." In *Theologies of Land: Contested Land, Spatial Justice, and Identity*, edited by K. K. Yeo and Gene L. Green, 98–131. Eugene, Ore.: Cascade, 2021.

Rambo, Shelly. "Exiling in America: The American Myth and the Spectral Christ." In *Interpreting Exile: Displacement and Deportation in Biblical and Modern Contexts*, edited by Brad E. Kelle, Frank Ames, and Jacob Wright, 409–22. Atlanta: SBL, 2011.

Ramesh, Sanjay. "State Hegemony and Ethnicity: Fiji's Problematic Colonial Past." In *The Palgrave Handbook of Ethnicity*, edited by Steven Ratuva, 247–63. Singapore: Palgrave Macmillan, 2019.

Randall, Rebecca. "Spirituality, Global Warming, and Grief: How Clergy Can Tackle Climate Anxiety." Grist, June 9, 2023. https://grist.org/culture/spirituality-global-warming-and-grief-how-clergy-can-help-tackle-climate-anxiety/.

Rasoqosogo, Losalini. "Renewed Symbols for Church." *Fiji Sun*, August 22, 2013. https://fijisun.com.fj/2013/08/22/renewed-symbols-for-church/.

Rawalai, Luke. "TC Yasa: Father's Will to Survive." *Fiji Times*, December 31, 2020. https://www.fijitimes.com/tc-yasa-fathers-will-to-survive/.

Rhyder, Julia. "Gates and Entrances in Ezekiel 40–48: The Social Utopia and the Temple Vision." *Vetus Testamentum* 73 (2023): 752–65.

Ricoeur, Paul. *The Conflict of Interpretation*. Evanston, Ill.: Northwestern University Press, 1974.

———. *Interpretation Theory: Discourse and the Surplus of Meaning*. Fort Worth: Texas Christian University Press, 1976.

Rienstra, Debra. *Refugia Faith: Seeking Hidden Shelters, Ordinary Wonders, and the Healing of the Earth*. Minneapolis: Fortress, 2022.

Ritchie, Hannah, Max Roser, and Pablo Rosado. "CO_2 and Greenhouse Gas Emissions." Accessed January 19, 2023. Our World in Data. https://ourworldindata.org/co2-and-other-greenhouse-gas-emissions.

Robertson, Dylan C. "Kiev Protestors Hold Mirrors to Police—Literally—as New Confrontation Looms." *Christian Science Monitor*, January 17, 2014. https://www.csmonitor.com/World/Europe/2014/0117/Kiev-protesters-hold-mirrors-to-police-literally-as-new-confrontation-looms.

Rossing, Barbara R. "Lost Land: Visualizing Deforestation and Eschatology in the Apocalypse of John and the Column of Trajan in Rome." In *People and Land: Decolonizing Theologies*, edited by Jione Havea, 159–74. Minneapolis: Fortress Academic, 2019.

———. "Reimagining Eschatology: Toward Healing and Hope for a World at the Eschatos." In *Planetary Solidarity: Global Women's Voices on Christian Doctrine and Climate Justice*, edited by Grace Ji-Sun Kim and Hilda Koster, 325–47. Minneapolis: Fortress, 2017.

———. "Waters Cry Out: Water Protectors, Watershed Justice, and the Voice of the Waters." In *Decolonizing Ecotheology: Indigenous and Subaltern Challenges*, edited by S. Lily Mendoza and George Zachariah, 39–57. Eugene, Ore.: Wipf and Stock, 2022.

Roundtable on Sustainable Palm Oil. "Driving Change with Sustainable Palm Oil." Accessed November 17, 2023. https://rspo.org/who-we-are/.

Rovoi, Christine. "Fiji Should Be Ashamed, Says Ali." RNZ News, December 30, 2021. https://www.rnz.co.nz/international/pacific-news/458794/fiji-should-be-ashamed-says-ali.

Royster, Jacqueline Jones. "When the First Voice You Hear Is Not Your Own (1996)." In *Landmark Essays on Rhetoric and Feminism, 1973–2000*, edited by Cheryl Glenn and Andrea Lunsford, 122–31. New York: Routledge, 2015.

Rubow, Cecilie, and Cliff Bird. "Eco-theological Responses to Climate Change in Oceania." *Worldviews: Global Religions, Culture, and Ecology* 20 (2016): 150–68.

Rudiak-Gould, Peter. "Promiscuous Corroboration and Climate Change Translation: A Case Study from the Marshall Islands." *Global Environmental Change* 22 (2012): 46–54.

Rush, Elizabeth. *Rising: Dispatches from the New American Shore*. Minneapolis: Milkweed Editions, 2018.

Ryle, Jacqueline. *My God, My Land: Interwoven Paths of Christianity and Tradition in Fiji*. Surrey: Ashgate, 2010.

———. "Religious Freedom." In *Christianity in Oceania*, edited by Kenneth Ross, Katalina Tahaafe-Williams, and Todd Johnson, 309–20. Edinburgh: Edinburgh University Press, 2021.

Said, Edward. "The Mind of Winter: Reflections on Life in Exile." *Harpers Magazine*, September 1984, 49–55.

Salazar, Noel B. *Envisioning Eden: Mobilizing Imaginaries in Tourism and Beyond*. New York: Berghahn Books, 2010.

Salmond, Amiria. "Re-building Ships at Sea: Ontological Innovation in Action." In *Environmental Transformations and Cultural Responses: Ontologies, Discourses, and Practices in Oceania*, edited by Eveline Dürr and Arno Pascht, 215–26. New York: Palgrave Macmillan, 2017.

———. "Transforming Translations (Part 2): Addressing Ontological Alterity." *HAU: Journal of Ethnographic Theory* 4, no. 1 (2014): 155–87.

Salmond, Dame Anne. "Think Like a Fish: Pacific Philosophies and Climate Change." In *Pacific Climate Cultures: Living Climate Change in Oceania*, edited by Peter Rudiak-Gould and Tony Crook, 155–59. Warsaw: De Gruyter, 2018.

Salusalu, Jone. "Kia Island Rewrites History 214 Years On." *Fiji Sun*, November 30, 2023. https://www.pressreader.com/fiji/fiji-sun/20231130/281616720126556.

Samson, Jane. "Fijian and Tongan Methodism." In *The Oxford History of Protestant Dissenting Traditions*, vol. 4, *The Twentieth Century Traditions in a Global Context*, edited by Jehu Hanciles, 409–32. Oxford: Oxford University Press, 2019.

Samuwai, Jale. Personal interview, Cogea, Fiji, July 5, 2023.

Sawyer, John F. A. "Ezekiel in the History of Christianity." In *After Ezekiel: Essays on the Reception of a Difficult Prophet*, edited by Paul Joyce and Andrew Mein, 1–10. New York: T&T Clark, 2011.

Schaafsma, Katrina. "Follow the Body: How Play Therapy Makes Sense of Ezekiel's River (47:1–7)." Unpublished paper presented to the Theological Perspectives on the Book of Ezekiel section, Society of Biblical Literature Annual Meeting, Boston, 2017.

Schade, Leah. "Building the Eco-ethical Ark in the Age of Climate Disruption." In *Rooted and Rising: Voices of Courage in a Time of Climate Crisis*, edited by Leah Schade and Margaret Bullitt-Jonas, 17–22. New York: Rowman and Littlefield, 2019.

———. *Creation-Crisis Preaching: Ecology, Theology, and the Pulpit*. St. Louis: Chalice Press, 2015.

Schade, Leah, and Margaret Bullit-Jonas, eds. *Rooted and Rising: Voices of Courage in a Time of Climate Crisis*. New York: Rowman and Littlefield, 2019.

Scharen, Christian. "Theological Ethnography and World Christianity: A Response." *Journal of World Christianity* 10, no. 1 (2020): 109–19.

Schwartz, Baruch. "A Priest Out of Place: Reconsidering Ezekiel's Role in the History of the Israelite Priesthood." In *Ezekiel's Hierarchical World: Wrestling with a Tiered Reality*, edited by Stephen L. Cook and Corrine L. Patton, 61–72. Atlanta: SBL, 2004.

Scott, James. *Domination and the Arts of Resistance: Hidden Transcripts*. New Haven: Yale University Press, 1990.

———. *Weapons of the Weak: Everyday Forms of Peasant Resistance*. New Haven: Yale University Press, 1985.

Scranton, Roy. *Learning to Die in the Anthropocene: Reflections on the End of a Civilization*. San Francisco: City Lights Books, 2015.

Sekek, Sam. "Police Again Detain Fiji Church Leaders." ABC News, July 22, 2009. http://www.abc.net.au/news/2009-07-22/police-again-detain-fiji-church-leaders/1363198.

Sen, Amrita, and Harini Nagendra. "The Role of Environmental Placemaking in Shaping Contemporary Environmentalism and Understanding Land Change." *Journal of Land Use Science* 14, nos. 4–6 (2019): 410–24.

Sharp, Carolyn J. "'Be Strong and Resolute!' Reading Joshua in the Contemporary Church." *Anglican Theological Review* 97, no. 1 (2015): 19–32.

———. "Sites of Conflict: Textual Engagements of Dislocation and Diaspora in the Hebrew Bible." In *Interpreting Exile: Displacement and Deportation in Biblical and Modern Contexts*, edited by Brad E. Kelle, Frank Ames, and Jacob Wright, 365–76. Atlanta: SBL, 2011.

Shelton-Ormond, Charlie, and Frank Stasio. "Planting Inequality: How Durham's Trees Show a History of Marginalization." WUNC, July 4, 2017. https://www.wunc.org/environment/2017-07-04/planting-inequality-how-durhams-trees-show-a-history-of-marginalization.

Singh, J. *Unthinking Mastery: Dehumanism and Decolonial Entanglements*. Durham, N.C.: Duke University Press, 2018.

Singh, Priyatma, Tammy Tabe, and Tess Martin. "The Role of Women in Community Resilience to Climate Change: A Case Study of an Indigenous Fijian Community." *Women's Studies International Forum* 90 (2022): 1–13.

Siwatibau, Suliana, and David Williams, *A Call to a New Exodus: An Anti-Nuclear Primer for Pacific People*. Suva, Fiji: Pacific Council of Churches, 1982.

Slabodsky, Santiago. "Empty Land: Righteous Theology, Sneaky Coloniality." In *People and Land: Decolonizing Theologies*, edited by Jione Havea, 51–64. Minneapolis: Fortress Academic, 2019.

Smith, Jonathan Z. *To Take Place: Toward Theory in Ritual*. Chicago: University of Chicago Press, 1987.

Smith, Tracy K. "Wade in the Water." In *Wade in the Water*, 15–16. Minneapolis: Graywolf Press, 2018.

Smith, Tui Nuku. "Reading Jeremiah 29:1–23 in Light of the *Daunivucu ni Vanua* in the Fijian Methodist Context." Bachelor of theology thesis, Davuilevu Theological College, Fiji, 2017.

Smith-Christopher, Daniel. *A Biblical Theology of Exile*. Minneapolis: Augsburg Fortress, 2002.

———. "Ezekiel in Abu Ghraib: Rereading Ezekiel 16:37–39 in the Context of Imperial Conquest." In *Ezekiel's Hierarchical World: Wrestling with a Tiered Reality*, edited by Stephen L. Cook and Corrine L. Patton, 141–58. Atlanta: SBL, 2004.

———. "Reading Exile Then: Reconsidering the Methodological Debates for Biblical Analysis in Dialogue with Sociological and Literary Analysis." In *By the Irrigation Canals of Babylon: Approaches to the Study of the Exile*, edited by John Ahn and Jill Middlemas, 139–57. London: T&T Clark, 2012.

———. "Reading War and Trauma: Suggestions Toward a Social-Psychological Exegesis of Exile and War in Biblical Texts." In *Interpreting Exile: Displacement and Deportation in Biblical and Modern Contexts*, edited by Brad E. Kelle, Frank Ames, and Jacob Wright, 253–74. Atlanta: SBL, 2011.

———. *The Religion of the Landless: The Social Context of the Babylonian Exile*. Eugene, Ore.: Wipf and Stock, 2015.

Sovaraki, Ana. "Act of God." *Fiji Sun*, March 11, 2015. https://fijisun.com.fj/2015/03/11/act-of-god/.

Spafford, Horatio. "It Is Well with My Soul, #377." In *The United Methodist Hymnal*. Nashville: United Methodist Publishing House, 1989.

Stalker, D. M. *Ezekiel*. London: SCM Press, 1968.

Stanfill, Jeff. *Disastrous Preaching: Preaching in the Aftermath of a Natural Environmental Disaster*. Eugene, Ore.: Wipf and Stock, 2022.

Stark, David. "Whiteness and the So-Called Problem of Preaching the Old Testament." *Theology Today* 79, no. 2 (2022): 115–26.

Stevenson, Kalinda Rose. *The Vision of Transformation: The Territorial Rhetoric of Ezekiel 40–48*. Atlanta: SBL, 1996.

Stewart, Anthony. "Tuvalu PM Slams Kevin Rudd's Proposal to Offer Australian Citizenship for Pacific Resources as Neo-colonialism." ABC News, February 17, 2019. https://www.abc.net.au/news/2019-02-18/tuvalu-pm-slams-kevin-rudd-suggestion-as-neo-colonialism/10820176.

Stockwell, Mary. *The Other Trail of Tears: The Removal of the Ohio Indians*. Yardley, Pa.: Westholme, 2016.

Strine, C. A. "Chaoskampf against Empire: YHWH's Battle against Gog (Ezekiel 38–39) as Resistance Literature." In *Divination, Politics, and Ancient Near East Empires*, edited by Alan Lenzi and Jonathan Stökl, 87–108. Atlanta: SBL, 2014.

Strong, John T. “Cosmic Re-creation and Ezekiel’s Vocabulary.” In *Ezekiel: Current Debates and Future Directions*, edited by William A. Tooman and Penelope Barter, 245–84. Tübingen: Mohr Siebeck, 2017.

Sweeney, Marvin A. “The Ezekiel That G-d Creates.” In *The God Ezekiel Creates*, edited by Paul Joyce and Dalit Rom-Shiloni, 150–61. London: T&T Clark, 2015.

Tarabe, Akanisi Sobusobu. Personal interview, Suva, December 29, 2022.

———. “Re-Assessing Gender Roles: A Study of Indigenous Fijian Women’s Role in Funerals.” Master’s thesis, School of Social Science, University of the South Pacific, Suva, September 2015.

Teikabua, Terikata. “The True and Righteous Faith: Unwrapping the Oceanic Word of ‘Tabu/Tapu,’ Romans 4:13–25.” Sermon preached at Davuilevu Theological College, Nausori, Fiji, March 2015.

Teves, Stephanie Nohelani, Andrea Smith, and Michelle H. Raheja. *Native Studies Keywords*. Tucson: University of Arizona Press, 2015.

Thompson, Lisa. “A Different Renewal of Sorts: Black Women’s Lives, Praxis, and Disciplinary Futures.” Keynote address at the Academy of Homiletics Virtual Meeting, December 2021.

———. *Ingenuity: Preaching as an Outsider*. Nashville: Abingdon, 2018.

———. “Struggles for Belonging: Environment, Land, and Borders.” In *Preaching the Headlines: Possibilities and Pitfalls*, 65–84. Minneapolis: Fortress, 2021.

Thornley, Andrew. *The Inheritance of Hope: John Hunt, Apostle to Fiji*. Suva: Institute of Pacific Studies, 2000.

———. “‘Through a Glass Darkly’: Ownership of Fijian Methodism, 1850–80.” In *Vision and Reality in Pacific Religion*, edited by Phyllis Herda, Michael Reilly, and David Hilliard, 132–50. Christchurch: Pandanus Books, 2005.

Thrupp, Sylvia L. “Millennial Dreams in Action: A Report on the Conference Discussion.” In vol. 2 of *Millennial Dreams in Action*, edited by Sylvia L. Thrupp, 11–27. The Hague: Moutnon, 1962.

Thunberg, Greta. “‘Our House Is on Fire’: Greta Thunberg, 16, Urges Leaders to Act on Climate.” *Guardian*, January 25, 2019. https://www.theguardian.com/environment/2019/jan/25/our-house-is-on-fire-greta-thunberg16-urges-leaders-to-act-on-climate.

Tinaimana, Peniana. Personal interview, Cogea, Fiji, July 5, 2023.

Tinker, George “Tink.” “The Full Circle of Liberation: An American Indian Theology of Place.” *Sojourners* 21, no. 8 (1992): 12–17.

Tomlinson, Matt. “Compelling Replication: Genesis 1:26, John 3:16 and Biblical Politics in Fiji.” *Journal of the Royal Anthropological Institute* 16 (2010): 743–60.

———. “A Consuming Tradition: Kava Drinking in Fiji.” *Expedition* 48, no. 3 (2006): 8–17.

———. “Gender in a Land-Based Theology.” *Oceania* 85, no. 1 (2015): 79–91.

———. "The Generation of the Now: Denominational Politics in Fijian Christianity." In *Christian Politics in Oceania*, edited by Matt Tomlinson and Debra McDougall, 78–102. New York: Berghahn Books, 2013.

———. *God Is Samoan: Dialogues between Culture and Theology in the Pacific*. Honolulu: University of Hawaii Press, 2020.

———. *In God's Image: The Metaculture of Fijian Christianity*. Berkeley: University of California Press, 2009.

———. "Ritual Risk and Danger: Chain Prayers in Fiji." *American Anthropologist* 106, no. 1 (2004): 6–16.

———. "Sacred Soil in Kadavu, Fiji." *Oceania* 72, no. 4 (2002): 237–57.

Tomlinson, Matt, and Sekove Bigitibau. "Theologies of *Mana* and *Sau* in Fiji." In *New Mana: Transformations of a Classic Concept in Pacific Languages and Cultures*, edited by Matt Tomlinson and T. P. K. Tengan, 239–57. Canberra: ANU Press, 2016.

Toney, Heather McTeer. *Before the Streetlights Come On: Black America's Urgent Call for Climate Solutions*. Minneapolis: Broadleaf Books, 2023.

———. "Collards Are Just as Good as Kale." In *All We Can Save: Truth, Courage, and Solutions for the Climate Crisis*, edited by Ayana E. Johnson and Katharine Wilkinson, 75–83. New York: Random House, 2020.

Trnka, Susanna. *State of Suffering: Political Violence and Community Survival in Fiji*. Ithaca, N.Y.: Cornell University Press, 2008.

Tuan, Yi-Fu. *Space and Place: The Perspective of Experience*. Minneapolis: University of Minnesota Press, 2001.

———. *Topophilia: A Study of Environmental Perceptions, Attitudes, and Values*. New York: Columbia University Press, 1974.

Tuck, Eve, and C. Ree. "A Glossary of Haunting." In *Handbook of Autoethnography*, edited by Stacy Holman Jones, Tony Adams, and Carolyn Ellis, 639–58. Philadelphia: Routledge, 2016.

Tuck, Eve, and K. Wayne Yang. "Decolonization Is Not a Metaphor." *Decolonization: Indigeneity, Education and Society* 1, no. 1 (2012): 1–40.

Tuck, Eve, Marcia McKenzie, and Kate McCoy. "Land Education: Indigenous, Post-colonial, and Decolonizing Perspectives on Place and Environmental Education Research." *Environmental Education Research* 20, no. 1 (2014): 1–23.

Tuell, Steven S. "Divine Presence and Absence in Ezekiel's Prophecy." In *The Book of Ezekiel: Theological and Anthropological Perspectives*, edited by Margaret Odell and John Strong, 97–116. Atlanta: SBL, 2000.

———. "The Temple Vision of Ezekiel 40–48: A Program of Restoration?" *Proceedings of the Eastern Great Lakes Biblical Society* 2 (1982): 96–103.

Tuhiwai Smith, Linda. *Decolonizing Methodologies: Research and Indigenous Peoples*. 2nd ed. London: Zed Books, 2012.

Tuidrakulu, Rusiate. "Fiji." In *Christianity in Oceania*, edited by Kenneth Ross, Katalina Tahaafe-Williams, and Todd Johnson, 91–99. Edinburgh: Edinburgh University Press, 2021.

Tuiuruone, Inoke. "Weakening Relationship Is One of the Major Causes of Climate Change: John 15:1–8." Sermon given for the Oneata Circuit, Fiji, September 2021.

Tuwere, Ilaitia S. "At Home: A *Were-Kalou* Response to Epistemology." Paper presented at Navigating Knowledge: Pacific Epistemologies Conference, University of the South Pacific, Suva, July 6, 2006.

———. "The Church-State Relation in Fiji—a Reflection." Paper presented at the 1997 MCIF Annual Conference, Centenary Church, Suva.

———. *Vanua: A Fijian Theology of Place*. Suva: Institute of Pacific Studies Press, 2001.

Uehlinger, Christoph. "The Cry of the Earth? Biblical Perspectives on Ecology and Violence." In *Ecology and Poverty: Cry of the Earth, Cry of the Poor*, edited by Virgil Elizondo and Leonardo Boff, 41–54. London: SCM Press, 1995.

Uiti, Jacob. "Behind the Meaning of the Classic Gospel Song, 'Wade in the Water.'" *American Songwriter: The Craft of Music*, February 23, 2023. https://americansongwriter.com/behind-the-meaning-of-the-classic-gospel-song-wade-in-the-water/.

United Nations' Secretariat of the Permanent Forum on Indigenous Issues. "The Concept of Indigenous Peoples." Workshop on Data Collection and Disaggregation for Indigenous Peoples. New York, January 19–21, 2004.

University of Delaware. "Placemaking." Delaware Complete Communities Planning Toolbox. Accessed October 24, 2021. https://www.completecommunitiesde.org/getting-started/.

Vaai, Upolu Lumā. "A Dirtified God: A Dirt Theology from the Pacific Dirt Communities." In *Theologies from the Pacific*, edited by Jione Havea, 15–30. Cham, Switzerland: Palgrave Macmillan, 2021.

———. "*Motu ma le Taula*: Towards an Island 'Let Be' Hermeneutics." *Pacific Journal of Theology* 2, no. 53 (2015): 25–40.

———. "'Others Are Enjoying Life from Our Death': Eco-relational Theology and a Methodist Ecological Revolution in Oceania." In *Methodist Revolutions: Evangelical Engagements of Church and World*, edited by Joerg Rieger and Upolu Lumā Vaai, 173–96. Nashville: Wesley's Foundery Books, 2021.

———. "Relational Hermeneutics: A Return to the Relationality of the Pacific Itulagi as a Lens for Understanding and Interpreting Life." In *Relational Hermeneutics: Decolonising the Mindset and the Pacific Itulagi*, edited by Upolu Lumā Vaai and Aisake Casimira, 17–41. Suva: University of the South Pacific Press, 2017.

———. "Va'atapalagi: De-heavening Trinitarian Theology in the Islands." In *Colonial Contexts and Postcolonial Theologies: Storyweaving in the Asia-Pacific*, edited by Jione Havea and Mark Brett, 41–53. Melbourne: Palgrave, 2014.

———. "We Are Because We Don't Have." In *The Relational Self: Decolonizing Personhood in the Pacific*, edited by Upolu Lumā Vaai and Unaisi Nabobo-Baba, 283–84. Suva: University of the South Pacific Press, 2017.

Vanderhooft, David Stephen. *The Neo-Babylonian Empire and Babylon in the Latter Prophets*. Harvard Semitic Monographs 59. Atlanta: Scholars Press, 2000.

Van Houtan, Kyle. "Response: Good Work." In *Diversity and Dominion: Dialogues in Ecology, Ethics, and Theology*, edited by Kyle Van Houtan and Michael Northcott, 64–68. Eugene, Ore.: Cascade, 2010.

Veland, Siri. "Transcending Ontological Schisms in Relationships with Water, Air, and Ice." *Weather, Climate, and Society* 9, no. 3 (2017): 607–19.

Veracini, Lorenzo. "'Settler Colonialism': Career of a Concept." *Journal of Imperial and Commonwealth History* 41, no. 2 (2013): 313–33.

Voelz, Richard. "Wrestling with Whiteness in Homiletic Pedagogy: A Reflection on Teaching 'Proclaiming Justice in the Church and Public Square.'" In *Unmasking White Preaching: Racial Hegemony, Resistance, and Possibilities in Homiletics*, edited by Lis Valle-Ruiz and Andrew Wymer, 141–54. Lanham, Md.: Lexington Books, 2022.

Von Rad, Gerhard. *Old Testament Theology*. Translated by D. Stalker. Edinburgh: Oliver and Boyd, 1962.

Wadsworth, JoAnne. "Ecological Racism and Deep Sea Mining in the Pacific." G20 Interfaith Forum, February 11–12, 2022. https://www.g20interfaith.org/2022-webinar-deep-sea-mining-and-ecological-racism/.

Walker-Jones, Arthur. "Psalm 104: A Celebration of the *Vanua*." In *The Earth Story in the Psalms and the Prophets*, edited by Norman Habel and Shirley Wurst, 84–97. Cleveland, Ohio: Pilgrim Press, 2001.

Walmsley, Claire, producer. "Paradise in Peril." *Everyman*, BBC TV, December 20, 1987. https://www.youtube.com/watch?v=23A4S1ZBmgM.

Walzer, Michael. *Exodus and Revolution*. New York: Basic Books, 1986.

Waqainabete, Jeremaia. *Christian Environmentalism: An Ecotheological Approach to Earthkeeping*. Suva: WWF Pacific, 2018. https://wwfasia.awsassets.panda.org/downloads/theological_mindset_english_2018_sep.pdf.

———. "Sermon for the Commemoration of National Fisheries Day, Genesis 1:26–28." Sermon preached at Epworth Hall, Suva, November 21, 2016.

Waqairatu, Tuikilakila. "Na Lako Yani [the Exodus]." Unpublished sermon notes, Centenary Church, Suva, November 4, 2012.

Waqalerua, Apenisa. "The Testimony of Dravuwalu Village." Sermon preached at Davuilevu Theological College, Nausori, Fiji, September 2016.

Warrior, Robert Allen. "A Native American Perspective: Canaanites, Cowboys, and Indians." In *Voices from the Margin: Interpreting the Bible in the Third World*, edited by R. S. Sugirtharajah, 283–90. 25th anniversary ed. Maryknoll, N.Y.: Orbis Books, 2016.

Watikaicolo, Makelesi. "Climate Change: Genesis 3, 2 Chronicles 7:14, Revelation 3:20." Sermon preached at Nakauvadra High School, Rakiraki, Fiji, September 2021.

Watson, Irene. "Buried Alive." *Law and Critique* 13 (2002): 253–69.

Watts, Vanessa. "Indigenous Place-Thought and Agency amongst Humans and Non-Humans (First Woman and Sky Woman Go on a European World Tour!)." *Decolonization: Indigeneity, Education and Society* 2, no. 1 (2013): 20–34.

Weaver, Jace. "Misfit Messengers: Indigenous Religious Traditions and Climate Change." *Journal of the American Academy of Religion* 83, no. 2 (2015): 320–35.

Wedel, Pailin. "'When We Lost the Forest, We Lost Everything': Oil Palm Plantations and Rights Violations in Indonesia." Human Rights Watch, September 22, 2019. https://www.hrw.org/report/2019/09/23/when-we-lost-forest-we-lost-everything/oil-palm-plantations-and-rights-violations.

Weeber, Christine. "Why Capitalize 'Indigenous'?" *Sapiens Anthropology Magazine*, May 19, 2020. https://www.sapiens.org/language/capitalize-indigenous/.

Wellhausen, J. *Prolegomena to the History of Ancient Israel.* New York: Meridian Books, 1957.

Wells, Lisa. *Believers: Making a Life at the End of the World.* New York: Farrar, Straus and Giroux, 2021.

Wesley, Howard-John. "I Can't Breathe." Sermon preached at Alfred Street Baptist Church, Alexandria, Va., August 11, 2019. https://www.youtube.com/watch?v=fc0KaXUhwK8.

Wheeling, Kate. "How the Methodist Church in Fiji Is Helping to Lead on Climate Change." *Pacific Standard*, November 13, 2017. https://psmag.com/environment/cop23-methodist-church-helping-lead-climate-action-fiji.

White, Lynn, Jr. "The Historical Roots of Our Ecologic Crisis." *Science* 155, no. 3767 (1967): 1203–7.

Whiteman, Darrell. "The Legacy of Allan R. Tippett." *International Bulletin of Missionary Research* 16, no. 4 (1992): 163–66.

Whitney, Elspeth. "Lynn White's 'The Historical Roots of Our Ecologic Crisis' After Fifty Years." *History Compass* 13, no. 8 (2015): 396–410.

Whyte, Kyle. "Settler Colonialism, Ecology, and Environmental Injustice." *Environment and Society: Advances in Research* 9, no. 1 (2018): 125–44.

Wilcock, Deirdre, Gary Brierley, and Richard Howitt. "Ethnogeomorphology." *Progress in Physical Geography* 37, no. 5 (2013): 573–600.

Wilson, Brittany. *The Embodied God: Seeing the Divine in Luke-Acts and the Early Church.* Oxford: Oxford University Press, 2021.

Wirzba, Norman. "Dramas of Love and Dirt: Soil and the Salvation of the Earth." *The Cresset* 77, no. 3 (2014): 6–15.

———. "The Soil of Spirituality: What Agrarians Can Teach Us about the Life of Faith." *Laudato Si' Research Institute* 2 (2023): 1–11.

———. "The Song of the Meadow Lark: A Conversation with Norman Wirzba." By Mari Joerstad. *Facing the Anthropocene* webinar series, Kenan Institute for Ethics, April 15, 2021. https://kenan.ethics.duke.edu/the-song-of-the-meadow-lark/.

——. *This Sacred Life: Humanity's Place in a Wounded World*. Cambridge: Cambridge University Press, 2021.

Woodley, Randy. *Shalom and the Community of Creation: An Indigenous Vision*. Grand Rapids: Eerdmans, 2012.

Woodruff, Jonathan, Jennifer Irish, and Suzana Camargo. "Coastal Flooding by Tropical Cyclones and Sea Level Rise." *Nature* 505 (2013): 44–52.

Worland, Justin. "The Leaders of These Sinking Countries Are Fighting to Stop Climate Change. Here's What the Rest of the World Can Learn." *Time*, June 13, 2019. https://time.com/longform/sinking-islands-climate-change/.

World Council of Churches. "Pacific Conference of Churches Calls for 'Less Talk and More Action' as COP27 Begins." November 7, 2022. https://www.oikoumene.org/news/pacific-conference-of-churches-calls-for-less-talk-more-action-as-cop27-begins.

World Wildlife Federation. "8 Things to Know about Palm Oil." Accessed November 17, 2023. https://www.wwf.org.uk/updates/8-things-know-about-palm-oil.

Wright, Christopher. *Old Testament Ethics for the People of God*. Leicester: InterVarsity Press, 2004.

Wright, Jacob L. "Warfare and Wanton Destruction: A Reexamination of Deuteronomy 20:19–20 in Relation to Ancient Siegecraft." *Journal of Biblical Literature* 127 (2008): 423–58.

Yadin, Yagail. *The Art of Warfare in Biblical Lands in the Light of Archaeological Study*. London: Weidenfeld and Nicolson, 1963.

Yee, Merewalesi, Karen McNamara, and Annah Piggott-McKellar. "The Role of *Vanua* in Climate-Related Voluntary Immobility in Fiji." *Frontiers in Climate* 4 (2022): 1–19.

Yeo, K. K. "Theologies of Land: Contested Land, Spatial Justice, and Covenantal Identity." In *Theologies of Land: Contested Land, Spatial Justice, and Identity*, edited by K. K. Yeo and Gene L. Green, 132–56. Eugene, Ore.: Cascade, 2021.

Zachariah, George. "Colonization of the Watersheds and the Green Politics of Hagar." In *People and Land: Decolonizing Theologies*, edited by Jione Havea, 141–57. Minneapolis: Fortress, 2019.

——. "Whose *Oikos* Is It Anyway? Towards a *Poromboke* Ecotheology of 'Commoning.'" In *Decolonizing Ecotheology: Indigenous and Subaltern Challenges*, edited by S. Lily Mendoza and George Zachariah, 201–18. Eugene, Ore.: Wipf and Stock, 2022.

Zacharias, H. Daniel. "The Land Takes Care of Us: Recovering Creator's Relational Design." In *Theologies of Land: Contested Land, Spatial Justice, and Identity*, edited by K. K. Yeo and Gene L. Green, 69–97. Eugene, Ore.: Cascade, 2021.

Zimmerli, Walther. *Ezekiel*. Translated by Ronald E. Clements. 2 vols. Philadelphia: Fortress, 1979.

Index of Authors

Index of Subjects

Index of Scripture

Joel

Matthew

Mark

Luke

John

Romans

Galatians

Ephesians

Colossians

Hebrews

1 Peter

Revelation